A CIVIC BIOLOGY

GEORGE W. HUNTER

A facsimile of the original 1914 edition

as published by the American Book Company

Reprinted by Suzeteo Enterprises, 2020.

ISBN: 978-1-64594-044-9

Compare the unfavorable artificial environment of a crowded city with the more favorable environment of the country.

A CIVIC BIOLOGY

Presented in Problems

BY

GEORGE WILLIAM HUNTER, Ph.D.

PROFESSOR OF BIOLOGY, KNOX COLLEGE, GALESBURG, ILLINOIS; FORMERLY
HEAD OF THE DEPARTMENT OF BIOLOGY, DE WITT CLINTON
HIGH SCHOOL, CITY OF NEW YORK.
AUTHOR OF "ELEMENTS OF BIOLOGY," "ESSENTIALS OF
BIOLOGY," ETC.

AMERICAN BOOK COMPANY

NEW YORK   CINCINNATI   CHICAGO
BOSTON   ATLANTA

Dedicated

TO MY

FELLOW TEACHERS

OF THE DEPARTMENT OF BIOLOGY

IN THE DE WITT CLINTON HIGH SCHOOL

WHOSE CAPABLE, EARNEST, UNSELFISH

AND INSPIRING AID HAS MADE

THIS BOOK POSSIBLE

FOREWORD TO TEACHERS

A course in biology given to beginners in the secondary school should have certain aims. These aims must be determined to a degree, first, by the capabilities of the pupils, second, by their native interests, and, third, by the environment of the pupils.

The boy or girl of average ability upon admission to the secondary school is not a thinking individual. The training given up to this time, with but rare exceptions, has been in the forming of simple concepts. These concepts have been reached didactically and empirically. Drill and memory work have been the pedagogic vehicles. Even the elementary science work given has resulted at the best in an interpretation of some of the common factors in the pupil's environment, and a widening of the meaning of some of his concepts. Therefore, the first science of the secondary school, elementary biology, should be primarily the vehicle by which the child is taught to solve problems and to think straight in so doing. No other subject is more capable of logical development. No subject is more vital because of its relation to the vital things in the life of the child. A series of experiments and demonstrations, discussed and applied as definite concrete problems which have arisen within the child's horizon, will develop power in thinking more surely than any other subject in the first year of the secondary school.

But in our eagerness to develop the power of logical thinking we must not lose sight of the previous training of our pupil. Up to this time the method of induction, that handmaiden of logical thought, has been almost unknown. Concepts have been formed deductively by a series of comparisons. All concepts have been handed down by the authority of the teacher or the text; the inductive search for the unknown is as yet a closed book. It is unwise, then, to directly introduce the pupil to the method of induction with a series of printed directions which, though definite in the mind of the teacher because of his wider horizon, mean

7

little or nothing as a definite problem to the pupil. The child must be brought to the appreciation of the problem through the deductive method, by a comparison of the future problem with some definite concrete experience within his own field of vision. Then by the inductive experiment, still led by a series of oral questions, he comes to the real end of the experiment, the conclusion, with the true spirit of the investigator. The result is tested in the light of past experiment and a generalization is formed which means something to the pupil.

For the above reason the laboratory problems, which naturally precede the textbook work, should be separated from the subject matter of the text. A textbook in biology should serve to verify the student's observations made in the laboratory, it should round out his concept or generalization by adding such material as he cannot readily observe and it should give the student directly such information as he cannot be expected to gain directly or indirectly through his laboratory experience. For these reasons the laboratory manual has been separated from the text.

"The laboratory method was such an emancipation from the old-time bookish slavery of pre-laboratory days that we may have been inclined to overdo it and to subject ourselves to a new slavery. It should never be forgotten that the laboratory is simply a means to the end; that the dominant thing should be a consistent chain of ideas which the laboratory may serve to elucidate. When, however, the laboratory assumes the first place and other phases of the course are made explanatory to it, we have taken, in my mind, an attitude fundamentally wrong. The question is, not what *types* may be taken up in the laboratory to be fitted into the general scheme afterwards, but what *ideas* are most worth while to be worked out and developed in the laboratory, if that happens to be the best way of doing it, or if not, some other way to be adopted with perfect freedom. Too often our course of study of an animal or plant takes the easiest rather than the most illuminating path. What is easier, for instance, particularly with large classes of restless pupils who apparently need to be kept in a condition of uniform occupation, than to kill a supply of animals, preferably as near alike as possible, and set the pupils to work drawing the dead remains? This method is usually supplemented by a series of questions concerning the remains which are sure to keep the pupils busy a while longer, perhaps until the bell strikes, and which usually are so planned as to anticipate any ideas that might naturally crop up in the pupil's mind during the drawing exercise.

" Such an abuse of the laboratory idea is all wrong and should be avoided. The ideal laboratory ought to be a retreat for rainy days; a substitute for out of doors; a clearing house of ideas brought in from the outside. Any course in biology which can be confined within four walls, even if these walls be of a modern, well-equipped laboratory, is in some measure a failure. Living things, to be appreciated and correctly interpreted, must be seen and studied in the open where they will be encountered throughout life. *The place where an animal or plant is found is just as important a characteristic as its shape or function.* Impossible field excursions with large classes within school hours, which only bring confusion to *inflexible* school programs, are not necessary to accomplish this result. Properly administered, it is without doubt one of our most efficient devices for developing biological ideas, but the laboratory should be kept in its proper relation to the other means at our disposal and never be allowed to degenerate either into a place for vacuous drawing exercises or a biological morgue where dead remains are viewed." — *Dr. H. E. Walter.*

For the sake of the pupil the number of technical and scientific terms has been reduced to a minimum. The language has been made as simple as possible and the problems made to hinge upon material already known, by hearsay at least, to the pupil. So far as consistent with a well-rounded course in the essentials of biological science, the interests of the children have been kept in the foreground. In a recent questionnaire sent out by the author and answered by over three thousand children studying biology in the secondary schools of Connecticut, Massachusetts, New Jersey, and New York by far the greatest number gave as the most interesting topics those relating to the care and functions of the human body and the control and betterment of the environment. As would be expected, boys have different biological interests from girls, and children in rural schools wish to study different topics from those in congested districts in large communities. The time has come when we must frankly recognize these interests and adapt the content of our courses in biology to interpret the *immediate* world of the pupil.

With this end in view the following pages have been written. This book shows boys and girls living in an urban community how they may best live within their own environment and how they may coöperate with the civic authorities for the betterment of their environment. A logical course is built up around the topics which appeal to the average normal boy or girl, topics given in a logical sequence so as to work out the solution of problems bearing on the ultimate problem of the entire course, that of preparation for citizenship in the largest sense.

Seasonal use of materials has been kept in mind in outlining this course. Field trips, when properly organized and later used as a basis for discussion in the classroom, make a firm foundation on which to build the superstructure of a course in biology. The normal environment, its relation to the artificial environment of the city, the relations of mutual give and take existing between plants and animals, are better shown by means of field trips than in any other way. Field and museum trips are enjoyed by the pupils as well. These result in interest and in better work. The course is worked up around certain great biological principles; hence insects may be studied when abundant in the fall in connection with their relations to green plants and especially in their relation to flowers. In the winter months material available for the laboratory is used. Saprophytic and parasitic organisms, wild plants in the household, are studied in their relations to mankind, both as destroyers of food, property and life and as man's invaluable friends. The economic phase of biology may well be taken up during the winter months, thus gaining variety in subject matter and in method of treatment. The apparent emphasis placed upon economic material in the following pages is not real. It has been found that material so given makes for variety, as it may be assigned as a topical reading lesson or simply used as reference when needed. Cyclic work in the study of life phenomena and of the needs of organisms for oxygen, food, and reproduction culminates, as it rightly should, in the study of life-processes of man and man's relation to his environment.

In a course in biology the difficulty comes not so much in knowing what to teach as in knowing what *not* to teach. The author believes that he has made a selection of the topics most vital in a well-rounded course in elementary biology directed toward civic betterment. The physiological functions of plants and animals, the hygiene of the individual within the community, conservation and the betterment of existing plant and animal products, the big underlying biological concepts on which society is built, have all been used to the end that the pupil will become a better, stronger and more unselfish citizen. The " spiral " or cyclic method of treatment has been used throughout, the purpose being to ultimately build up a number of well-rounded concepts by constant repetition but with constantly varied viewpoint.

The sincere thanks of the author is extended to all who have helped make this book possible, and especially to the members of the Department of Biology in the De Witt Clinton High School. Most of the men there have directly or indirectly contributed their time and ideas to help make this book worth more to teachers and pupils. The following have read the manuscript in its entirety and have offered much valuable constructive criticism : Dr. Herbert E. Walter, Professor of Zoölogy in Brown University; Miss Elsie Kupfer, Head of the Department of Biology in Wadleigh High School; George C. Wood, of the Department of Biology in the Boys' High School, Brooklyn; Edgar A. Bedford, Head of Department of Biology in the Stuyvesant High School; George E. Hewitt, George T. Hastings, John D. McCarthy, and Frank M. Wheat, all of the Department of Biology in the De Witt Clinton High School.

Thanks are due, also, to Professor E. B. Wilson, Professor G. N. Calkins, Mr. William C. Barbour, Dr. John A. Sampson, W. C. Stevens, and C. W. Beebe, Dr. Alvin Davison, and Dr. Frank Overton; to the United States Department of Agriculture; the New York Aquarium; the Charity Organization Society; and the American Museum of Natural History, for permission to copy and use certain photographs and cuts which have been found useful in teaching. Dr. Charles H. Morse and Dr. Lucius J. Mason, of the De Witt Clinton High School, prepared the hygiene outline in the appendix. Frank M. Wheat and my former pupil, John W. Teitz, now a teacher in the school, made many of the line drawings and took several of the photographs of experiments prepared for this book. To them especially I wish to express my thanks.

At the end of each of the following chapters is a list of books which have proved their use either as reference reading for students or as aids to the teacher. Most of the books mentioned are within the means of the small school. Two sets are expensive: one, *The Natural History of Plants*, by Kerner, translated by Oliver, published by Henry Holt and Company, in two volumes, at $11; the other, *Plant Geography upon a Physiological Basis*, by Schimper, published by the Clarendon Press, $12; but both works are invaluable for reference.

For a general introduction to physiological biology, Parker, *Elementary Biology*, The Macmillan Company; Sedgwick and Wilson, *General Biology*, Henry Holt and Company; Verworn, *General Physiology*, The Macmillan Company; and Needham, *General Biology*, Comstock Publishing Company, are most useful and inspiring books.

Two books stand out from the pedagogical standpoint as by far the most helpful of their kind on the market. No teacher of botany or zoölogy can afford to be without them. They are: Lloyd and Bigelow, *The Teaching of Biology*, Longmans, Green, and Company, and C. F. Hodge, *Nature Study and Life*, Ginn and Company. Other books of value from the teacher's standpoint are: Ganong, *The Teaching Botanist*, The Macmillan Company; L. H. Bailey, *The Nature Study Idea*, Doubleday, Page, and Company; and McMurry's *How to Study*, Houghton Mifflin Company.

CONTENTS

A CIVIC BIOLOGY

I. THE GENERAL PROBLEM — SOME REASONS FOR THE STUDY OF BIOLOGY

What is Biology? — *Biology is the study of living beings, both plant and animal.* Inasmuch as man is an animal, the study of biology includes the study of man in his relations to the plants and the animals which surround him. Most important of all is that branch of biology which treats of the mechanism we call the human body, — of its parts and their uses, and its repair. This subject we call *human physiology.*

Why study Biology? — Although biology is a very modern science, it has found its way into most high schools; and an increasingly large number of girls and boys are yearly engaged in its study. These questions might well be asked by any of the students: Why do I take up the study of biology? Of what practical value is it to me? Besides the discipline it gives me, is there anything that I can take away which will help me in my future life?

Human Physiology. — The answer to this question is plain. If the study of biology will give us a better understanding of our own bodies and their care, then it certainly is of use to us. That phase of biology known as *physiology* deals with the uses of the parts of a plant or animal; human physiology and hygiene deal with the uses and care of the parts of the human animal. The prevention of sickness is due in a large part to the study of hygiene. It is estimated that over twenty-five per cent of the deaths that occur yearly in this country could be averted if *all* people lived in a hygienic manner. In its application to the lives of each of us, as a member of our family, as a member of the school we attend, and as a future citizen, a knowledge of hygiene is of the greatest importance.

Relations of Plants to Animals. — But there are other reasons why an educated person should know something about biology.

We do not always realize that if it were not for the green plants, there would be no animals on the earth. Green plants furnish food to animals. Even the meat-eating animals feed upon those that feed upon plants. How the plants manufacture this food and the relation they bear to animals will be discussed in later chapters. Plants furnish man with the greater part ef his food in the form of grains and cereals, fruits and nuts, edible roots and leaves; they provide his domesticated animals with food; they give him timber for his houses and wood and coal for his fires; they provide him with pulp wood, from which he makes his paper, and oak galls, from which he may make ink. Much of man's clothing and the thread with which it is sewed together come from fiber-producing plants. Most medicines, beverages, flavoring extracts, and spices are plant products, while plants are made use of in hundreds of ways in the useful arts and trades, producing varnishes, dyestuffs, rubber, and other products.

Bacteria in their Relation to Man. — In still another way, certain plants vitally affect mankind. Tiny plants, called *bacteria*, so small that millions can exist in a single drop of fluid, exist almost everywhere about us, — in water, soil, food, and the air. They play a tremendous part in shaping the destiny of man on the earth. They help him in that they act as scavengers, causing things to decay; thus they remove the dead bodies of plants and animals from the surface of the earth, and turn this material back to the ground; they assist the tanner; they help make cheese and butter; they improve the soil for crop growing; so the farmer cannot do without them. But they likewise sometimes spoil our meat and fish, and our vegetables and fruits; they sour our milk, and may make our canned goods spoil. Worst of all, they cause diseases, among others tuberculosis, a disease so harmful as to be called the "white plague." Fully one half of all yearly deaths are caused by these plants. So important are the bacteria that a subdivision of biology, called *bacteriology*, has been named after them, and hundreds of scientists are devoting their lives to the study of bacteria and their control. The greatest of all bacteriologists, Louis Pasteur, once said, "It is within the power of man to cause all parasitic diseases (diseases mostly caused by bacteria) to disappear from the world." His prophecy is gradually being fulfilled, and it may be the lot of some boys or girls who read this book to do their share in helping to bring this condition of affairs about.

The Relation of Animals to Man. — Animals also play an important part in the world in causing and carrying disease. Animals that cause disease are usually tiny, and live in other animals as *parasites;* that is, they get their living from their hosts on which they feed. Among the diseases caused by parasitic animals are malaria, yellow fever, the sleeping sickness, and the hookworm disease. Animals also *carry* disease, especially the flies and mosquitoes; rats and other animals are also well known as spreaders of disease.

From a money standpoint, animals called insects do much harm. It is estimated that in this country alone they are annually responsible for $800,000,000 worth of damage by eating crops, forest trees, stored food, and other material wealth.

The Uses of Animals to Man. — We all know the uses man has made of the domesticated animals for food and as beasts of burden. But many other uses are found for animal products, and materials made from animals. Wool, furs, leather, hides, feathers, and silk are examples. The arts make use of ivory, tortoise shell, corals, and mother-of-pearl; from animals come perfumes and oils, glue, lard, and butter; animals produce honey, wax, milk, eggs, and various other commodities.

The Conservation of our Natural Resources. — Still another reason why we should study biology is that we may work understandingly for the conservation of our natural resources, especially of our forests. The forest, aside from its beauty and its health-giving properties, holds water in the earth. It keeps the water from drying out of the earth on hot days and from running off on rainy days. Thus a more even supply of water is given to our rivers, and thus freshets are prevented. Countries that have been deforested, such as China, Italy, and parts of France, are now subject to floods, and are in many places barren. On the forests depend our supply of timber, our future water power, and the future commercial importance of cities which, like New York, are located at the mouths of our navigable rivers.

HUNTER, CIV. BI. — 2

Plants and Animals mutually Helpful. — Most plants and animals stand in an attitude of mutual helpfulness to one another, plants providing food and shelter for animals; animals giving off waste materials useful to plants in the making of food. We also learn that plants and animals need the same conditions in their surroundings in order to live : water, air, food, a favorable temperature, and usually light. The life processes of both plants and animals are essentially the same, and the living matter of a tree is as much alive as is the living matter in a fish, a dog, or a man.

Biology in its Relation to Society. — Again, the study of biology should be part of the education of every boy and girl, because society itself is founded upon the principles which biology teaches. Plants and animals are living things, taking what they can from their surroundings; they enter into competition with one another, and those which are the best fitted for life outstrip the others. Animals and plants tend to vary each from its nearest relative in all details of structure. The strong may thus hand down to their offspring the characteristics which make them the winners. Health and strength of body and mind are factors which tell in winning.

Man has made use of this message of nature, and has developed improved breeds of horses, cattle, and other domestic animals. Plant breeders have likewise selected the plants or seeds that have varied toward better plants, and thus have stocked the earth with hardier and more fruitful domesticated plants. Man's dominion over the living things of the earth is tremendous. This is due to his understanding the principles which underlie the science of biology.

Finally the study of biology ought to make us better men and women by teaching us that unselfishness exists in the natural world as well as among the highest members of society. Animals, lowly and complex, sacrifice their comfort and their very lives for their young. In the insect communities the welfare of the individual is given up for the best interests of the community. The law of mutual give and take, of sacrifice for the common good, is seen everywhere. This should teach us, as we come to take our places in society, to be willing to give up our individual pleasure or selfish gain for the good of the community in which we live. Thus the application of biological principles will benefit society.

II. THE ENVIRONMENT OF PLANTS AND ANIMALS

Problem. — *To discover some of the factors of the environment of plants and animals.*

(a) *Environment of a plant.*

(b) *Environment of an animal.*

(c) *Home environment of a girl or boy.*

LABORATORY SUGGESTIONS

Laboratory demonstrations. — Factors of the environment of a living plant or animal in the vivarium.

Home exercise. — The study of the factors making up my own environment and how I can aid in their control.

Environment. — Each one of us, no matter where he lives, comes in contact with certain surroundings. Air is everywhere around us; light is necessary to us, so much so that we use artificial light at night. The city street, with its dirty and hard paving stones, has come to take the place of the soil of the village or farm. Water and food are a necessary part of our surroundings. Our clothing, useful to maintain a certain temperature, must also be included. All these things air, light, heat, water, food — together make up our *environment*.

An unfavorable city environment.

All other animals, and all plants as well, are surrounded by and use practically the same things from their environment as we do. The potted plant in the window, the goldfish in the aquarium, your pet dog at home, all use, as we will later prove, the factors of their environ-

19

ment in the same manner.   Air, water, light, a certain amount of
heat, soil to live in or on, and food form parts of the surroundings
of *every* living thing.

**The Same Elements found in Plants
and Animals as in their Environment.**
— It has been found by chemists that
the plants and animals as well as their
environment may be reduced to about
eighty very simple substances known
as *chemical elements.*   For example,
the air is made up largely of two elements, *oxygen* and *nitrogen.*   Water,
by means of an electric current, may
be broken up into two elements, *oxygen*
and *hydrogen.*   The elements in water
are combined to make a *chemical compound.*   The oxygen and nitrogen of
the air are not so united, but exist as
separate gases.   If we were to study

An experiment that shows the
air contains about four fifths
nitrogen.

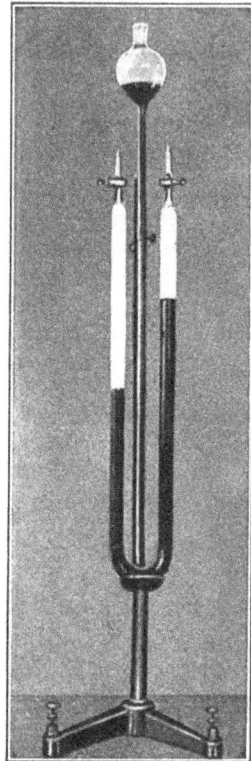

Apparatus for separating
water by means of an
electric current into the
two elements, hydrogen
and oxygen.

the chemistry of the bodies of plants and animals and of their
foods, we would find them to be made up of certain chemical
elements combined in various complex compounds.   These elements are principally *carbon, hydrogen, oxygen, nitrogen,* and
perhaps a dozen others in very minute proportions.   But the
same elements present in the living things might also be found

in the environment, for example, water, food, the air, and the soil. It is logical to believe that living things use the chemical elements in their surroundings and in some wonderful manner build up their own bodies from the materials found in their en-
vironment. How this is done we will learn in later chapters.

What Plants and Animals take from their Environment. Air. — It is a self-evident fact that animals need air. Even those living in the water use the air dissolved in the water. A fish placed in an air-tight jar will soon die. It will be proven later that plants also need air in order to live.

Water. — We all know that water must form part of the environment of plants and animals. It is a matter of common knowledge that pets need water to drink; so do other animals. Every one knows we must water a potted plant if we expect it to grow. Water is of so much importance to man that from the time of the Cæsars until

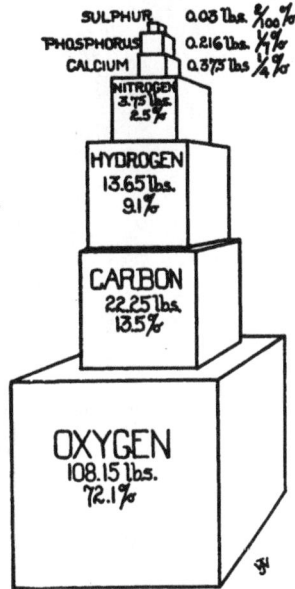

Chart to show the percentage of chemical elements in the human body.

now he has spent enormous sums of money to bring pure water to his cities. The United States government is spending millions of dollars at the present time to bring by irrigation the water needed to support life in the western desert lands.

Light as Condition of the Environment. — Light is another important factor of the environment. A study of the leaves on any green plant growing near a window will convince one that such plants grow toward the light. All green plants are thus influenced by the sun. Other plants which are not green seem either indifferent or are negatively influenced (move away from) the source of light. Animals may or may not be attracted by light. A moth, for example, will fly toward a flame, an earthworm will move away from light. Some animals prefer a moderate or

The effect of water upon the growth of trees. These trees were all planted at the same time in soil that is sandy and uniform. They are watered by a small stream which runs from left to right in the picture. Most of the water soaks into the ground before reaching the last trees.

weak intensity of light and live in shady forests or jungles, prowling about at night. Others seem to need much and strong light. And man himself enjoys only moderate intensity of light and heat. Look at the shady side of a city street on any hot day to prove this statement.

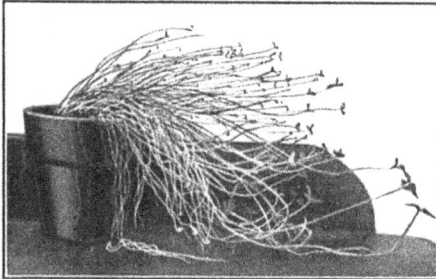

The effect of light upon a growing plant.

Heat. — Animals and plants are both affected by heat or the absence of it. In cold weather green plants either die or their life activities are temporarily suspended, — the plant becomes *dormant*. Likewise small animals, such as insects, may be killed by cold or they may *hibernate* under stones or boards. Their life activities are stilled until the coming of warm weather. Bears and other large animals go to sleep during the winter and awake thin and active at the approach of warm weather. Animals or plants used to certain temperatures are killed if removed from those temperatures. Even man, the most adaptable of all animals, cannot stand great changes without discomfort and sometimes death. He heats his houses in winter and cools them in summer so as to have the amount of heat most acceptable to him, *i.e.* about 70° Fahrenheit.

The Environment determines the Kind of Animals and Plants within It. — In our study of geography we learned that certain

Vegetation in Northern Russia. The trees in this picture are nearly one hundred years old. They live under conditions of extreme cold most of the year.

luxuriant growths of trees and climbing plants were characteristic of the tropics with its moist, warm climate. No one would expect to find living there the hardy stunted plants of the arctic region. Nor would we expect to find the same kinds of animal life in warm regions as in cold. The surroundings determine the kind of living things there. Plants or animals *fitted to live* in a given locality will probably be found there if they have had an opportunity to reach that locality. If, for example, temperate forms of life were introduced by man into the tropics, they would either die or they would gradually change so as to become fitted to live in their new environment. Sheep with long wool fitted to live in England, when removed to Cuba, where conditions of greater heat exist,

Plant life in a moist tropical forest. Notice the air plants to the left and the resurrection ferns on the tree trunk.

soon died because they were not fitted or *adapted* to live in their changed environment.

Adaptations. — Plants and animals are not only fitted to live under certain conditions, but each part of the body may be fitted to do certain work. I notice that as I write these words the fingers of my right hand grasp the pen firmly and the hand and arm execute some very complicated movements. This they are able to do because of the free movement given through the arrangement of the delicate bones of the wrist and fingers, their attachment to the bones of the arm, a wonderful complex of muscles which move the bones, and a directing nervous system which plans the work. Because of the peculiar fitness in the structure of the hand for this work we say it is adapted to its function of grasping objects.   Each part of a plant or animal is usually fitted for some particular work.   The root of a green plant, for example, is fitted to take in water by having tiny absorbing organs growing from it, the stems have pipes or tubes to convey liquids up and down and are strong enough to support the leafy part of the plant.   Each part of a plant does work, and is fitted, by means of certain structures, to do that work.   It is because of these adaptations that living things are able to do their work within their particular environment.

Plants and Animals and their Natural Environment. — Those of us who have tried to keep potted plants in the schoolroom know how difficult it is to keep them healthy.   Dust, foreign gases in the air, lack of moisture, and other causes make the artificial environment in which they are placed unsuitable for them.

A goldfish placed in a small glass jar with no food or no green water plants soon seeks the surface of the water, and if the water is not changed frequently so as to supply air the fish will die.   Again the artificial environment lacks something that the fish needs. Each plant and animal is limited to a certain environment because of certain individual needs which make the surroundings fit for it to live in.

Changes in Environment. — Most plants and animals do not change their environment.   Trees, green plants of all kinds, and some animals remain

A natural barrier on a stream. No trout would be found above this fall. Why not?

fixed in one spot practically all their lives. Certain tiny plants and most animals move from place to place, either in air, water, on the earth or in the earth, but they maintain relatively the same conditions in environment. Birds are perhaps the most striking exception, for some may fly thousands of miles from their summer homes to winter in the south. Other animals, too, migrate from place to place, but not usually where there are great changes in the surroundings. A high mountain chain with intense cold at the upper altitudes would be a barrier over which, for example, a bear, a deer, or a snail could not travel. Fish like trout will migrate up a stream until they come to a fall too high for them to jump. There they must stop because their environment limits them.

Man in his Environment. — Man, while he is like other animals in requiring heat, light, water, and food, differs from them in that

A new apartment house, with out-of-door sleeping porch.

he has come to live in a more or less artificial environment. Men who lived on the earth thousands of year ago did not wear clothes or have elaborate homes of wood or brick or stone. They did not use fire, nor did they eat cooked foods. In short, by slow degrees, civilized man has come to live in a changed environment from that of other animals. The living together of men in communities has caused certain needs to develop. Many things can be supplied in common, as water, milk, foods. Wastes of all kinds have to be disposed of in a town or city. Houses have come to be placed close together, or piled on top of each other, as in the modern apartment.   Fields and trees, all outdoor life, has practically disappeared.   Man has come to live in an artificial environment.

Care and Improvement of One's Environment.—Man can modify or change his surroundings by making this artificial environment favorable to live in.   He may heat his dwellings in winter and cool them in summer so as to maintain a moderate and nearly constant temperature.   He may see that his dwellings have windows so as to let light and air pass in and out.   He may have light at night and shade by day from intense light.   He may have a system of pure water supply and may see that drains or sewers carry away his wastes.   He may see to it that people ill with " catching " or *infectious* diseases are isolated or *quarantined* from others.   This care of the artificial environment is known as *sanitation*, while the care of the *individual* for himself within the environment is known as *hygiene*.   It will be the chief end of this book to show girls and boys how they may become good citizens through the proper control of personal hygiene and sanitation.

REFERENCE BOOKS

ELEMENTARY

Hunter, *Laboratory Problems in Civic Biology*.   American Book Company.
Hough and Sedgwick, *Elements of Hygiene and Sanitation*.   Ginn and Company.
Jordan and Kellogg, *Animal Life*.   Appleton.
Sharpe, *A Laboratory Manual for the Solution of Problems in Biology, p. 95*.   American Book Company.
Tolman, *Hygiene for the Worker*.   American Book Company.

ADVANCED

Allen, *Civics and Health*.   Ginn and Company.

III. THE INTERRELATIONS OF PLANTS AND ANIMALS

Problem. — *To discover the general interrelations of green plants and animals.*

(a) *Plants as homes for insects.*

(b) *Plants as food for insects.*

(c) *Insects as pollinating agents.*

LABORATORY SUGGESTIONS

A field trip: — Object : to collect common insects and study their general characteristics ; to study the food and shelter relation of plant and insects. The pollination of flowers should also be carefully studied so as to give the pupil a general viewpoint as an introduction to the study of biology.

Laboratory exercise. — Examination of simple insect, identification of parts — drawing. Examination and identification of some orders of insects.

Laboratory demonstration. — Life history of monarch and some other butterflies or moths.

Laboratory exercise. — Study of simple flower — emphasis on work of essential organs, drawing.

Laboratory exercise. — Study of mutual adaptations in a given insect and a given flower, *e.g.* butter and eggs and bumble bee.

Demonstration of examples of insect pollination.

The Object of a Field Trip. — Many of us live in the city, where the crowded streets, the closely packed apartments, and the city playgrounds form our environment. It is very artificial at best. To understand better the *normal environment* of plants or animals we should go into the country. Failing in this, an overgrown city lot or a park will give us much more closely the environment as it touches some animals lower than man. We must then remember that in learning something of the natural environment of other living creatures we may better understand our own environment and our relation to it.

28

On any bright warm day in the fall we will find insects swarming everywhere in any vacant lot or the less cultivated parts of a city park. Grasshoppers, butterflies alighting now and then on the flowers, brightly marked hornets, bees busily working over the purple asters or golden rod, and many other forms hidden away on the leaves or stems of plants may be seen. If we were to select for observation some partially decayed tree, we would find it also inhabited. Beetles would be found boring through its bark and wood, while caterpillars (the young stages of butterflies or moths) are feeding on its leaves or building homes in its branches. Everywhere above, on, and under ground may be noticed small forms of life, many of them insects. Let us first see how we would go to work to identify some of the common forms we would be likely to find on plants. Then a little later we will find out what they are doing on these plants.

How to tell an Insect. — A bee is a good example of the group of animals we call *insects*. If we examine its body carefully, we notice that it has three regions, a front part or *head*, a middle part called the *thorax*, and a hind portion, jointed and hairy, the *abdomen*. We cannot escape noting the fact that this insect has wings with which it flies and that it also has legs. The three pairs of legs, which are jointed and provided with tiny hooks at the end, are attached to the thorax. Two pairs of delicate wings are attached to the upper or *dorsal* side of the thorax. The thorax and indeed the entire body, is covered with a hard shell of material similar to a cow's horn, there being no skeleton inside for the attachment of muscles. If we carefully watch the abdomen of a living bee, we notice it move up and down quite regularly. The animal is breathing through tiny breathing holes called *spiracles.*

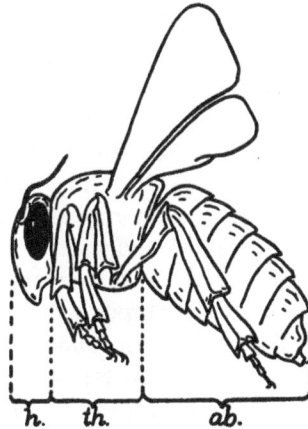

An insect viewed from the side. Notice the head, thorax, and abdomen. What other characters do you find?

h.   th.   ab.

placed along the side of the thorax and abdomen. Bees also have
compound eyes. Wings are not found on all insects, but all the
other characters just given are marks of the
great group of animals we call *insects*.

Forms to be looked for on a Field Trip. —
Inasmuch as there are over 360,000 different
species or kinds of insects, it is evident that it
would be a hopeless task for us even to think of
recognizing all of them. But we can learn to
recognize a few examples of the common forms
that might be met on a field trip. In the fields,
on grass, or on flowering plants we may count on
finding members from six groups or *orders* of insects. These may
be known by the following characters.

Part of the com-
pound eye of an
insect (highly mag-
nified.

The order *Hymenoptera* (membrane wing) to which the bees,
wasps, and ants belong is the only insect group the members of
which are provided with true stings. This sting is placed in a
sheath at the extreme hind end of the abdomen. Other characteristics, which show them to be insects, have been given above.

Butterflies or moths will be found hovering over flowers. They
belong to the order *Lepidoptera* (scale wings). This name is
given to them because their wings are covered with tiny scales,
which fit into little sockets on the wing much as shingles are placed
on a roof. The dust which comes off on the fingers when one
catches a butterfly is composed of these scales. The wings are
always large and usually brightly colored, the legs small, and one
pair is often inconspicuous. These insects may be seen to take
liquid food through a long tubelike organ, called the *proboscis*,
which they keep rolled up under the head when not in use. The
young of the butterfly or moth are known as *caterpillars* and feed
on plants by means of a pair of hard jaws.

Grasshoppers, found almost everywhere, and crickets, black
grasshopper-like insects often found under stones, belong to the
order *Orthoptera* (straight wings). Members of this group may
usually be distinguished by their strong, jumping hind legs, by
their chewing or biting mouth parts, and by the fact that the hind
wings are folded up under the somewhat stiffer front wings.

Another group of insects sometimes found on flowers in the fall are flies. They belong to the order *Diptera* (two wings). These insects are usually rather small and have a single pair of gauzy

Forms of life to be met on a field trip. *A*, The red-legged locust, one of the *Orthoptera; o*, the egg-layer, about natural size. *B*, the honey bee, one of the *Hymenoptera*, about natural size. *C*, a bug, one of the *Hemiptera*, about natural size. *D*, a butterfly, an example of the *Lepidoptera*, slightly reduced. *E*, a house fly, an example of the *Diptera*, about twice natural size. *F*, an orb-weaving spider, about half natural size. (This is not an insect, note the number of legs.) *G*, a beetle, slightly reduced, one of the *Coleoptera*.

wings. Flies are of much importance to man because certain of their number are disease carriers.

Bugs, members of the order *Hemiptera* (half wings), have a jointed proboscis which points backward between the front legs. They are usually small and may or may not have wings.

The beetles or *Coleoptera* (sheath wings), often mistaken for bugs by the uneducated, have the first pair of hardened wings meeting in a straight line in the middle of the back, the second pair of wings being covered by them. Beetles are frequently found on goldenrod blossoms in the fall.

Other forms of life, especially *spiders*, which have four pairs of walking legs, *centipedes* and *millepedes*, both of which are worm-like and have many pairs of legs, may be found.

Try to discover members of the six different orders named above. Collect specimens and bring them to the laboratory for identification.

Why do Insects live on Plants? — We have found insect life abundant on living green plants, some visiting flowers, others hidden away on the stalks or leaves of the plants. Let us next try to find out *why* insects live among and upon flowering green plants.

The Life History of the Milkweed Butterfly. — If it is possible to find on our trip some growing milkweed, we are quite likely to find hovering near, a golden brown and black butterfly, the monarch or milkweed butterfly (*Anosia plexippus*). Its body, as in all insects, is composed of three regions. The monarch frequents the milkweed in order to lay eggs there. This she may be found doing at almost any time from June until September.

Egg and Larva. — The eggs, tiny hat-shaped dots a twentieth of an inch in length, are fastened singly to the underside of milkweed leaves. Some wonderful instinct leads the animal to deposit the eggs on the milkweed, for the young feed upon no other plant. The eggs hatch out in four or five days into rapid-growing worm-like caterpillars, each of which will shed its skin several times before it becomes full size. These caterpillars possess, in addition to the three pairs of true legs, additional pairs of *prolegs* or caterpillar legs. The animal at this stage is known as a *larva*.

Formation of Pupa. — After a life of a few weeks at most, the caterpillar stops eating and begins to spin a tiny mat of silk upon a leaf or stem. It attaches itself to this web by the last pair of prolegs, and there hangs in the dormant stage known as the *chrysalis* or *pupa*. This is a resting stage during which the body changes from a caterpillar to a butterfly.

The Adult. — After a week or more of inactivity in the pupa state, the outer skin is split along the back, and the adult butterfly emerges. At first the wings are soft and much smaller than in the adult. Within fifteen minutes to half an hour after the butterfly emerges, however, the wings are full-sized, having been pumped full of blood and air, and the little insect is ready after her wedding flight to follow her instinct to deposit her eggs on a milkweed plant.

Monarch butterfly: adults, larvæ, and pupa on their food plant, the milkweed. (From a photograph loaned by the American Museum of Natural History.)

Plants furnish Insects with Food. — Food is the most important factor of any animal's environment. The insects which we have seen on our field trip feed on the green plants among which they live. Each insect has its own particular favorite food plant or plants, and in many cases the eggs of the insect are laid on the food plant so that the young may have food close at hand. Some insects prefer the rotted wood of trees. An American zoölogist, Packard, has estimated that over 450 kinds of insects live upon oak trees alone. Everywhere animals are engaged in taking their nourishment from plants, and millions of dollars of damage is done every year to gardens, fruits, and cereal crops by insects.

Damage done by insects. These trees have been killed by boring insects.

All Animals depend on Green Plants. — But insects in their turn are the food of birds; cats and dogs may kill birds; lions or tigers live on still larger defenseless animals as deer or cattle. And finally comes man, who eats the bodies of both plants and animals. But if we reduce this search after food to its final limit, we see that green plants provide *all* the food for animals. For the lion or tiger eats the deer which feeds upon grass or green shoots of young trees, or the cat eats the bird that lives on weed seeds. Green plants supply the food of the world. Later by experiment we will prove this.

Homes and Shelter. — After a field trip no one can escape the knowledge that plants often give animals a home. The grass shelters millions of grasshoppers and countless hordes of other small insects which can be obtained by sweeping through the grass with an insect net. Some insects build their homes in the trees or bushes on which they feed, while others tunnel through the wood, making homes there. Spiders build webs on plants, often using the leaves for shelter. Birds nest in trees, and many other wild animals use the forest as their home. Man has come to use all kinds of plant products to aid him in making his home, wood and various fibers being the most important of these.

What do Animals do for Plants? — So far it has seemed that green plants benefit animals and receive nothing in return. We will later see that plants and animals *together* form a balance of life on the earth and that one is necessary for the other. Certain substances found in the body wastes from animals are necessary to the life of a green plant.

Insects and Flowers. — Certain other problems can be worked out in the fall of the year. One of these is the biological interrelations between insects and flowers. It is easy on a field trip to find insects lighting upon flowers. They evidently have a reason for doing this. To find out why they go there and what they do when there, it will be first necessary for us to study flowers with the idea of finding out what the insects get from them, and what the flowers get from the insects.

The Use and Structure of a Flower. — It is a matter of common knowledge that flowers form fruits and that fruits contain seeds. They are, then, very important parts of certain plants. Our field trip shows us that flowers are of various shapes, colors, and sizes. It will now be our problem first to learn to know the parts of a flower, and then find out how they are fitted to attract and receive insect visitors.

The Floral Envelope. — In a flower the expanded portion of the flower stalk, which holds the parts of the flower, is called the *receptacle*. *The green leaflike parts covering the unopened flower are called the sepals.* Together they form the *calyx*.

The more brightly colored structures are the petals. Together they form

A section of a flower, cut lengthwise. In the center find the pistil with the ovary containing a number of ovules. Around this organ notice a circle of stalked structures, the stamens; the knobs at the end contain pollen. The outer circles of parts are called the petals and sepals, as we go from the inside outward.

the *corolla*. The corolla is of importance, as we shall see later, in making the flower conspicuous. Frequently the petals or corolla have bright marks or dots which lead down to the base of the cup of the flower, where a sweet fluid called *nectar* is made and

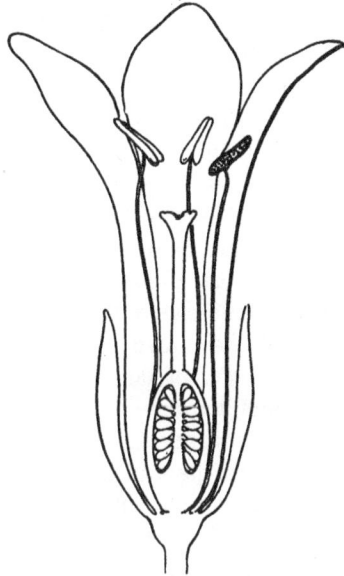

secreted. It is principally this food substance, later made into honey by bees, that makes flowers attractive to insects.

The Essential Organs. — A flower, however, could live without sepals or petals and still do the work for which it exists. Certain *essential organs* of the flower are within the so-called floral envelope. They consist of the *stamens* and *pistil*, the latter being in the center of the flower. The structures with the knobbed ends are called *stamens*. In a single stamen the boxlike part at the end is the *anther;* the stalk which holds the anther is called the *filament*. The anther is in reality a hollow box which produces a large number of little grains called *pollen*. Each pistil is composed of a rather stout base called the *ovary*, and a more or less lengthened portion rising from the ovary called the *style*. The upper end of the style, which in some cases is somewhat broadened, is called the *stigma*. The free end of the stigma usually secretes a sweet fluid in which grains of pollen from flowers of the same kind can grow.

Insects as Pollinating Agents. — Insects often visit flowers to obtain pollen as well as nectar. In so doing they may transfer some of the pollen from one flower to another of the same kind. This transfer of pollen, called *pollination,* is of the greatest use to the plant, as we will later prove. No one who sees a hive of bees with their wonderful communal life can fail to see that these insects play a great part in the life of the flowers near the hive. A famous observer named Sir John Lubbock tested bees and wasps to see how many trips they made daily from their homes to the flowers, and found that the wasp went out on 116 visits during a working day of 16 hours, while the bee made but a few less visits, and worked only a little less time than the wasp worked. It is evident that in the course of so many trips to the fields a bee must light on hundreds of flowers.

Adaptations in a Bee. — If we look closely at the bee, we find the body and legs more or less covered with tiny hairs; especially are these hairs found on the legs. *When a plant or animal structure is fitted to do a certain kind of work, we say it is adapted to do that work*. The joints in the leg of the bee adapt it for complicated movements; the arrangement of stiff hairs along the edge of a concavity in one of the joints of the leg forms a structure well adapted to hold pollen. In this way pollen is collected by the bee and taken to the hive to be used as food. But while gathering pollen for itself, the dust is caught on the hairs and other pro-

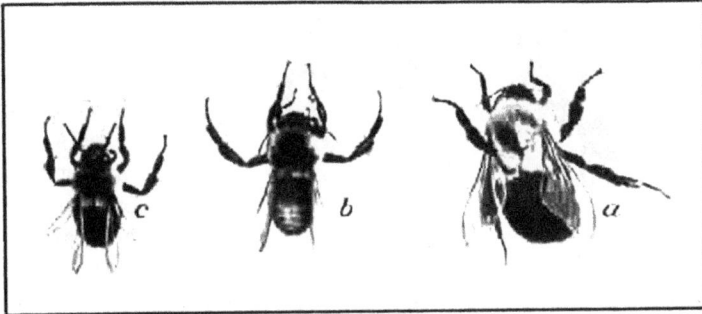

Bumblebees. *a*, queen; *b*, worker; *c*, drone.

jections on the body or legs and is thus carried from flower to flower. The value of this to a flower we will see later.

Field Work. — Is Color or Odor in a Flower an Attraction to an Insect? — Sir John Lubbock tried an experiment which it would pay a number of careful pupils to repeat. He placed a few drops of honey on glass slips and placed them over papers of various colors. In this way he found that the honeybee, for example, could evidently distinguish different colors. Bees seemed to prefer blue to any other color. Flowers of a yellow or flesh color were preferred by flies. It would be of considerable interest for some student to work out this problem with our native bees and with other insects by using paper flowers and honey or sirup. Test the keenness of sight in insects by placing a white object (a white golf ball will do) in the grass and see how many insects will alight on it. Try to work out some method by which you can decide whether a given insect is attracted to a flower by odor alone.

The Sight of the Bumblebee. — The large eyes located on the sides of the head are made up of a large number of little units, each of which is considered to be a very simple eye. The large eyes are therefore called the *compound eyes*. All insects are provided with compound eyes, with simple eyes, or in most cases with both. The simple eyes of the bee may be found by a careful observer between and above the compound eyes.

Insects can, as we have already learned, distinguish differences in color at some distance; they can see *moving* objects, but they do not seem to be able to make out form well. To make up for this, they appear to have an extremely well-developed sense of smell. Insects can distinguish at a great distance odors which to the human nose are indistinguishable. Night-flying insects, especially, find the flowers by the odor rather than by color.

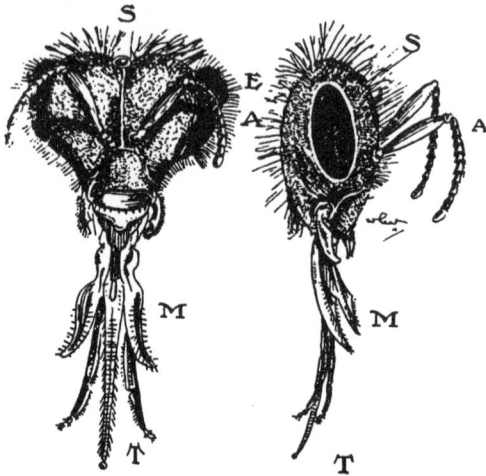

The head of a bee. *A*, antennæ or "feelers"; *E*, compound eye; *S*, simple eye; *M*, mouth parts; *T*, tongue.

Mouth Parts of the Bee. — The mouth of the bee is adapted to take in the foods we have mentioned, and is used for the purposes for which man would use the hands and fingers. The honeybee laps or sucks nectar from flowers, it chews the pollen, and it uses part of the mouth as a trowel in making the honeycomb. The uses of the mouth parts may be made out by watching a bee on a well-opened flower.

Suggestions for Field Work. — In any locality where flowers are abundant, try to answer the following questions: How many bees visit the locality in ten minutes? How many other insects alight on the flowers? Do bees visit flowers of the same kinds in succession, or fly from one flower on a given plant to another on a plant of a different kind? If the bee lights on a flower cluster, does it visit more than one flower in the same cluster? How does a bee alight? Exactly what does the bee do when it alights?

Butter and Eggs (*Linaria vulgaris*). — From July to October this very abundant weed may be found especially along roadsides and in sunny fields. The flower cluster forms a tall and conspicuous cluster of orange and yellow flowers.

The corolla projects into a spur on the lower side; an upper two-parted lip shuts down upon a lower three-parted lip. The four stamens are in pairs, two long and two short.

Certain parts of the corolla are more brightly colored than the rest of the

Flower cluster of "butter and eggs."

flower. This color is a guide to insects. Butter and eggs is visited most by bumblebees, which are guided by the orange lip to alight just where they can push their way into the flower. The bee, seeking the nectar secreted in the spur, brushes its head and shoulders against the stamens. It may then, as it pushes down after nectar, leave some pollen upon the pistil, thus assisting in *self-pollination*. Visiting another flower of the cluster, it would be an easy matter accidentally to transfer

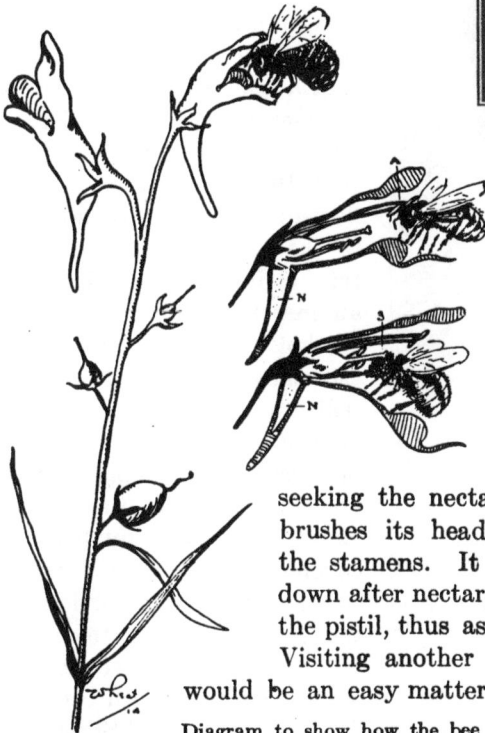

Diagram to show how the bee pollinates "butter and eggs." The bumblebee, upon entering the flower, rubs its head against the long pair of anthers (a), then continuing to press into the flower so as to reach the nectar at (N) it brushes against the stigma (S), thus pollinating the flower. Inasmuch as bees visit other flowers in the same cluster, cross-pollination would also be likely. Why?

this pollen to the stigma of another flower. In this way pollen is carried by the insect to another flower of the same kind. This is known as *cross-pollination*. *By pollination we mean the transfer of pollen from an anther to the stigma of a flower. Self-pollination is the transfer of pollen from the anther to the stigma of the same flower; cross-pollination is the transfer of pollen from the anthers of one flower to the stigma of another flower on the same or another plant of the same kind.*

History of the Discoveries regarding Pollination of Flowers. — Although the ancient Greek and Roman naturalists had some vague ideas on the subject of pollination, it was not until the first part of the nineteenth century that a book appeared in which a German named Conrad Sprengel worked out the facts that the structure of certain flowers seemed to be adapted to the visits of insects. Certain facilities were offered to an insect in the way of easy foothold, sweet odor, and especially food in the shape of pollen and nectar, the latter a sweet-tasting substance manufactured by certain parts of the flower known as the nectar glands. Sprengel further discovered the fact that pollen could be and was carried by the insect visitors from the anthers of the flower to its

A wild orchid, a flower of the type from which Charles Darwin worked out his theory of cross-pollination by insects.

stigma. It was not until the middle of the nineteenth century, however, that an Englishman, Charles Darwin, applied Sprengel's discoveries on the relation of insects to flowers by his investigations upon cross-pollination. The growth of the pollen on the stigma of the flower results eventually in the production of seeds, and thus new plants. Many species of flowers are self-pollinated and do not do so well in seed production if cross-pollinated, but Charles Darwin found that some flowers which were self-pollinated did not produce so many seeds, and that the plants which grew from their seeds were smaller and weaker than plants from seeds produced by cross-pollinated flowers of the same kind. He also found that plants grown from cross-pollinated seeds tended to *vary* more than those grown from self-pollinated seed. This has an important bearing, as we shall see later, in the production of new varieties of plants. Microscopic examination of the stigma at the time of pollination also shows that the pollen from another flower usually germinates before the pollen which has fallen from the anthers of the same flower. This latter fact alone in most cases renders it unlikely for a flower to produce seeds by its own pollen. Darwin worked for years on the pollination of many insect-visited flowers, and discovered in almost every case that showy, sweet-scented, or otherwise attractive flowers were adapted or fitted to be cross-pollinated by insects. He also found that, in the case of flowers that were inconspicuous in appearance, often a compensation appeared in the odor which rendered them attractive to certain insects. The so-called carrion flowers, pollinated by flies, are examples, the odor in this case being like decayed flesh. Other flowers open at night, are white, and provided with a powerful scent. Thus they attract night-flying moths and other insects.

Other Examples of Mutual Aid between Flowers and Insects. — Many other examples of adaptations to secure cross-pollination by means of the visits of insects might be given. The mountain laurel, which makes our hillsides so beautiful in late spring, shows a remarkable adaptation in having the anthers of the stamens caught in little pockets of the corolla. The weight of the visiting insect on the corolla releases the anther from the pocket in which it rests so that it springs up, dusting the body of the visitor with pollen.

In some flowers, as shown by the primroses or primula of our hothouses, the stamens and pistils are of different lengths in different flowers. Short styles and long or high-placed filaments are found in one flower, and long styles with short or low-placed filaments in the other. Pollination will be effected only when some of the pollen from a low-placed anther reaches the stigma of a short-styled flower, or when the pollen from a high anther is placed upon

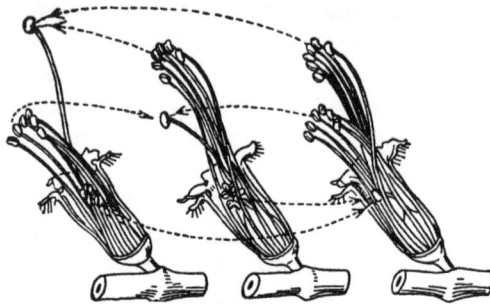

The condition of stamens and pistils on the spiked loosestrife (*Lythrum salicaria*).

a long-styled pistil. There are, as in the case of the loosestrife, flowers having pistils and stamens of three lengths. Pollen only grows on pistils of the same length as the stamens from which it came.

The milkweed or butterfly weed already mentioned is another example of a flower adapted to insect pollination.[1]

A very remarkable instance of insect help is found in the pollination of the yucca, a semitropical lily which lives in deserts (to be seen in most botanic gardens). In this flower the stigmatic surface is above the anther, and the pollen is sticky and cannot be transferred except by insect aid. This is accomplished in a remarkable manner. A little moth, called the *pronuba*, after gathering pollen from an anther, deposits an egg in the ovary of the pistil, and then rubs its load of pollen over the stigma of the flower. The young hatch out

The pronuba moth within the yucca flower.

and feed on the young seeds which have grown because of the pollen placed on the stigma by the mother. The baby cater-

[1] For an excellent account of cross-pollination of this flower, the reader is referred to W. C. Stevens, *Introduction to Botany*. Orchids are well known to botanists as showing some very wonderful adaptations. A classic easily read is Darwin, *On the Fertilization of Orchids*.

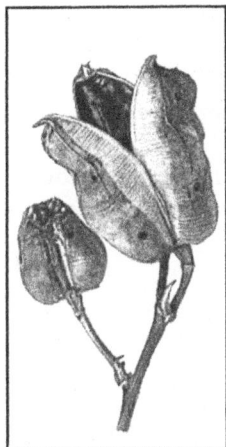

Pod of yucca showing where the young pronubas escaped.

pillars eat some of the developing seeds and later bore out of the seed pod and escape to the ground, leaving the plant to develop the remaining seeds without further molestation.

The fig insect (*Blastophaga grossorum*) is another member of the insect tribe that is of considerable economic importance. It is only in recent years that the fruit growers of California have

The pronuba pollinating the pistil of the yucca.

discovered that the fertilization of the female flowers is brought about by a gallfly which bores into the young fruit. By importing the gallflies it has been possible to grow figs where for many years it was believed that the climate prevented figs from ripening.

Other Flower Visitors. — Other insects besides those already mentioned are pollen carriers for flowers. Among the most useful are moths and butterflies. Projecting from each side of the head of a butterfly is a fluffy structure, the palp. This collects and carries a large amount of pollen, which is deposited upon the stigmas of other flowers when the butterfly pushes its head down into the flower tube after nectar. The scales and hairs on the wings, legs, and body also carry pollen.

Flies and some other insects are agents in cross-pollination. Humming birds are also active agents in

A humming bird about to cross-pollinate a lily.

some flowers. Snails are said in rare instances to carry pollen. Man and the domesticated animals undoubtedly frequently pollinate flowers by brushing past them through the fields.

Pollination by the Wind. — Not all flowers are dependent upon insects or other animals for cross-pollination. Many of the earliest of spring flowers appear almost before the insects do. Such flowers are dependent upon the wind for carrying pollen from the stamens

A cornfield showing staminate and pistillate flowers, the latter having become grains of corn. An ear of corn is a bunch of ripened fruits.

of one flower to the pistil of another. Most of our common trees, oak, poplar, maple, and others, are cross-pollinated almost entirely by the wind.

Flowers pollinated by the wind are generally inconspicuous and often lack a corolla. The anthers are exposed to the wind and provided with much pollen, while the surface of the stigma may be long and feathery. Such flowers may also lack odor, nectar, and bright color. Can you tell why?

Imperfect Flowers. — Some flowers, the wind-pollinated ones in particular, are imperfect; that is, they lack either stamens or pistils. Again, in some cases, imperfect flowers having stamens only are alone found on one plant, while those flowers having pistils only are found on another plant of the same kind. In such flowers, cross-pollination must of necessity follow. Many of our common trees are examples.

Other Cases. — The stamens and pistil ripen at different times in some flowers. The " Lady Washington " geranium, a common

A   B

The flower of "Lady Washington" geranium, in which stamens and pistil ripen at different times, thus insuring cross-pollination. *A*, flower with ripe stamens; *B*, flower with stamens withered and ripe pistil.

house plant, shows this condition. Here also cross-pollination must take place if seeds are to be formed.

Summary. — If we now collect our observations upon flowers with a view to making a summary of the different devices flowers have assumed to prevent self-pollination and to secure cross-pollination, we find that they are as follows : —

(1) *The stamens and pistils may be found in separate flowers, either on the same or on different plants.*

(2) *The stamens may produce pollen before the pistil is ready to receive it, or vice versa.*

(3) *The stamens and pistils may be so placed with reference to each other that pollination can be brought about only by outside assistance.*

Artificial Cross-pollination and its Practical Benefits to Man. — Artificial cross-pollination is practiced by plant breeders and can easily be tried in the laboratory or at home. First the anthers must be carefully removed from the bud of the flower so as to eliminate all possibility of self-pollination. The flower must then be covered so as to prevent access of pollen from without; when the ovary is sufficiently developed, pollen from another flower, having the characters desired, is placed on the stigma and the flower again covered to prevent any other pollen reaching the flower. The seeds from this flower when planted *may* give rise to plants with the best characters of each of the plants which contributed to the making of the seeds.

REFERENCE BOOKS

ELEMENTARY

Hunter, *Laboratory Problems in Civic Biology.* American Book Company.
Andrews, *A Practical Course in Botany*, pages 214–249. American Book Company.
Atkinson, *First Studies of Plant Life*, Chaps. XXV–XXVI. Ginn and Company.
Coulter, *Plant Life and Plant Uses*, pages 301–322. American Book Company.
Dana, *Plants and their Children*, pages 187–255. American Book Company.
Lubbock, *Flowers, Fruits, and Leaves*, Part I. The Macmillan Company.
Needham, *General Biology*, pages 1–50. The Comstock Publishing Company.
Newell, *A Reader in Botany*, Part II, pages 1–96. Ginn and Company.
Sharpe, *A Laboratory Manual in Biology*, pages 43–48. American Book Company.

ADVANCED

Bailey, *Plant Breeding.* The Macmillan Company.
Campbell, *Lectures on the Evolution of Plants.* The Macmillan Company.
Coulter, Barnes, and Cowles, *A Textbook of Botany*, Part II. American Book Company.
Darwin, *Different Forms of Flowers on Plants of the Same Species.* D. Appleton and Company.
Darwin, *Fertilization in the Vegetable Kingdom*, Chaps. I and II. D. Appleton and Company.
Darwin, *Orchids Fertilized by Insects.* D. Appleton and Company.
Lubbock, *British Wild Flowers.* The Macmillan Company.
Müller, *The Fertilization of Flowers.* The Macmillan Company.

IV. THE FUNCTIONS AND COMPOSITION OF LIVING THINGS

Problems. — *To discover the functions of living matter.*
(a) *In a living plant.*
(b) *In a living animal.*

LABORATORY SUGGESTIONS

Laboratory study of a living plant. — Any whole plant may be used ; a weed is preferable.

Laboratory demonstration or home study. — The functions of a living animal.

Demonstration. — The growth of pollen tubes.

Laboratory exercise. — The growth of the mature ovary into the fruit, *e.g.* bean or pea pod.

A Living Plant and a Living Animal Compared. — A walk into the fields or any vacant lot on a day in the early fall will give us first-hand acquaintance with many common plants which, because of their ability to grow under somewhat unfavorable conditions, are called *weeds*. Such plants — the dandelion, butter and eggs, the shepherd's purse — are particularly well fitted by nature to produce many of their kind, and by this means drive out other plants which cannot do this so well. On these or other plants we find feeding several kinds of animals, usually insects.

If we attempt to compare, for example, a grasshopper with the plant on which it feeds, we see several points of likeness and difference at once. Both plant and insect are made up of parts, each of which, as the stem of the plant or the leg of the insect, appears to be distinct, but which is a part of the whole living plant or animal. Each part of the living plant or animal which has a separate work to do is called an *organ*. Thus plants and animals are spoken of as living *organisms*.

47

Functions of the Parts of a Plant. — We are all familiar with the parts of a plant, — the root, stem, leaves, flowers, and fruit. But we may not know so much about their uses to the plant. Each of these structures differs from every other part, and each has a separate work or function to perform for the plant. *The root holds the plant firmly in the ground and takes in water and mineral matter from the soil; the stem holds the leaves up to the light and acts as a pathway for fluids between the root and leaves; the leaves, under certain conditions, manufacture food for the plant and breathe; the flowers form the fruits; the fruits hold the seeds, which in turn hold young plants which are capable of reproducing adult plants of the same kind.*

A weed — notice the unfavorable environment.

The Functions of an Animal. — As we have already seen, the grasshopper has a head, a jointed body composed of a middle and a hind part, three pairs of jointed legs, and two pairs of wings. Obviously, the wings and legs are used for movement; a careful watching of the hind part of the animal shows us that breathing movements are taking place; a bit of grass placed before it may be eaten, the tiny black jaws biting little pieces out of the grass. If disturbed, the insect hops away, and if we try to get it, it jumps or flies away, evidently seeing us before we can grasp it. Hundreds of little grasshoppers on the grass indicate that the grasshopper can reproduce its own kind, but in other respects the animal seems quite unlike the plant. The animal moves, breathes, feeds, and has sensation, while *apparently* the plant does none of these. It will be the purpose of later chapters to prove that the functions of plants and animals are in many respects similar and that *both plants* and *animals breathe, feed,* and *reproduce.*

Organs. — If we look carefully at the organ of a plant called a leaf, we find that the materials of which it is composed do not ap-

pear to be everywhere the same.
The leaf is much thinner and
more delicate in some parts
than in others. Holding the
flat, expanded blade away from
the branch is a little stalk,
which extends into the blade of
the leaf. Here it splits up into
a network of tiny " veins "
which evidently form a framework for the flat blade somewhat as the sticks of a kite
hold the paper in place. If we
examine under the compound
microscope a thin section cut
across the leaf, we shall find
that the veins as well as the

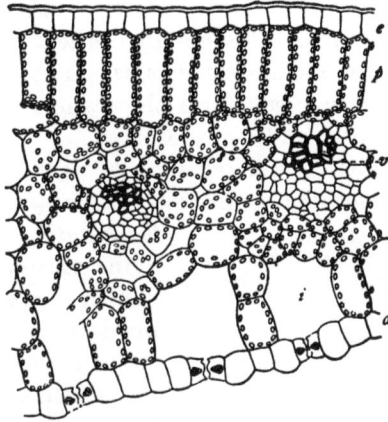

Section through the blade of a leaf. *e*,
cells of the upper surface; *d*, cells of the
lower surface; *i*, air spaces in the leaf;
v, vein in cross sections; *p*, green cells.

other parts are made up of many tiny boxlike units of various
sizes and shapes. These smallest units of building material of the
plant or animal disclosed by the
compound microscope are called
cells. The organs of a plant or
animal are built of these tiny
structures.

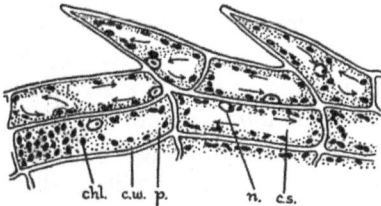

Several cells of *Elodea*, a water plant.
chl., chlorophyll bodies; *c.s.*, cell sap;
c.w., cell wall; *n.*, nucleus; *p.* protoplasm. The arrows show the direction of the protoplasmic movement.

Tissues.[1] — The cells which
form certain parts of the veins,
the flat blade, or other portions
of the plant, are often found in
groups or collections, the cells
of which are more or less alike

[1] *To the Teacher.* — Any simple plant or animal tissue can be used to demonstrate the cell. Epidermal cells may be stripped from the body of the frog or
obtained by scraping the inside of one's mouth. The thin skin from an onion
stained with tincture of iodine shows well, as do thin sections of a young stem, as
the bean or pea. One of the best places to study a tissue and the cells of which
it is composed is in the leaf of a green water plant, *Elodea*. In this plant the cells
are large, and not only their outline, but the movement of the living matter within
the cells, may easily be seen, and the parts described in the next paragraph can
be demonstrated.

HUNTER, CIV. BI. — **4**

in size and shape. Such a collection of cells is called a *tissue*. Examples of tissues are the cells covering the outside of the human body, the muscle cells, which collectively allow of movement, bony tissues which form the framework to which the muscles are attached, and many others.

Cells. — *A cell may be defined as a tiny mass of living matter containing a nucleus, either living alone or forming a unit of the building material of a living thing.* The living matter of which all cells are formed is known as *protoplasm* (formed from two Greek words meaning *first form*). If we examine under a compound microscope a small bit of the water plant *Elodea*, we see a number of structures resembling bricks in a wall. Each " brick," however, is really a plant cell bounded by a thin wall. If we look carefully, we can see that the material inside of this wall is slowly moving and is carrying around in its substance a number of little green bodies. This moving substance is living matter, the protoplasm of the cell. The green bodies (the *chlorophyll* bodies) we shall learn more about later ; they are found only in plant cells. All plant and animal cells appear to be alike in the fact that every living cell possesses a structure known as the *nucleus* (pl. *nuclei*), which is found within the body of the cell. This nucleus is not easy to find in the cells of *Elodea*. Within the nucleus of all cells are found certain bodies called *chromosomes*. These chromosomes in a given plant or animal are always constant in number. These chromosomes are supposed to be the bearers of the qualities which we believe can be handed down from plant to plant and from animal to animal, in other words, the inheritable qualities which make the offspring like its parents.

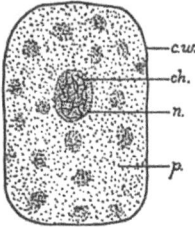

A cell. *ch.*, chromosomes; *c.w.*, cell wall; *n.*, nucleus; *p.*, protoplasm.

How Cells form Others. — Cells grow to a certain size and then split into two new cells. In this process, which is of very great importance in the growth of both plants and animals, the nucleus divides first. The chromosomes also divide, each splitting lengthwise and the parts going in equal numbers to each of the two cells formed from the old cell. In this way the matter in the chromosomes is divided equally between the two new cells. Then the rest of the protoplasm separates, and two new cells are formed. This process is known as *fis-sion*. It is the usual method of growth found in the tissues of plants and animals.

Cells of Various Sizes and Shapes. — Plant cells and animal cells are of very diverse shapes and sizes. There are cells so large that they can easily be seen with the unaided eye; for example, the root hairs

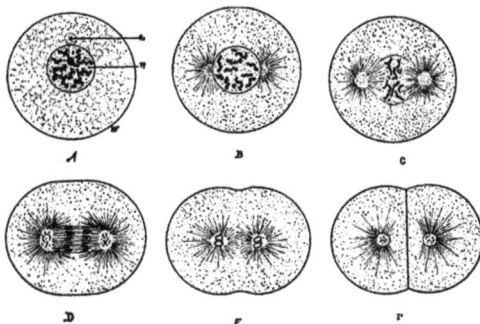

Stages in the division of one cell to form two. Which part of the cell divides first? What seems to become of the chromosomes?

of plants and eggs of some animals. On the other hand, cells may be so minute, as in the case of the plant cells named bacteria, that several million might be present in a few drops of milk. The forms of cells may be extremely varied in different tissues; they may assume the form of cubes, columns, spheres, flat plates, or may be extremely irregular in shape. One kind of tissue cell, found in man, has a body so small as to be quite invisible to the naked eye, although it has a prolongation several feet in length. Such are some of the cells of the nervous system of man and other large animals, as the ox, elephant, and whale.

Varying Sizes of Living Things. — Plant cells and animal cells may live alone, or they may form collections of cells. Some plants are so simple in structure as to be formed of only one kind of cells. Usually living organisms are composed of several groups of different kinds of cells. It is only necessary to call attention to the fact that such collections of cells may form organisms so tiny as to be barely visible to the eye; as, for instance, some of the small flowerless plants or many of the tiny animals living in fresh water or salt water. On the other hand, among animals, the bulk of the elephant and whale, and among plants the big trees of California, stand out as notable examples. The large plants and ani‑ mals are made up of *more*, not necessarily larger, cells.

What Protoplasm can Do. — It responds to influences or stimu‑ lation from without its own substance. Both plants and animals are sensitive to touch or stimulation by light, heat or cold, certain chemical substances, gravity, and electricity. Green plants turn toward the source of light. Some animals are attracted to light and others repelled by it; the earthworm is an example of the latter. *Protoplasm is thus said to be irritable.*

Protoplasm has the power to contract and to move. Muscular movement is a familiar instance of this power. Movement may also take place in plants. Some plants fold up their leaves at night; others, like the sensitive plant, fold their leaflets when touched.

Protoplasm can form new living matter out of food. To do this, food materials must be absorbed into the cells of the living organism. To make protoplasm, it is evident that the same chem‑ ical elements must enter into the composition of the food sub‑ stances as are found in living matter. The simplest plants and animals have this wonderful power as certainly developed as the most complex forms of life.

Protoplasm, be it in plant or animal, breathes and throws off waste materials. When a living thing does work oxygen unites with food in the body; the food is burned or *oxidized* and work is done by means of the energy released from the food. The waste materials are *excreted* or passed out. Plants and animals alike pass off the carbon dioxide which results from the oxidation of food and of parts of their own bodies. Animals eliminate wastes containing nitrogen through the skin and the kidneys.

Protoplasm can reproduce, that is, form other matter like itself. New plants are constantly appearing to take the places of those that die. The supply of living things upon the earth is not de‑ creasing; reproduction is constantly taking place. In a general way it is possible to say that plants and animals reproduce in a very similar manner.

The Importance of Reproduction. — Reproduction is the final process that plants and animals are called upon to perform.

Without the formation of *new* living things no progress would be possible on the earth. We have found that insects help flowering plants in this process. Let us now see exactly what happens when pollen is placed by the bee on the stigma of another flower of the same kind. To understand this process of reproduction in flowers, we must first study carefully pollen grains from the anther of some growing flower.

Pollen. — Pollen grains of various flowers, when seen under the microscope, differ greatly in form and appearance. Some are relatively large, some small, some rough, others smooth, some spherical,

Pollen grains of different shapes and sizes.

and others angular. They all agree, however, in having a thick wall, with a thin membrane under it, the whole inclosing a mass of protoplasm. At an early stage the pollen grain contains but a single cell. A little later, however, two nuclei may be found in the protoplasm. Hence we know that at least two cells exist there, one of which is called the sperm cell; its nucleus is the sperm nucleus.

Growth of Pollen Grains. — Under certain conditions a pollen grain will grow or germinate. This growth can be artificially produced in the laboratory by sprinkling pollen from well-opened flowers of sweet pea or nasturtium on a solution of 15 parts of sugar to 100 of water. Left for a few hours in a warm and moist place and then examined under the microscope, the grains of pollen will be found to have germinated, a long, threadlike mass of protoplasm growing from it into the sugar solution.

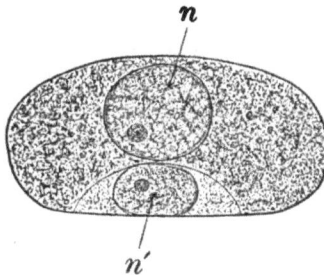

A pollen grain greatly magnified. Two nuclei are found (n, n') at this stage of its growth.

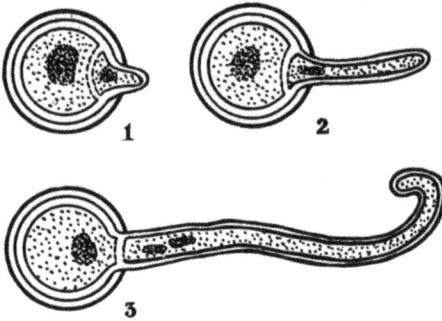

Three stages in the germination of the pollen grain. The nuclei in the tube in (3) are the sperm nuclei. Drawn under the compound microscope.

The presence of this sugar solution was sufficient to induce growth. When the pollen grain germinates, the nuclei enter the thread-like growth (this growth is called the pollen tube; see Figure). One of the nuclei which grows into the pollen tube is known as the *sperm nucleus*.

Fertilization of the Flower. — If we cut the pistil of a large flower (as a lily) lengthwise, we notice that the style appears to be composed of rather spongy material in the interior; the ovary is hollow and is seen to contain a number of rounded structures which appear to grow out from the wall of the ovary. These are the *ovules*. The ovules, under certain conditions, will become *seeds*. An explanation of these conditions may be had if we examine, under the microscope, a very thin section of a pistil, on which pollen has begun to germinate. The central part of the style is found to be either hollow or composed of a soft tissue through which the pollen tube can easily grow. Upon germination, the pollen tube grows downward through the spongy center of the style, follows the path of least resistance to the space within the ovary, and there enters the ovule. It is believed that some chemical influ-

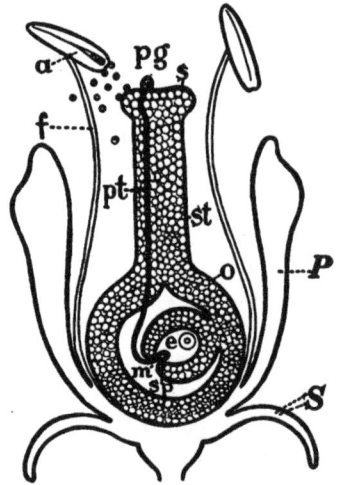

Fertilization of the ovule. A flower cut down lengthwise (only one side shown). The pollen tube is seen entering the ovule. *a*, anther; *f*, filament; *pg*, pollen grain; *s*, stigmatic surface; *pt*, pollen tube; *st*, style; *o*, ovary; *m*, micropyle; *sp*, space within ovary; *e*, egg cell; *P*, petal; *S*, sepal.

ence thus attracts the pollen tube. When it reaches the ovary, the sperm cell penetrates an ovule by making its way through a little hole called the *micropyle*. It then grows toward a clear bit of protoplasm known as the *embryo sac*. The embryo sac is an ovoid space, microscopic in size, filled with semifluid protoplasm containing several nuclei. (See Figure.) *One of the nuclei, with the protoplasm immediately surrounding it, is called the egg cell.* It is this cell that the sperm nucleus of the pollen tube grows toward; ultimately the sperm nucleus reaches the egg nucleus and unites with it. *The two nuclei, after coming together, unite to form a single cell. This process is known as fertilization.* This single cell formed by the union of the pollen tube cell or sperm and the egg cell is now called a *fertilized egg*.

Development of Ovule into Seed. — *The primary reason for the existence of a flower is that it may produce seeds from which future plants will grow. After fertilization the ovule grows into a seed.* The first beginning of the growth of the seed takes place at the moment of fertilization. From that time on there is a growth of the fertilized egg within the ovule which makes a baby plant called the *embryo*. *The embryo will give rise to the adult plant.*

A Typical Fruit, — the Pea or Bean Pod. — If a withered flower of any one of the pea or bean family is examined carefully, it will be found that the pistil of the flower continues to grow after the rest of the flower withers. If we remove the pistil from such a flower and examine it carefully, we find that it is the ovary that has enlarged. The space within the ovary has become nearly filled with a number of nearly ovoid bodies, attached along one edge of the inner wall. These we recognize as the young seeds.

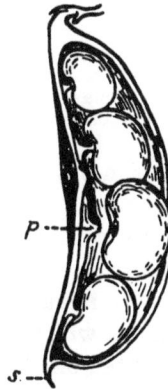

The fruit of the locust, a bean-like fruit. *p*, the attachment to the placenta; *s*, the stigma.

The pod of a bean, pea, or locust illustrates well the growth from the flower. The pod, which is in reality a ripened ovary with other parts of the pistil attached to it, is considered as a *fruit*. By definition, *a fruit is a ripened ovary and its contents together with any parts of the flower that may be attached to it.* The chief use of the fruit to the flower is to hold and to protect the seeds; it may ultimately distribute them where they can reproduce young plants.

The Necessity of Fruit and Seed Dispersal to a Plant. — We have seen that the chief reason for flowers, from the plant's standpoint, is to produce fruits which contain seeds. Reproduction and the ultimate scattering of fruits and seeds are absolutely neces-

The development of an apple. Notice that in this fruit additional parts besides the ovary (*o*) become part of the fruit. Certain outer parts of the flower, the sepals (*s*) and receptacle, become the fleshy part of the fruit, while the ovary becomes the core. Stages numbered 1 to 7 are in the order of development.

sary in order that colonies of plants may reach new localities. It is evident that plants best fitted to scatter their seeds, or place fruits containing the seeds some little distance from the parent plants, are the ones which will spread most rapidly. A plant, if it is to advance into new territory, must get its seeds there first. Plants which are best fitted to do this are the most widely distributed on the earth.

How Seeds and Fruits are Scattered. — Seed dispersal is accomplished in many different ways. Some plants produce enormous numbers of seeds which may or may not have special devices to aid in their scattering. Most weeds are thus started " in pastures new." Some prolific plants, like the milkweed, have *seeds* with a little tuft of hairlike down which allows them to be carried by the wind. Others, as the omnipresent dandelion, have their *fruits* provided with a similar structure, the pappus. Some plants, as the burdock and clotbur, have fruits provided with tiny hooks which stick to the hair of animals, thus proving a means of transportation. Most fleshy fruits contain indigestible seeds, so that when the fruits are eaten by animals the seeds are passed off from the body unharmed and may, if favorably placed, grow. Nuts of various kinds are often carried off by animals, buried, and forgotten, to grow later. Such are a few of the ways in which seeds are scattered. All other things being equal, the plants best equipped to scatter seeds or fruits are those which will drive out other plants in a given locality. Because of their adaptations they are likely to be very numerous, and when unfavorable conditions come, for that reason, if for no other, are likely to survive. Such plants are best exemplified in the weeds of the grassplots and gardens.

REFERENCE BOOKS

ELEMENTARY

Hunter, *Laboratory Problems in Civic Biology.* American Book Company.
Andrews, *A Practical Course in Botany*, pages 250–270. American Book Company.
Atkinson, *First Studies of Plant Life*, Chaps. XXV–XXVI. Ginn and Company.
Bailey, *Lessons with Plants*, Part III, pages 131–250. The Macmillan Company.
Coulter, *Plant Life and Plant Uses.* American Book Company.
Dana, *Plants and their Children*, pages 187–255. American Book Company.
Lubbock, *Flowers, Fruit, and Leaves*, Part I. The Macmillan Company.
Newell, *A Reader in Botany*, Part II, pages 1–96. Ginn and Company.

ADVANCED

Bailey, *Plant Breeding.* The Macmillan Company.
Campbell, *Lectures on the Evolution of Plants.* The Macmillan Company.
Coulter, Barnes, and Cowles, *A Textbook of Botany*, Part II. American Book Company.
Darwin, *Different Forms of Flowers on Plants of the Same Species.* Appleton.
Darwin, *Fertilization in the Vegetable Kingdom*, Chaps. I and II. Appleton.
Darwin, *Orchids Fertilized by Insects.* D. Appleton and Company.
Müller, *The Fertilization of Flowers.* The Macmillan Company.

V. PLANT GROWTH AND NUTRITION. CAUSES OF GROWTH

Problem. — *What causes a young plant to grow?*

(a) *The relation of the young plant to its food supply.*

(b) *The outside conditions necessary for germination.*

(c) *What the young plant does with its food supply.*

(d) *How a plant or animal is able to use its food supply.*

(e) *How a plant or animal prepares food to use in various parts of the body.*

LABORATORY SUGGESTIONS

Laboratory exercise. — Examination of bean in pod. Examination and identification of parts of bean seed.

Laboratory demonstration. — Tests for the nutrients: starch, fats or oils, protein.

Laboratory demonstration. — Proof that such foods exist in bean.

Home work. — Test of various common foods for nutrients. Tabulate results.

Extra home work by selected pupils. — Factors necessary for germination of bean. Demonstration of experiments to class.

Demonstration. — Oxidation of candle in closed jar. Test with lime water for products of oxidation.

Demonstration. — Proof that materials are oxidized within the human body.

Demonstration. — Oxidation takes place in growing seeds. Test for oxidation products. Oxygen necessary for germination.

Laboratory exercise. — Examination of corn on cob, the corn grain, longitudinal sections of corn grain stained with iodine to show that embryo is distinct from food supply.

Demonstration. — Test for grape sugar.

Demonstration. — Grape sugar present in growing corn grain.

Demonstration. — The action of diastase on starch. Conditions necessary for action of diastase.

What makes a Seed Grow. — The general problem of the pages that follow will be to explain how the baby plant, or *embryo,* formed in the seed as the result of the fertilization of the egg cell, is able to grow into an adult plant.   Two sets of factors are necessary for its growth : first, the presence of food to give the young plant a start; second, certain stimulating factors outside the young plant, such as water and heat.

If we open a bean pod, we find the seeds lying along one edge of the pod, each attached by a little stalk to the inner wall of the ovary.   If we pull a single bean from its attachment, we find that the stalk leaves a scar on the coat of the bean; this scar is called the *hilum*.   The tiny hole near the hilum is called the *micropyle*.   Turn back to the figure (page 54) showing the ovule in the ovary.   Find there the little hole through which the pollen tube reached the embryo sac.   This hole is identical with the micropyle in the seed.   The thick outer coat (the *testa*) is easily removed from a soaked bean, the delicate coat under it easily escaping notice.   The seed separates into two parts; these are called the *cotyledons*. If you pull apart the coty-

Three views of a kidney bean, the lower one having one cotyledon removed to show the hypocotyl and plumule.

ledons very carefully, you find certain other structures between them.   The rodlike part is called the *hypocotyl* (meaning *under the cotyledons*).   This will later form the root (and part of the stem) of the young bean plant.   The first true leaves, very tiny structures, are folded together between the cotyledons.   That part of the plant above the cotyledons is known as the *plumule* or *epicotyl* (meaning *above the cotyledons*).   All the parts of the seed within the seed coats together form the *embryo* or young plant.   A bean seed contains, then, a tiny *plant* protected by a tough coat.

Food in the Cotyledons. — The problem now before us is to find out how the embryo of the bean is adapted to grow into an adult plant. Up to this stage of its existence it has had the advantage of food and protection from the parent plant. Now it must begin the battle of life alone. We shall find in all our work with plants and animals that the problem of food supply is always the most important problem to be solved by the growing organism. Let us see if the embryo is able to get a start in life (which many animals get in the egg) from food provided for it within its own body.

Organic Nutrients. — Organic foods (those which come from living sources) are made up of two kinds of substances, the *nutrients* or food substances and *wastes* or *refuse*. An egg, for example, contains the white and the yolk, composed of nutrients, and the shell, which is waste. The organic nutrients are classed in three groups.

Carbohydrates, foods which contain carbon, hydrogen, and oxygen in a certain fixed proportion ($C_6H_{10}O_5$ is an example). They are the simplest of these very complex chemical compounds we call organic nutrients. Starch and sugar are common examples of carbohydrates.

Fats and Oils. — These foods are also composed of carbon, hydrogen, and oxygen in a proportion which enables them to unite readily with oxygen.

Proteins. — A third group of organic foods, *proteins*, are the most complex of all in their composition, and have, besides carbon, oxygen, and hydrogen, the element nitrogen and minute quantities of other elements.

Starch grains in the cells of a potato tuber.

Test for Starch. — If we boil water with a piece of laundry starch in a test tube, then cool it and add to the mixture two or three drops of iodine solution,[1] we find that the mixture in the test tube

---

[1] Iodine solution is made by simply adding a few crystals of the element iodine to 95 per cent alcohol; or, better, take by weight 1 gram of iodine crystals, ⅓ gram of iodide of potassium, and dilute to a dark brown color in weak alcohol (35 per cent) or distilled water.

turns purple or deep blue. It has been discovered by experiment that starch, and no *other known substance*, will be turned purple or dark blue by iodine. Therefore, iodine solution has come to be used as a test for the presence of starch.

Starch in the Bean. — If we mash up a little piece of a bean cotyledon which has been previously soaked in water, and test for starch with iodine solution, the characteristic blue-black color appears, showing the presence of the starch. If a little of the stained material is mounted in water on a glass slide under the compound microscope, you will find that the starch is in the

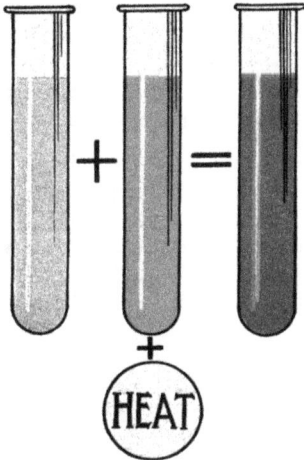

Test for starch.

form of little ovoid bodies called *starch grains*. The starch grains and other food products are made use of by the growing plant.

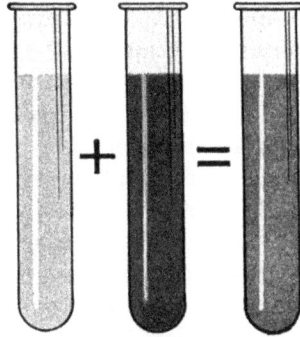

Test for protein.

Test for Oils. — If the substance believed to contain oil is rubbed on brown paper or is placed on paper and then heated in an oven, the presence of oil will be known by a translucent spot on the paper.

Protein in the Bean. — Another nutrient present in the bean cotyledon is *protein*. Several tests are used to detect the presence of this nutrient. The following is one of the best known : —

Place in a test tube the substance to be tested; for example, a bit of hard-boiled egg. Pour over it a little strong (60 per cent) nitric acid and heat gently. Note the color that appears —a lemon yellow. If the egg is washed in water and a little ammonium hydrate added, the color changes to a deep orange, showing that a protein is present.

If the protein is in a liquid state, its presence may be proved by heating, for when it coagulates or thickens, as does the white of an egg when boiled, protein in the form of an *albumin* is present.

Another characteristic protein test easily made at home is burning the substance. If it burns with the odor of burning feathers or leather, then protein forms part of its composition.[1]

A test of the cotyledon of a bean for protein food with nitric acid and ammonium hydrate shows us the presence of this food. Beans are found by actual test to contain about 23 per cent of protein, 59 per cent of carbohydrates, and about 2 per cent oils. The young plant within a pea or bean is thus shown to be well supplied with nourishment until it is able to take care of itself. In this respect it is somewhat like a young animal within the egg, a bird or fish, for example.

Beans and Peas as Food for Man. — So much food is stored in legumes (as beans and peas) that man has come to consider them a very valuable and cheap source of food. Study carefully the following table : —

NUTRIENTS FURNISHED FOR TEN CENTS IN BEANS AND PEAS AT CERTAIN PRICES PER POUND

FOOD MATERIALS AS PURCHASED	PRICES PER POUND	TEN CENTS WILL PAY FOR —			
		Total Food Material	Proteid	Fat	Carbo-hydrates
	Cents	*Pounds*	*Pounds*	*Pounds*	*Pounds*
Kidney beans, dried	5	2.00	0.45	0.04	1.19
Lima beans, fresh, shelled . . .	8	1.25	.04	——	.12
Lima beans, dried	6	1.67	.30	.03	1.10
String beans, fresh, 30 cents per peck	3	3.33	.07	.01	.23
Beans, baked, canned	5	2.00	.14	.05	.39
Lentils, dried	10	1.00	.26	.01	.59
Peas, green, in pod, 30 cents per peck	3	3.33	.12	.01	.33
Peas, dried	4	2.50	.62	.03	1.55

[1] Other tests somewhat more reliable, but much more delicate, are the biuret test and test with Millon's reagent.

Germination of the Bean. — If dry seeds are planted in sawdust or earth, they will not grow.   A moderate supply of water must be

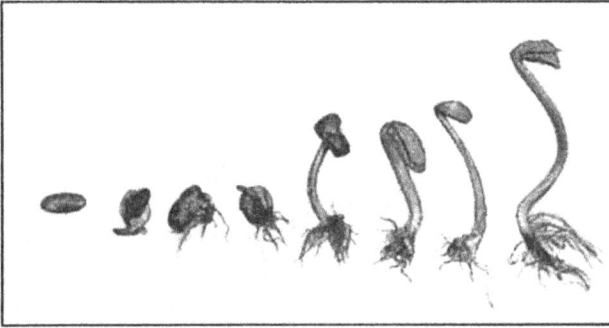

A series of early stages in the germination of the kidney bean.

given to them.   If seeds were to be kept in a freezing temperature or at a very high temperature, no growth would take place. A moderate temperature and a moderate water supply are most favorable for their development.

If some beans were planted so that we might make a record of their growth, we would find the first signs of germination to be the breaking of the testa and the pushing outward of the hypocotyl to form the first root.   A little later the hypocotyl begins to curve downward.   A later stage shows the hypocotyl lifting the cotyledon upward.   In consequence the hypocotyl forms an arch, dragging after it the bulky cotyledons.   The stem, as soon as it is released from the

Bean seedlings.   The older seedlings at the left have used up all of the food supply in the cotyledons.

ground, straightens out. From between the cotyledons the bud-like plumule or epicotyl grows upward, forming the first true leaves and all of the stem above the cotyledons. As growth continues, we notice that the cotyledons become smaller and smaller, until their food contents are completely absorbed into the young plant. The young plant is now able to care for itself and may be said to have passed through the stages of germination.

What makes an Engine Go. — If we examine the sawdust or soil in which the seeds are growing, we find it forced up by the growing seed. Evidently work was done; in other words, *energy* was released by the seeds. A familiar example of release of energy is seen in an engine. Coal is placed in the firebox and lighted, the lower door of the furnace is then opened so as to make a draft of air which will reach the coal. You know the result. The coal burns, heat is given off, causing the water in the boiler to make steam, the engine wheels to turn, and work to be done. Let us see what happens from the chemical standpoint.

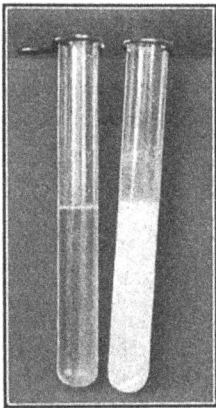

Coal, Organic Matter. — Coal is made largely from dead plants, long since pressed into its present hard form. It contains a large amount of a chemical element called carbon, the presence of which is characteristic of all organic material.

Oxidation, its Results. — When things containing carbon are lighted, they burn. If we place a lighted candle which contains carbon in a closed glass jar, the candle soon goes out. If we then carefully test the air in the jar with a substance known as *limewater*,[1] the latter, when shaken up with the air in the jar, turns milky. This test proves the presence in the jar of a gas, known as *carbon dioxide*. This gas is formed by the carbon of the candle uniting with the oxygen in

The limewater test. The tube at the right shows the effect of the carbon dioxide.

[1] Limewater can be made by shaking up a piece of quicklime the size of your fist in about two quarts of water. Filter or strain the limewater into bottles and it is ready for use.

the air. When the oxygen of the air in the jar was used up, the flame went out, showing that oxygen is necessary to make a thing burn. This uniting of oxygen with some other substance is called *oxidation*.

Oxidation possible without a Flame. — But a flame is not necessary for oxidation. Iron, if left in a damp place, becomes rusty. A union between the oxygen in the water or air and the iron makes what is known as iron oxide or rust. This is an example of *slow oxidation*.

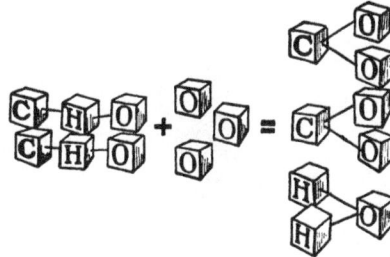

Diagram to show that when a piece of wood is burned it forms water and carbon dioxide.

Oxidation in our Bodies. — If we expel the air from our lungs through a tube into a bottle of limewater, we notice the limewater becomes milky. Evidently carbon dioxide is formed in our own bodies and oxidation takes place there. Is it fair to believe that the heat of our body (for example, 98.6° Fahrenheit under the tongue) is due to oxidation within the body, and that the work we do results from this chemical process. If so, what is oxidized?

Energy comes from Foods. — From the foregoing experiment it is evident that food is oxidized within the human body to release energy for our daily work. Is it not logical to suppose that all living things, both plant and animal, release energy as the result of oxidation of foods within their cells? Let us see if this is true in the case of the pea.

Food oxidized in Germinating Seeds. — If we take equal numbers of soaked peas, placed in two bottles, one tightly stoppered, the other having no stopper, both bottles being exposed to identical conditions of light, temperature, and moisture, we find that the seeds in both bottles start to germinate, but that those in the closed bottle soon stop, while those in the open jar continue to grow almost as well as similar seeds placed in an open dish would.

Why did not the seeds in the covered jar germinate? To answer this question, let us carefully remove the stopper from the stoppered jar and insert a lighted candle. The candle goes out at once. The surer test of limewater shows the presence of carbon dioxide in the jar. The carbon of the foodstuffs of the pea united with the oxygen of the air, forming carbon dioxide. Growth stopped as soon as the oxygen was exhausted. The presence of carbon dioxide in the jar is an indication that a very important process which we associate with animals rather than plants, that of *respiration*, is taking place. The seed, in order to release the energy locked up in its food supply, must have oxygen, so that the oxidation of the food may take

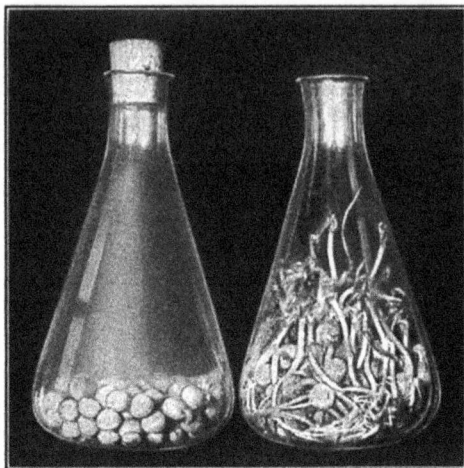

Experiment that shows the necessity for air in germination.

place. *Hence a constant supply of fresh air is an important factor in germination.* It is important that air should penetrate between the grains of soil around a seed. The frequent stirring of the soil enables the air to reach the seed. Air also acts upon some materials in the soil and puts them in a form that the germinating seed can use. This necessity for oxygen shows us at least one reason why the farmer plows and harrows a field and one important use of the earthworm. Explain.

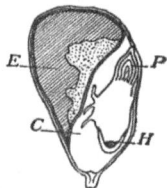

A grain of corn cut lengthwise. *C*, cotyledon; *E*, endosperm; *H*, hypocotyl; *P*, plumule.

Structure of a Grain of Corn. — Examination of a well-soaked grain of corn discloses a difference in the two flat sides of the grain. A light-colored area found on one surface marks the position of the embryo; the rest of the grain contains the food supply. The interesting thing to remember here is that the food supply is *outside* of the embryo.

A grain cut lengthwise perpendicular to the flat side and then dipped in weak iodine shows two distinct parts, an area containing considerable starch, the *endosperm*, and the embryo or young plant. Careful inspection shows the hypo-cotyl and plumule (the latter pointing toward the free end of the grain) and a part surrounding them, the *single* cotyledon (see Figure). Here again we have an example of a fitting for future needs, for in this fruit the one seed has at hand all the food material necessary for rapid growth, although the food is here outside the embryo.

Endosperm the Food Supply of Corn. — We find that the one cotyledon of the corn grain does not serve the same purpose to the young plant as do the two cotyledons of the bean. Although we find a little starch in the corn cotyledon, still it is evident from our tests that the endosperm is the chief source of food supply. The study of a thin section of the corn grain under the compound micro-scope shows us that the starch grains in the endosperm are large and regular in size. When the grain has begun to grow, examina-tion shows that the starch grains near the edge of the cotyledon are much smaller and quite irregular, having large holes in them. We know that the germinating grain has a much sweeter taste than that which is not growing. This is noticed in sprouting barley or malt. We shall later find that, in order to make use of starchy food, a plant or animal must in some manner change it over to sugar. This change is necessary, because starch will

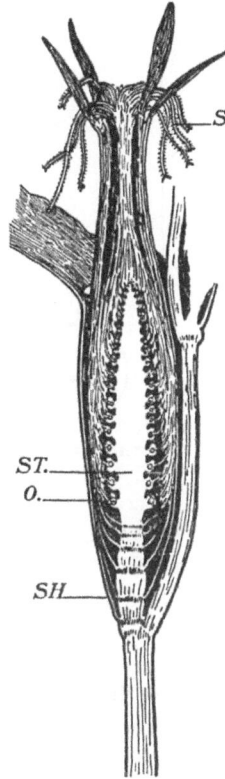

Longitudinal section of young ear of corn. *O*, the fruits; *S*, the stigmas; *SH*, the sheath-like leaves; *ST*, the flower stalk. (After Sargent.)

not dissolve in water, while sugar will; in this form substances can pass from cell to cell in the plant and thus distribute the food where it is needed.

A Test for Grape Sugar. — Place in a test tube the substance to be tested and heat it in a little water so as to dissolve the sugar.

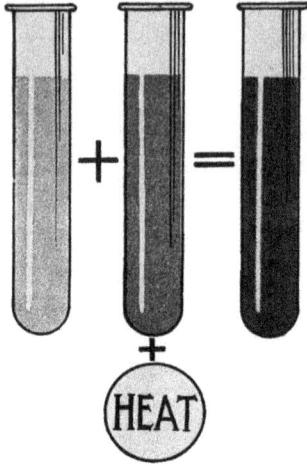

Add to the fluid twice its bulk of Fehling's solution,[1] which has been previously prepared. Heat the mixture, which should now have a blue color, in the test tube. If grape sugar is present in considerable quantity, the contents of the tube will turn first a greenish, then yellow, and finally a brick-red color. Smaller amounts will show less decided red. No other substance than sugar will give this reaction. If Benedict's test [1] is used, a colored precipitate will appear in the test tube after boiling.

Test for grape sugar.

Starch changed to Grape Sugar in the Corn. — That starch is being changed to grape sugar in the germinating corn grain can easily be shown if we cut lengthwise through the embryos of half a dozen grains of corn that have just begun to germinate, place them in a test tube with some Fehling's solution, and heat almost to the boiling point. They will be found to give a reaction showing the presence of sugar along the edge of the cotyledon and between it and the endosperm.

Digestion. — This change of starch to grape sugar in the corn is a process of *digestion*. If you chew a bit of unsweetened cracker in the mouth for a little time, it will begin to taste sweet, and if the chewed cracker, which we know contains starch, is tested with Fehling's solution, some of the starch will be found to have changed to grape sugar. Here, again, a process of digestion has taken place. In both the corn and in the mouth, the change is brought about by the action of peculiar substances known as digestive ferments, or *enzymes*. Such substances have the power under certain conditions to change insoluble foods — solids — into

[1] Directions for making these solutions will be found in Hunter's *Laboratory Problems in Civic Biology*.

soluble substances — liquids. The result is that substances which before digestion would not dissolve in water now will dissolve.

The Action of Diastase on Starch. — The enzyme found in the cotyledon of the corn, which changes starch to grape sugar, is called *diastase.* It may be separated from the cotyledon and used in the form of a powder.

To a little starch in half a cup of water we add a very little (1 gram) of diastase and put the vessel containing the mixture in a warm place, where the temperature will remain nearly constant at about 98° Fahrenheit. On testing part of the contents at the end of half an hour, and the remainder the next morning, for starch and for grape sugar, we find from the morning test that the starch has been almost completely changed to grape sugar. Starch and warm water alone under similar conditions will not react to the test for grape sugar.

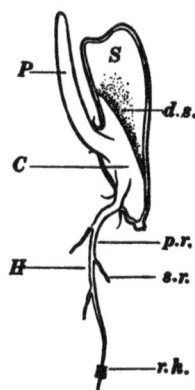

A germinating corn grain. *C,* cotyledon; *H,* growing root (*hypocotyl*); *P,* growing stem (*plumule*); *S,* endosperm; *d.s.,* digested starch; *p.r.,* primary root; *s.r.,* secondary root; *r.h.,* root hairs.

Digestion has the Same Purpose in Plants and Animals. — In our own bodies we know that solid foods taken into the mouth are broken up by the teeth and moistened by saliva. If we could follow that food, we would find that eventually it became part of the blood. It was made soluble by digestion, and in a liquid form was able to reach the blood. Once a part of the body, the food is used either to release energy or to build up the body.

Summary. — We have seen:

1. That seeds, in order to grow, must possess a food supply either in or around their bodies.

2. That this food supply must be oxidized before energy is released.

3. That in cases where the food is not stored at the point where it is to be oxidized the food must be digested so that it may be transported from one part to another in the same plant.

The life processes of plants and animals, so far, may be considered as alike; they both feed, breathe (oxidize their food), do work, and grow.

REFERENCE BOOKS

ELEMENTARY

Hunter, *Laboratory Problems in Civic Biology*. American Book Company.
Andrews, *A Practical Course in Botany*, pages 1–21. American Book Company.
Atkinson, *First Studies of Plant Life*, Chap. XXX. Ginn and Company.
Bailey, *Botany*, Chaps. XX, XXX. The Macmillan Company.
Beal, *Seed Dispersal*. Ginn and Company.
Bergen and Davis, *Principles of Botany*, Chaps. XX, XXX. Ginn and Company
Coulter, *Plant Life and Plant Uses*. American Book Company.
Dana, *Plants and their Children*. American Book Company.
Mayne and Hatch, *High School Agriculture*. American Book Company.
Lubbock, *Flowers, Fruits, and Leaves*. The Macmillan Company.
Newell, *Reader in Botany*, pages 24–49. Ginn and Company.
Sharpe, *A Laboratory Manual in Biology*, pages 55–65. American Book Company

ADVANCED

Bailey, *The Evolution of our Native Fruits*. The Macmillan Company.
Bailey, *Plant Breeding*. The Macmillan Company.
Coulter, Barnes, and Cowles, *A Textbook of Botany*, Vol. I. American Book Company.
De Candolle, *Origin of Cultivated Plants*. D. Appleton and Company.
Duggar, *Plant Physiology*. The Macmillan Company.
Farmers' Bulletins, Nos. 78, 86, 225, 344. U. S. Department of Agriculture.
Hodge, *Nature Study and Life*, Chaps. X, XX. Ginn and Company.
Kerner (translated by Oliver), *Natural History of Plants*. Henry Holt and Company, 4 vols. Vol. II, Part 2.
Sargent, *Corn Plants*. Houghton, Mifflin, and Company.

VI. THE ORGANS OF NUTRITION IN PLANTS — THE SOIL AND ITS RELATION TO THE ROOTS

Problem. — *What a plant takes from the soil and how it gets it.*

(a) *What determines the direction of growth of roots?*

(b) *How is the root built?*

(c) *How does a root absorb water?*

(d) *What is in the soil that a root might take out?*

(e) *Why is nitrogen necessary, and how is it obtained?*

LABORATORY SUGGESTIONS

Demonstration. — Roots of bean or pea.

Demonstration or home experiment. — Response of root to gravity and to water. What part of root is most responsive?

Laboratory work. — Root hairs, radish or corn, position on root, gross structure only. Drawing.

Demonstration. — Root hair under compound microscope.

Demonstration. — Apparatus illustrating osmosis.

Demonstration or a home experiment. — Organic matter present in soil.

Demonstration. — Root tubercles of legume.

Demonstration. — Nutrients present in some roots.

Uses of the Root. — If one of the seedlings of the bean spoken of in the last chapter is allowed to grow in sawdust and is given light, air, and water, sooner or later it will die. Soil is part of its natural environment, and the roots which come in contact with the soil are very important. It is the purpose of this chapter to find out just how the young plant is fitted to get what it needs from this part of its environment; namely, the soil.

The development of a bean seedling has shown us that the root grows first. *One of the most important functions of the root to a young seed plant is that of a holdfast, an anchor to fasten it in the place where it is to develop.* It has many other uses, as the taking in of water with the mineral and organic matter dissolved therein, the stor-

71

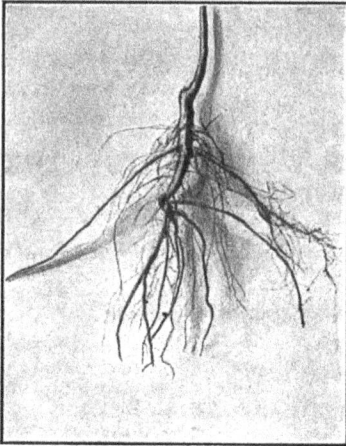

A root system, showing primary and secondary roots.

age of food, climbing, etc. All functions other than the first one stated arise after the young plant has begun to develop.

Root System. — If you dig up a young bean seedling and carefully wash the dirt from the roots, you will see that a long root is developed as a continuation of the hypocotyl. This root is called the *primary* root. Other smaller roots which grow from the primary root are called *secondary,* or *tertiary,* depending on their relation to the first root developed.

Downward Growth of Root. Influence of Gravity. — Most of the roots examined take a more or less downward direction. We are all familiar with the fact that the force we call gravity influences life upon this earth to a great degree. Does gravity act on the growing root? This question may be answered by a simple experiment.

Plant mustard or radish seeds in a pocket garden, place it on one edge and allow the seeds to germinate until the root has grown to a length of about half an inch. Then turn it at right angles to the first position and allow it to remain for one day undisturbed. The roots now will be found to have turned in response to the change in position, that part of the root near the growing point being the most sensitive to the change. This experiment seems to indicate that the roots are influenced to grow downward by the force of gravity.

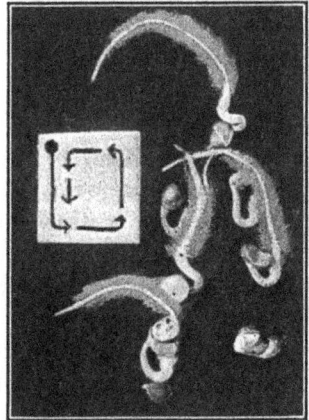

Revolve this figure in the direction of the arrows to see if the roots of the radish respond to gravity.

Experiments to determine the Influence of Moisture on a Growing Root. — The objection might well be interposed that possibly the roots in the pocket garden [1] grew downward after water. That moisture has an influence on the growing root is easily proved.

Plant bird seed, mustard or radish seed in the underside of a sponge, which should be kept wet, and may be suspended by a string under a bell jar in the schoolroom window. . Note whether the roots leave the sponge to grow downward, or if the moisture in the sponge is sufficient to counterbalance the force of gravity.

Water a Factor which determines the Course taken by Roots. — *Water, as well as the force of gravity, has much to do with the direction taken by roots.* Water is always found below the surface of the ground, but sometimes at a great depth. Most trees, and all grasses, have a greater area of surface exposed by the roots than by the branches. The roots of alfalfa, a cloverlike plant used for hay in the Western states, often penetrate the soil after water for a distance of ten to twenty feet below the surface of the ground.

Fine Structure of a Root. [2] — When we examine a delicate root in thin longitudinal section under the compound microscope, we find the entire root to be made up of cells, the walls of which are uniformly rather thin. Over the lower end of the root is found a collection of cells, most of which are dead, loosely arranged so as to form a cap over the growing tip. This is evidently an adaptation which protects the young and actively growing cells just under the root cap. In the body of the root a central cylinder can easily be distinguished from the surrounding cells. In a longitudinal section a series of tubelike structures may be found within the central cylinder. These structures are cells which have grown together at the small end, the long axis of the cells running

---

[1] *The Pocket Garden.* — A very convenient form of pocket germinator may be made as follows. Obtain two cleaned four by five negatives (window glass will do); place one flat on the table and place on this half a dozen pieces of colored blotting paper cut to a size a little less than the glass. Now cut four thin strips of wood to fit on the glass just outside of the paper. Next moisten the blotter, place on it some well-soaked radish, mustard seeds or barley grains, and cover with the other glass. The whole box thus made should be bound together with bicycle tape. Seeds will germinate in this box and with care may live for two weeks or more.

[2] Sections of tradescantia roots are excellent for demonstration of these structures.

the length of the main root.   In their development the cells mentioned have grown together in such a manner as to lose their small

ends, and now form continuous hollow tubes with rather strong walls.   Other cells have come to develop greatly thickened walls; these cells give mechanical support to the tubelike cells.   Collections of such tubes and supporting woody cells together make up what are known as *fibrovascular bundles.*

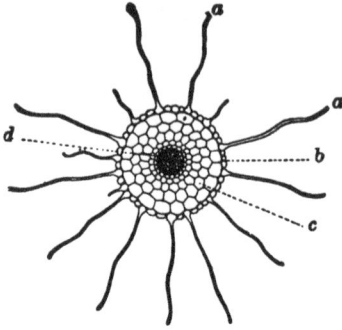

Cross section of a young taproot; *a, a,* root hairs; *b,* outer layer of bark; *c,* inner layer of bark; *d,* wood or central cylinder.

Root Hairs. — Careful examination of the root of one of the seedlings of mustard, radish, or barley grown in the pocket germinator shows a covering of tiny fuzzy structures.   These structures are very minute, at most 3 to 4 millimeters in length.   They vary in length according to their position on the root, the most and the longest root hairs being found near the point marked *R. H.* in the figure.   These structures are outgrowths of the outer layer of the root (the *epidermis*), and are of very great importance to the living plant.

Structure of a Root Hair. — A single root hair examined under a compound microscope will be found to be a long, round structure, almost colorless in appearance.   The wall, which is very flexible and thin, is made up of cellulose, a substance somewhat like wood in chemical composition, through which fluids may easily pass.   Clinging close to the cell wall is the protoplasm of the cell.

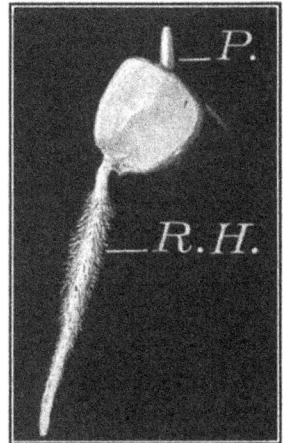

Young embryo of corn, showing root hairs (*R. H.*) and growing stem (*P.*).

The interior of the root hair is more or less filled with a fluid called *cell sap*. Forming a part of the living protoplasm of the root hair, sometimes in the hairlike prolongation and sometimes in that part of the cell which forms the epidermis, is found a *nucleus*. The protoplasm and nucleus are alive; the cell wall formed by the living matter in the cell is dead. *The root hair is a living plant cell* with a wall so delicate that water and mineral substances from the soil can pass through it into the interior of the root.

How the Root absorbs Water. — The process by which the root hair takes up soil water can better be understood if we make

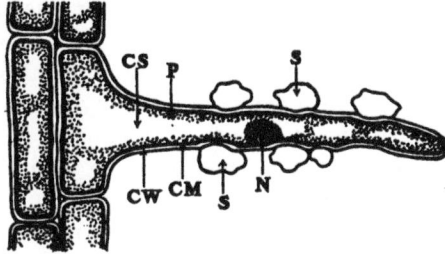

Diagram of a root hair; *CS*, cell sap; *CW*, cell wall; *P*, protoplasm; *N*, nucleus; *S*, particles of soil.

an artificial root hair large enough to be easily seen. An egg with part of the outer shell removed so as to expose the soft skinlike membrane underneath is an example. Better, an artificial root hair may be *made* in the following way. Pour some soft celloidin into a test tube; carefully revolve the test tube so that an even film of celloidin dries on the inside. This membrane is removed, filled with white of egg, and tied over the end of a rubber cork in which a glass tube has previously been inserted. When placed in water, it gives a very accurate picture of the root hair at work. After a short time water begins to rise in the tube, having passed through the film of celloidin. If grape sugar, salt, or some other substance which will dissolve in water were placed in the water outside the artificial root hair, it could soon be proved by test to pass through the wall and into the liquid inside.

Osmosis. — To explain this process we must remember that gases and liquids of different densities, when separated by a membrane, tend to flow toward each other and mingle, the greater flow always being in the direction of the denser medium. *The process by which two gases or fluids, separated by a membrane, tend to pass through the membrane and mingle with each other, is called osmosis.* The method by which the root hairs take up soil water is exactly the same process. It is by osmosis. The white of the egg is the best possible substitute for living matter; the celloidin membrane separating the egg from the water is much like the delicate membrane-like wall which separates the protoplasm of the root hair from the water in the soil surrounding it. The fluid in the root hair is denser than the soil water; hence the greater flow is toward the interior of the root hair.[1]

Passage of Soil Water within the Root. — We have already seen that in an exchange of fluids by osmosis the greater flow is always toward the denser fluid. Thus it is that the root hairs take in more fluid than they give up. The cell sap, which partly fills the interior of the root hair, is a fluid of greater density than the water outside in the soil. When the root hairs become filled with

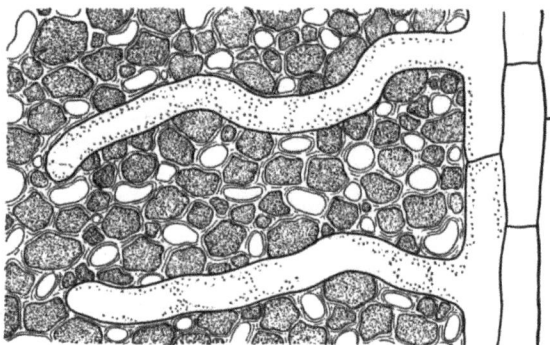

The soil particles are each surrounded with a delicate film of water.
How might the root hairs take up this water?

water, the density of the cell sap is lessened, and the cells of the epidermis are thus in a position to pass along their supply of water to the cells next to them and nearer to the center of the root. These cells, in turn, become less dense than their inside neighbors, and so the transfer of water goes on until the water at last reaches the central cylinder. Here it is passed over to the tubes of the woody bundles and started up the stem. The pressure created

[1] For an excellent elementary discussion of osmosis see Moore, *Physiology of Man and Other Animals*. Henry Holt and Company.

by this process of osmosis is sufficient to send water up the stem to a distance, in some plants, of 25 to 30 feet. Cases are on record of water having been raised in the birch a distance of 85 feet.

Physiological Importance of Osmosis. — It is not an exaggeration to say that osmosis is a process not only of great importance to a plant, but to an animal as well. Foods are digested in the food tube of an animal; that is, they are changed into a soluble form so that they may pass through the walls of the food tube and become part of the blood. The inner lining of part of the food tube is thrown into millions of little fingerlike projections which look somewhat, in size at least, like root hairs. These fingerlike processes are (unlike a root hair) made up of many cells. But they serve the same purpose as the root hairs, for they absorb liquid food into the blood. This process of absorption is largely by osmosis. Without the process of osmosis we should be unable to use much of the food we eat.

Composition of Soil. — If we examine a mass of ordinary loam carefully, we find that it is composed of numerous particles of varying size and weight. Between these particles, if the soil is not caked and hard packed, we can find tiny spaces. In well-tilled soil these spaces are constantly being formed and enlarged. They allow air and water to penetrate the soil. If we examine soil under the microscope, we find considerable water clinging to the soil particles and forming a delicate film around each particle. In this manner most of the water is held in the soil.

Inorganic soil is being formed by weathering.

How Water is held in Soil. — To understand what comes in with the soil water, it will be necessary to find out a little more about soil. Scientists who have made the subject of the composition of the earth a study, tell us that once upon a time at least a part of the earth was molten. Later, it cooled into solid rock.   Soil making began when the ice and frost, working alternately with the heat, chipped off pieces of rock.   These pieces in time became ground into fragments by action of ice, glaciers, running water, or the atmosphere.   This process is· called weathering. Weathering is aided by oxidation.   A glance at almost any crumbling stones will convince you of this, because of the yellow oxide of iron (rust) disclosed.   So by slow degrees this earth became covered with a coating of what we call inorganic soil.   Later, generation after generation of tiny plants and animals which lived in the soil died, and their remains formed the first organic materials of the soil.

This picture shows how the forests help to cover the inorganic soil with an organic coating. Explain how.

You are all familiar with the difference between the so-called rich soil and poor soil.   The dark soil contains more dead plant and animal matter, which forms the portion called *humus*.

Humus contains Organic Matter. — It is an

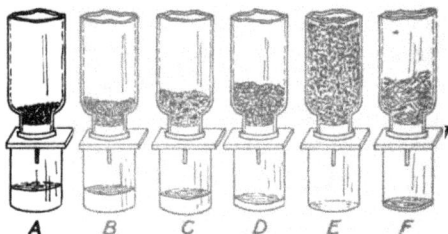

Apparatus for testing the capacity of soils to take in and retain moisture.

easy matter to prove that black soil contains organic matter, for if
an equal weight of carefully dried humus and soil from a sandy road
is heated red-hot for some time and
then reweighed, the humus will be
found to have lost considerably in
weight, and the sandy soil to have
lost very little.   The material left
after heating is inorganic material,
the organic matter having been
burned out.

Soil containing organic materials
holds water much more readily than
inorganic soil, as a glance at the
accompanying figure shows.   If we
fill each of the vessels with a given
weight (say 100 grams each) of
gravel, sand, barren soil, rich loam,
leaf mold, and 25 grams of dry,
pulverized leaves, then pour equal
amounts of water (100 c.c.) on each

Soil particles cling to root hairs.
Why?

and measure all that runs through, the water that has been retained will represent the water supply that plants could draw on
from such soil.

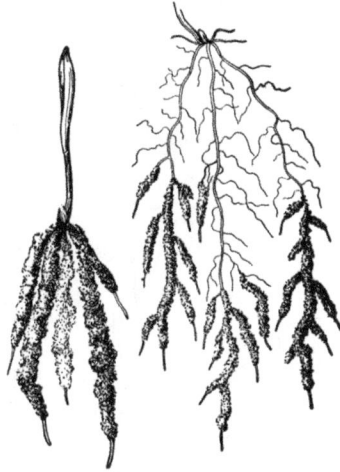

The Root Hairs take more than Water out of the Soil. — If a
root containing a fringe of root hairs is washed carefully, it will be
found to have little particles of soil still clinging to it.   Examined
under the microscope, these particles of soil seem to be cemented
to the sticky surface of the root hair.   The soil contains, besides
a number of chemical compounds of various mineral substances, —
lime, potash, iron, silica, and many others, — a considerable amount
of organic material.   Acids of various kinds are present in the soil.
These acids so act upon certain of the mineral substances that
they become dissolved in the water which is absorbed by the root
hairs.   Root hairs also give off small amounts of acid.   An interesting experiment may be shown (see Figure on page 80) to
prove this.   A solution of *phenolphthalein* loses its color when an
acid is added to it.   If a growing pea be placed in a tube contain-

ing some of this solution the latter will quickly change from a rose pink to a colorless solution.

A Plant needs Mineral Matter to Make Living Matter. — Living matter (protoplasm), besides containing the chemical elements carbon, hydrogen, oxygen, and nitrogen, contains a very minute proportion of various elements which make up the basis of certain minerals. These are calcium (lime), sulphur, iron, potassium, magnesium, phosphorus, sodium, and chlorine.

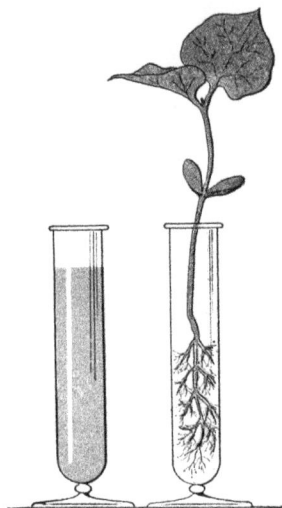

That plants will not grow well without certain of these mineral substances can be proved by the growth of seedlings in a so-called nutrient solution.[1] Such a solution contains all the mineral matter that a plant uses for food. If certain ingredients are left out of this solution, the plants placed in it will not live.

Effect of root hairs on phenolphthalein solution. The change of color indicates the presence of acid.

Nitrogen in a Usable Form necessary for Growth of Plants. — A chemical element needed by the plant to make protoplasm is *nitrogen*. The air can be proven by experiment to be made up of about four fifths nitrogen, but this element cannot be taken from either soil water or air in a pure state, but is usually obtained from the organic matter in the soil, where it exists with other substances in the form of *nitrates*. Ammonia and other organic compounds which contain nitrogen are changed by two groups of little plants called *bacteria*, first into nitrites and then nitrates.[2]

[1] See Hunter's *Laboratory Problems in Civic Biology* for list of ingredients.

[2] It has recently been discovered that under some conditions these bacteria are preyed upon by tiny one-celled animals (*protozoa*) living in the soil and are so reduced in numbers that they cannot do their work effectively. If, then, the soil is heated artificially or treated with antiseptics so as to kill the protozoa, the bacteria which escape multiply so rapidly as to make the land much richer than before.

Relation of Bacteria to Free Nitrogen. — It has been known since the time of the Romans that the growth of clover, peas, beans, and other legumes in soil causes it to become more favorable for growth of other plants. The reason for this has been discovered in late years. On the roots of the plants mentioned are found little swellings or nodules; in the nodules exist millions of bacteria, which take nitrogen from the atmosphere and fix it so that it can be used by the plant; that is, they assist in forming nitrates for the plants to use. Only these bacteria, of all the living plants, have the power to take the free nitrogen from the air and make it over into a form that can be used by the roots. As all the compounds of nitrogen are used over and over again, first by plants, then as food for animals, eventually returning to the soil again, or in part being turned into free nitrogen, it is evident that any *new* supply of usable nitrogen must come by means of these nitrogen-fixing bacteria.

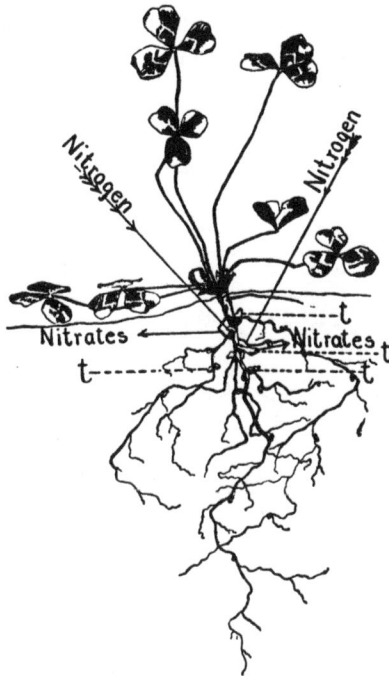

Diagram to show how the nitrogen-fixing bacteria prepare nitrogen for use by plants; *t*, tubercles.

Rotation of Crops. — The facts mentioned above are made use of by careful farmers who wish to make as much as possible from a given area of ground in a given time. Such plants as are hosts for the nitrogen-fixing bacteria are planted early in the season. Later these plants are plowed in and a second crop is planted. The latter grows quickly and luxuriantly because of the nitrates left in the soil by the bacteria which lived with the first crop. For this reason, clover is often grown on land in which it is proposed to plant corn, the nitrogen left in the soil thus giving nourishment to the young corn plants. In scientifically managed farms, different crops are planted in a given field on different years so that one crop may replace some of the elements taken from the soil by the previous crop. This is known as rotation of crops.[1] The annual yield of the average farm may thus be greatly increased.

Five of the elements necessary to the life of the plant which may be taken out of the soil by constant use are calcium, nitrogen, phosphorus, potassium, and sulphur. Several methods are used by the farmer to prevent the exhaustion of these and other raw food materials from the soil. One method known as *fallowing* is to allow the soil to remain idle until bacteria and oxidation have renewed the chemical materials used by the plants. This is an expensive method, if land is dear. The most common method of enriching soil is by means of fertilizing material rich in plant food. Manure is most frequently used, but many artificial fertilizers, most of which contain nitrogen in the form of some nitrate, are used,

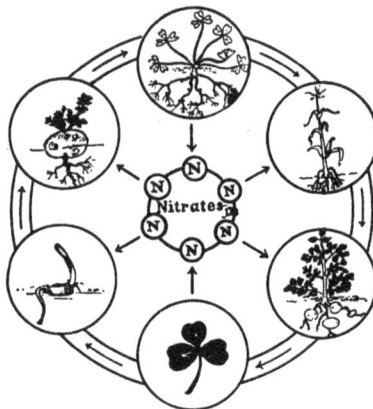

Nitrogen in the soil is necessary for plants. Explain from this diagram how nitrogen is put into the soil by some plants and taken out by others.

because they can be more easily transported and sold. Such are ground bone, guano (bird manure), nitrate of soda, and many others. These also contain other important raw food materials for plants, especially potash and phosphoric acid. Both of these substances are made soluble so as to be taken into the roots by the action of the carbon dioxide in the soil.

The Indirect Relation of this to the City Dweller. — All of us living in the city are aware of the importance of fresh vegetables,

[1] That crop rotation is not primarily a process to conserve the fertility of the soil, but is a sanitary measure to prevent infection of the soil, is the latest belief of the scientist.

brought in from the neighboring market gardens. But we sometimes forget that our great staple crops, wheat and other cereals, potatoes, fruits of all kinds, our cotton crop, and all plants we make use of grow directly in proportion to the amount of raw food materials they take in through the roots. When we also remember that many industries within the cities, as mills, bakeries, and the like, as well as the earnings of our railways and steamship lines, are largely dependent on the abundance of the crops, we may recognize the importance of what we have read in this chapter.

Food Storage in Roots of Commercial Importance. — Some plants, as the parsnip, carrot, and radish, produce no seed until the second year, storing food in the roots the first year and using it to get an early start the following spring, so as to be better able to produce seeds when the time comes. This food storage in roots is of much practical value to mankind. Many of our commonest garden vegetables, as those mentioned above, and the beet, turnip, oyster plant, sweet potato and many others, are of value because of the food stored. The sugar beet has, in Europe especially, become the basis of a great industry.

REFERENCE BOOKS

ELEMENTARY

Hunter, *Laboratory Problems in Civic Biology*. American Book Company.
Bigelow, *Applied Biology*. The Macmillan Company.
Coulter, *Plant Life and Plant Uses*, Chaps. III, IV. American Book Company.
Mayne and Hatch, *High School Agriculture*. American Book Company.
Moore, *The Physiology of Man and Other Animals*. Henry Holt and Company.
Sharpe, *Laboratory Manual in Biology*, pp. 73–87. American Book Company.

ADVANCED

Coulter, Barnes, and Cowles, *A Textbook of Botany*, Part II. Amer. Book Co
Duggar, *Plant Physiology*. The Macmillan Company.
Goodale, *Physiological Botany*. American Book Company.
Green, *Vegetable Physiology*, Chaps. V, VI. J. and A. Churchill.
Kerner-Oliver, *Natural History of Plants*. Henry Holt and Company.
MacDougal, *Plant Physiology*. Longmans, Green, and Company.

VII. PLANT GROWTH AND NUTRITION — PLANTS MAKE FOOD

Problem. — *Where, when, and how green plants make food?*

(a) *How and why is moisture given off from leaves?*

(b) *What is the reaction of leaves to light?*

(c) *What is made in green leaves in the sunlight?*

(d) *What by-products are given off in the above process?*

(e) *Other functions of leaves.*

LABORATORY SUGGESTIONS

Demonstration. — Water given off by plant in sunlight. Loss of weight due to transpiration measured.

Laboratory exercise. —

    (a) Gross structure of a leaf.

    (b) Study of stoma and lower epidermis under microscope.

    (c) Study of cross section to show cells and air spaces.

Demonstration. — Reaction of leaves to light.

Demonstration. — Light necessary to starch making.

Demonstration. — Air necessary to starch making.

Demonstration. — Oxygen a by-product of starch making.

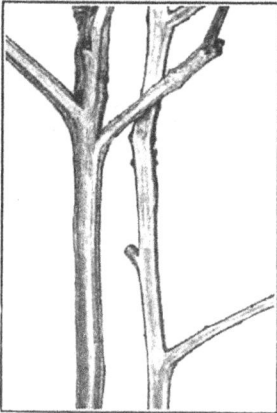

Apple twigs split to show the course of colored water up the stem.

What becomes of the Water taken in by the Roots? — We have seen that more than pure water has been absorbed through the root hairs into the roots. What becomes of this water and the other substances that have been absorbed? This question may be partly answered by the following experiments.

Passage of Fluids up the Stem. — If any young growing shoots (young seedlings of corn or pea, or the older stems of garden balsam, touch-me-not, or sunflower) are placed in red ink (eosin), and left in the sun for a few hours, the red ink will be found to have passed up the stem. If such stems were examined

84

carefully, it would be seen that the colored fluid is confined to collections of woody tubes immediately under the inner bark. Water evidently rises in that part of the stem we call the wood.

Water given off by Evaporation from Leaves. — Take some well-watered potted green plant, as a geranium or hydrangea, cover the pot with sheet rubber, fastening the rubber close to the stem of the plant. Next weigh the plant with the pot. Then cover it with a tall bell jar and place the apparatus in the sun. In a few minutes drops of moisture are seen to gather on the *inside* of the jar. If we now weigh the potted plant, we find it weighs less than before. Obviously the loss comes from the water lost, and evidently this water escapes as vapor from either the stem or leaves.

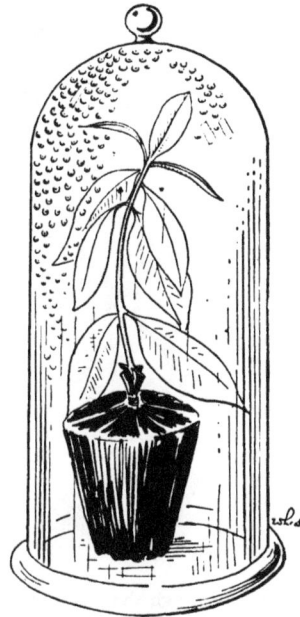

Experiment to prove that water is given off through the leaves of a green plant.

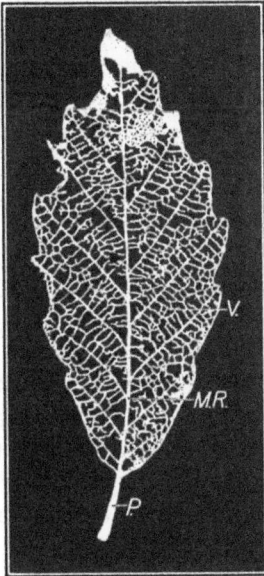

The skeleton of a leaf. *M.R.*, the midrib; *P.*, the leafstalk; *V.*, the veins.

The Structure of a Leaf. — In the experiment with the red ink mentioned above we will find that the fluid has gone out into the skeleton or framework of the leaf. Let us now examine a leaf more carefully. It shows usually (1) a flat, broad *blade*, which may take almost any conceivable shape; (2) a *stem* which spreads out in the blade (3) in a number of *veins*.

The Cell Structure of a Leaf. — The under surface of a leaf seen under the microscope usually shows numbers of tiny oval openings. These are called *stomata* (singular *stoma*). Two cells, usually kidney-shaped, are found, one on each side of the opening. These are the *guard cells*. By change in shape of these cells the opening of the stoma is made larger or smaller. Larger irregular cells form the *epidermis*, or outer covering of the leaf. Study of the leaf in cross section shows that these stomata open directly into air chambers which penetrate between and around the loosely arranged cells composing the underpart of the leaf. The upper surface of leaves sometimes contains stomata, but more often they are lacking. The under surface of an oak leaf of ordinary size contains about 2,000,000 stomata. Under the upper epidermis is a layer of green cells closely packed together (called collectively the *palisade layer*). These cells are more or less columnar in shape. Under these are several rows of rather loosely placed cells just mentioned. These are called collectively the *spongy tissue*. If we happen to have a section cut through a vein, we find this composed of a number of tubes made up of, and strengthened by, thick-walled cells. The veins are evidently a continuation of the tubes of the stem out into the blade of the leaf.

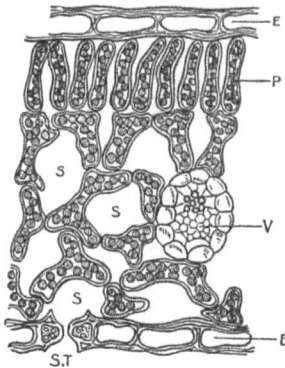

Section through the blade of a leaf as seen under the compound microscope. *S*, air spaces, which communicate with the outside air; *V*, vein in cross section; *S.T.*, breathing hole (stoma); *E*, outer layer of cells; *P*, green cells.

Evaporation of Water. — During the day an enormous amount of water is taken up by the roots and passed out through the leaves. So great is this excess at times that a small grass plant on a summer's day evaporates more than its own weight in water. This would make nearly half a ton of water delivered to the air during twenty-four hours by a grass plot twenty-five by one hundred feet, the size of the average city lot. According to Ward, an oak tree may pass off two hundred and twenty-six times its own weight in water during the season from June to October.

From which Surface of the Leaf is Water Lost? — In order to find out whether water is passed out from any particular part of the leaf, we may remove two leaves of the same size and weight from some large-leaved plant [1] — a mullein was used for the illustrations given below — and cover the upper surface of one leaf and the lower surface of the other with vaseline. The leaf stalks of each should be covered with wax or vaseline, and the two leaves exactly balanced on the pans of a balance which has previously been placed in a warm and sunny place. Within an hour the leaf which has the upper surface covered with vaseline will show a loss of

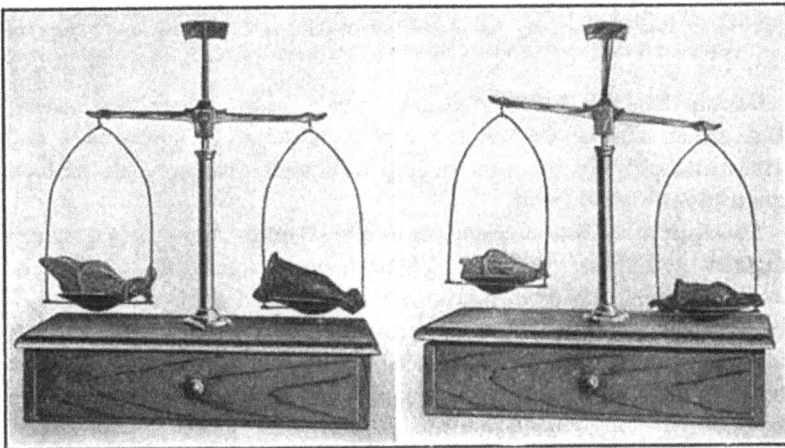

Experiment to show through which surface of a leaf water passes off.

weight. Examination of the surface of a mullein leaf shows us that the *lower surface of the leaf is provided with stomata.* It is through these organs, then, that water is passed out from the tissues of the leaf.

Factors in Transpiration. — The amount of water lost from a plant varies greatly under different conditions. The humidity of the air, its temperature, and the temperature of the plant all affect the rate of transpiration. The stomata also tend to close under some conditions, thus helping to prevent evaporation. But there seems to be no certain regulation of this water loss. Consequently plants droop or wilt on hot dry days because they cannot

[1] The "rubber plant" leaf is an easily obtainable and excellent demonstration.

obtain water rapidly enough from the soil to make up for the loss through the leaves.

Diagrams of a stoma. *a*, surface view of a closed stoma; *b*, the same stoma opened. (After Hanson.) *c*, diagrams of a transverse section through a stoma, dotted lines indicate the closed position of the guard cells, the heavy lines the open condition. (After Schwendener.)

Green Plants Food Makers. — We have previously stated that green plants are the great food makers for themselves and for animals. We are now ready to attack the problem of how green plants *make* food.

The Sun a Source of Energy. — We all know the sun is a source of most of the energy that is released on this earth in the form of heat or light. Every boy knows the power of a " burning glass." Solar engines have not come into any great use as yet, because fuel is cheaper, but some day we undoubtedly will directly harness the energy of the sun in everyday work. Actual experiments have shown that vast amounts of energy are given to the earth. When the sun is highest in the sky, energy equivalent to one hundred horse power is received by a plot of land twenty-five by one hundred feet, the size of a city lot. Plants receive and use much of this energy by means of their leaves.

Effect of Light on Plants. — In young plants which have been grown in total darkness, no green color is found in either stems or leaves, the latter often being reduced to mere scales. The stems are long and more or less reclining. We can explain the changed condition of the seedling grown in the dark only by assuming that light has some effect on the protoplasm of the seedling and induces the growth of the green part of the plant. If seedlings have been growing on a window sill, or where the light comes in from one side, you have doubtless noticed that the stem and leaves of the seedlings incline in the direction from which the light comes.

The experiment pictured shows this effect of light very plainly. A hole was cut in one end of a cigar box and barriers were,erected in the interior of the box so that the seeds planted in the sawdust received their light by an indirect course. The young seedling in this case responded to the influence of the stimulus of light so as to grow out finally through the hole in the box into the open

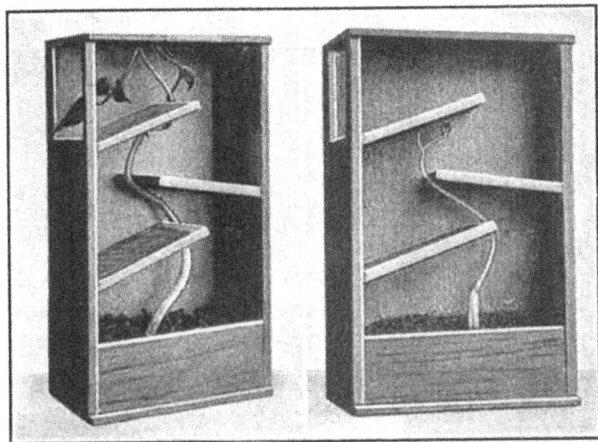

Two stages in an experiment to show that green plants grow toward the light.

air. This growth of the stem to the light is of very great importance to a growing plant, because, as we shall see later, food making depends largely on the amount of sunlight the leaves receive.

Effect of Light on Leaf Arrangement. — It is a matter of common knowledge that green leaves turn toward the light. Place growing pea seedlings, oxalis, or any other plants of rapid growth near a window which receives full sunlight. Within a short time the leaves are found to be in positions to receive the most sunlight possible. Careful observation of any plant growing outdoors shows us that in almost every case the leaves are so disposed as to get much sunlight. The ivy climbing up the wall, the morning-glory, the dandelion, and the burdock all show different arrangements of leaves, each presenting a large surface to the light. Leaves are often definitely arranged, fitting in between one another so as to present their upper surface to the sun. Such an arrangement is known as a *leaf mosaic*. In the case of the dandelion, a *rosette* or whorled cluster of leaves is found. In the horse-chestnut, where the leaves come out opposite each other, the older leaves have longer petioles than the young ones. In the mullein the entire plant forms a cone. The old leaves near the bottom have long stalks, and the little ones near the apex come out close

A lily, showing long narrow leaves.

The dandelion, showing a whorled arrangement of long irregular leaves.

to the main stalk. In every case each leaf receives a large amount of light. Other modifications of these forms may easily be found on any field trip.

Starch made by a Green Leaf. — If we examine the palisade layer of the leaf, we find cells which are almost cylindrical in form. In the protoplasm of such cells are found a number of little green-colored bodies, which are known as *chloroplasts* or *chlorophyll bodies*. If we place the leaf in wood alcohol, we find that the bodies still remain, but that the color is extracted, going into the alcohol and giving to it a beautiful green color. The chloroplasts are, indeed, simply part of the protoplasm of the cell colored green. These bodies are of the greatest importance directly to plants and indirectly to animals. *The chloroplasts, by means of the energy re-*

ceived from the sun, manufacture starch out of certain raw materials.
These raw materials are soil water, which is passed up through
the bundles of tubes into the veins of the leaf from the roots, and
carbon dioxide, which is taken in through the stomata or pores,
which dot the under surface of the leaf. A plant with variegated
leaves, as the *coleus*, makes starch only in the green part of the
leaf, even though these raw materials reach all parts of the leaf.

**Light and Air necessary for
Starch Making.** — If we pin strips
of black cloth, such as alpaca, over
some of the leaves of a growing
hydrangea which has previously
been placed in a dark room for a

An experiment to show the effect of excluding light (but not air) from the
leaves of a green plant. The result of
this experiment is seen in the next
picture. (Experiment performed by
C. Dobbins and A. Schwartz.)

Starchless area in a leaf caused
by excluding sunlight by
means of a strip of black
cloth.

few hours, and then put the plant in direct sunlight for an hour
or two, we are ready to test for starch. We then remove some of
the covered leaves and extract the chlorophyll with wood alcohol
(because the green color of the chlorophyll interferes with the blue
color of the starch test). A test then shows that starch is present
only in the portions of the leaves exposed to sunlight. From this
experiment we infer that the sun has something to do with starch
making in a leaf. The necessity of a part of the air (carbon
dioxide) for starch making may also easily be proved, for the

parts of leaves covered with vaseline will be found to contain no starch, while parts of the leaf without vaseline, but exposed to the sun and air, do contain starch.

Air is necessary for the process of starch making in a leaf, not only because carbon dioxide gas is absorbed (there are from three to four parts in ten thousand present in the atmosphere),

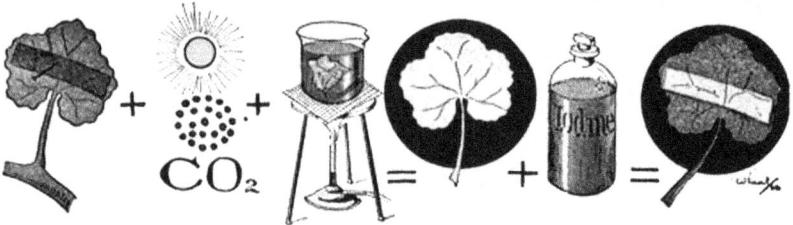

Diagram to show starch making. Read the text carefully and then explain this diagram.

but also because the leaf is alive and must have oxygen in order to do work. This oxygen it takes from the air around it.

Comparison of Starch Making and Milling. — The manufacture of starch by the green leaf is not easily understood. The process has been compared to the milling of grain. In this case the mill is the green part of the leaf. The sun furnishes the motive power, the chloroplasts constitute the machinery, and soil water and carbon dioxide are the raw products taken into the mill. The manufactured product is starch,[1] and a certain by-product (corresponding to the waste in a mill) is also given out. This by-product is oxygen. To

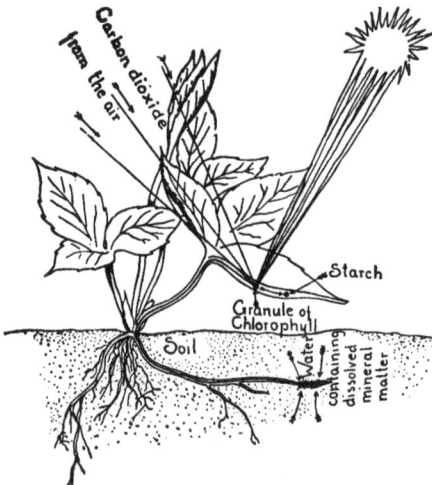

Diagram to illustrate the formation of starch in a leaf.

[1] Sugar is first manufactured and then transformed into starch.

understand the process fully, we must refer to a small portion
of the leaf shown below. Here we find that the cells of the green
layer of the leaf, under the upper epidermis, perform most of
the work. The carbon dioxide is taken in through the stomata
and reaches the green cells by way of the intercellular spaces and
by osmosis from cell to cell. Water reaches the green cells
through the veins. It then passes into the cells by osmosis, and
there becomes part of the cell sap. The light of the sun easily
penetrates to the cells of the palisade layer, giving the energy

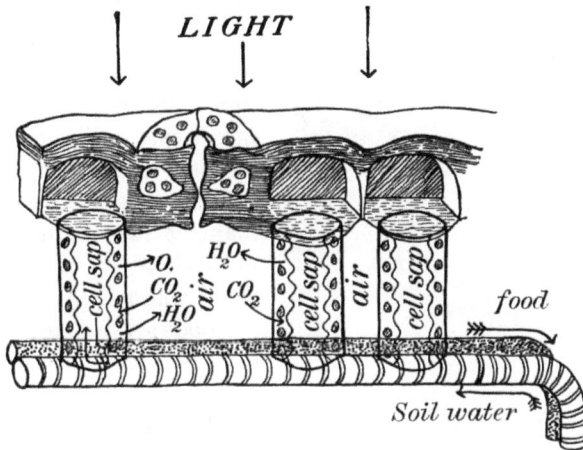

Diagram (after Stevens) to illustrate the processes of breathing and food
making in the cells of a green leaf in the sunlight.

needed to make the starch. This whole process is a very delicate
one, and will take place only when external conditions are favorable.
For example, too much heat or too little heat stops starch making
in the leaf. This building up of food and the release of oxygen
by the plant in the presence of sunlight is called *photosynthesis*.

Manufacture of Fats. — Inasmuch as tiny droplets of oil are
found *inside* the chlorophyll bodies in the leaf, we believe that fats,
too, are made there, probably by a transformation of the starch
already manufactured.

Protein Making and its Relation to the Making of Living Matter.
— Protein material is a food which is necessary to form protoplasm.

Protein food is present in the leaf, and is found in the stem or root as well. Proteins can apparently be manufactured in any of the cells of green plants, the presence of light not seeming to be a necessary factor. How it is manufactured is a matter of conjecture. The minerals brought up in the soil water form part of its composition, and starch or grape sugar give three elements (C, H, and O). The element nitrogen is taken up by the roots as a nitrate (nitrogen in combination with lime or potash). Proteins are probably not made directly into protoplasm in the leaf, but are stored by the cells of the plant and used when needed, either to form new cells in growth or to repair waste. While plants and animals obtain their food in different ways, they probably make it into living substance (*assimilate* it) in exactly the same manner.

An example of how a tree may exert energy. This rock has been split by the growing tree.

Foods serve exactly the same purposes in plants and in animals; they either build living matter or they are burned (oxidized) to furnish energy (power to do work). If you doubt that a plant exerts energy, note how the roots of a tree bore their way through the hardest soil, and how stems or roots of trees often split open the hardest rocks, as illustrated in the figure above.

Starch-Making and its Relation to Human Welfare. — Leaves which have been in darkness show starch to be present soon after exposure to light. A corn plant sends 10 to 15 grams of reserve material into the ears in a single day. The formation of fruit, and especially the growth of the grain fields, show the economic importance of this fact. Not only do plants make their own food and store it away, but they make food for animals as well. And the food is stored in such a stable form that it may be sent to all parts of the world in the form of grain or other fruits. Animals, herbivorous and flesh-eating, man himself, all are dependent upon the starch-making processes of the green plant for the ultimate source of their food. When we remember that in 1913 in the United States the total value of all farm crops was over $6,000,000,000, and when we realize that these products came from the air and soil through the energy of the sun, we may begin to realize why as city boys and girls the study of plant biology is of importance to us.

Green Plants give off Oxygen in Sunlight. — In still another way green plants are of direct use to us in the city. During this process of starch-making oxygen is given off as a by-product. This may easily be proven by the following experiment.[1] Place any green water plant in a battery jar partly filled with water, cover the plants with a glass funnel and mount a test tube full of water over the mouth of the funnel. Then place the apparatus in a warm sunny window. Bubbles of gas are seen to rise from the plant. After two or three hours of hot sun, enough of the gas can be obtained by displacement of the water to make the oxygen test.

That oxygen is given off as a by-product by green plants is a fact of far-reaching importance. City parks are true "breathing spaces." The green covering of the

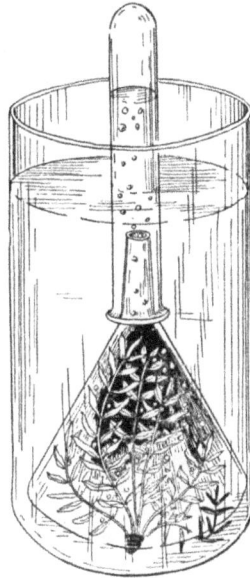

Experiment to show that oxygen is given off by green plants in the sunlight.

earth is giving to animals an element that they must have, while the animals in their turn are supplying to the plants carbon dioxide, a compound used in food-making. Thus a widespread relation of mutual helpfulness exists between plants and animals.

[1] Immediate success with this experiment will be obtained if the water has been previously charged with carbon dioxide.

Respiration by Leaves. — All living things require oxygen. It is by means of the oxidation of food materials within the plant's body that the energy used in growth and movement is released. A plant takes in oxygen largely through the stomata of the leaves, to a less extent through the *lenticels* or breathing holes in the stem, and through the roots. Thus rapidly growing tissues receive the oxygen necessary for them to perform their work. The products of oxidation in the form of carbon dioxide are also passed off through these same organs. It can be shown by experiment that a plant uses up oxygen in the darkness; in the light the amount of oxygen given off as a by-product in the process of starch-making is, of course, much greater than the amount used by the plant.

Summary. — From the above paragraphs it is seen that a leaf performs the following functions: (1) breathing, or the taking in of oxygen and passing off of carbon dioxide; (2) starch-making, with the incidental passing out of oxygen; (3) formation of proteins, with their digestion and assimilation to form new tissues; and (4) the transpiration of water.

REFERENCE BOOKS

ELEMENTARY

Hunter, *Laboratory Problems in Civic Biology*. American Book Company.
Andrews, *A Practical Course in Botany*, pages 160–177. American Book Company.
Coulter, *A Textbook of Botany*, pages 5–40. D. Appleton and Company.
Coulter, *Plant Life and Plant Uses*. American Book Company.
Dana, *Plants and their Children*, pages 135–185. American Book Company.
Sharpe, *A Laboratory Manual in Biology*, pages 90–102. American Book Company.
Stevens, *Introduction to Botany*, pages 81–99. D. C. Heath and Company.

ADVANCED

Clement, *Plant Physiology and Ecology*. Henry Holt and Company.
Coulter, Barnes, and Cowles, *A Textbook of Botany*, Part II, and Vol. II. American Book Company.
Darwin, *Insectivorous Plants*. D. Appleton and Company.
Duggar, *Plant Physiology*. The Macmillan Company.
Goodale, *Physiological Botany*, pages 337–353 and 409–424. American Book Company.
Green, *Vegetable Physiology*. J. and A. Churchill.
Lubbock, *Flowers, Fruits, and Leaves*, last part. The Macmillan Company.
MacDougal, *Practical Textbook of Plant Physiology*. Longmans, Green, and Company.
Report of the Division of Forestry, U. S. Department of Agriculture, 1899.
Ward, *The Oak*. D. Appleton and Company.

VIII. PLANT GROWTH AND NUTRITION—THE CIR-CULATION AND FINAL USES OF FOOD BY PLANTS

Problem.—*How green plants store and use the food they make.*

(a) *What are the organs of circulation?*
(b) *How and where does food circulate?*
(c) *How does the plant assimilate its food?*

LABORATORY SUGGESTIONS

Laboratory exercise. — The structure (cross section) of a woody stem.

Demonstration. — To show that food passes downward in the bark.

Demonstration. — To show the condition of food passing through the stem.

Demonstration. — Plants with special digestive organs.

The Circulation and Final Uses of Foods in Green Plants. — We have seen that cells of green plants make food and that such cells are mostly in the leaves. But *all* parts of the bodies of plants grow. Roots, stems, leaves, flowers, and fruits grow. Seeds are storehouses of food. We must now examine the stem of some plant in order to see how food is distributed, stored, and finally used in the various parts of the plant.

The Structure of a Woody Stem. — If we cut a cross section through a young willow or apple stem, we find it shows three distinct regions. The center is occupied by the spongy, soft *pith;* surrounding this is found the rather tough *wood,* while the outermost area is *bark.* More careful study of the bark reveals the presence of three layers — an outer layer, a middle green layer, and an inner fibrous layer, the latter usually brown in color. This layer is made up largely of tough fiberlike cells known as *bast* fibers. The most important parts of this inner bark, so far as the plant is concerned, are many tubelike structures known as *sieve tubes.* These are long rows of living cells, having perforated sievelike ends. Through these cells food materials pass downward from the upper part of the plant, where they are manufactured.

In the wood will be noticed (see Figure) a number of lines radiating outward from the pith toward the bark. These are thin plates of pith which separate the wood into a number of wedge-shaped masses. These masses of wood are composed of many elongated cells, which, placed end to end, form thousands of little tubes connecting the leaves with the roots. In addition to these are many thick-walled cells, which give strength to the mass of wood. The bundles of tubes with their surrounding hard walled cells are the continuation of the bundles of tubes which are found in the root. In sections of wood which have taken several years to grow, we find so-called *annual rings*. The distance between one ring and the next (see Figure) usually represents the amount of growth in one year. Growth takes place from an actively dividing layer of cells, known as the *cambium layer*. This layer forms wood cells from its inner surface and bark from its outer surface. Thus new wood is formed as a distinct ring around the old wood.

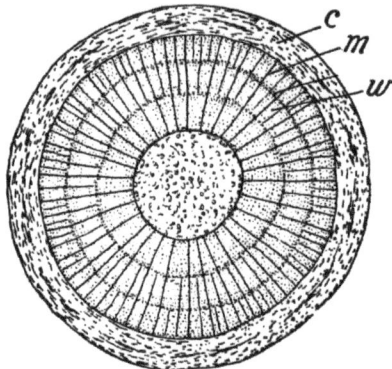

Section of a twig of box elder three years old, showing three annual growth rings. The radiating lines (*m*) which cross the wood (*w*) represent the pith rays, the principal ones extending from the pith in the center to the cortex or bark. (From Coulter's *Plant Relations*.)

Use of the Outer Bark. — The outer bark of a tree is protective. The cells are dead, the heavy woody skeletons serving to keep out cold and dryness, as well as prevent the evaporation of fluids from within. The bark also protects the tree from attack of other plants or animals which might harm it. Most trees are provided with a layer of corky cells. This layer in the cork oak is thick enough to be of commercial importance. The function of the corky layer in preventing evaporation is well seen in the case of the potato, which is a true stem, though found underground. If two potatoes of equal weight are balanced on the scales, the skin having been peeled from one, the peeled potato will be found to lose weight rapidly.   This is due to loss of water, which is held in by the skin of the unpeeled potato (see right hand figure below).

There are also small breathing holes known as *lenticels* scattered through the surface of the bark.   These can easily be seen in a young woody stem of apple, beech, or horse-chestnut.

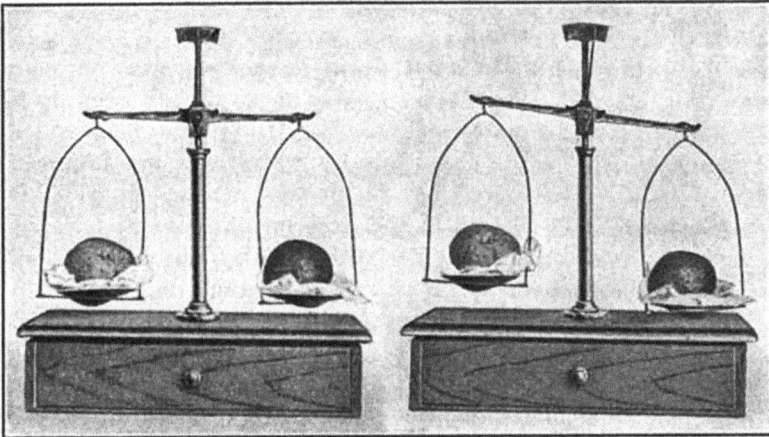

Experiment to show that the skin of the potato (a stem) retards evaporation.

Proof that Food passes down the Stem. — If freshly cut willow twigs are placed in water, roots soon begin to develop from that part of the stem which is under water.   If now the stem is girdled by removing the bark in a ring just above where the roots are growing, the latter will eventually die, and new roots will appear above the girdled area.   The food material necessary for the outgrowth of roots evidently comes from above, and the passage of food materials takes place in a downward direction just outside the wood in the layer of bark which contains the bast fibers and sieve tubes.   This experiment with the willow explains why it is that trees die when girdled so as to cut the sieve tubes of the inner bark.   The food supply is cut off from the protoplasm of the cells in the part of the tree below the cut area.   Many of the canoe birches of our Adirondack forest are thus killed, girdled by thoughtless visitors. In the same manner mice and other gnawing animals kill fruit trees. Food substances are also conducted to a much less extent in the wood itself, and food passes from the inner bark to the center of the tree by way of the pith plates. This can be proved by testing for starch in the pith plates of young stems. It is found that much starch is stored in this part of the tree trunk.

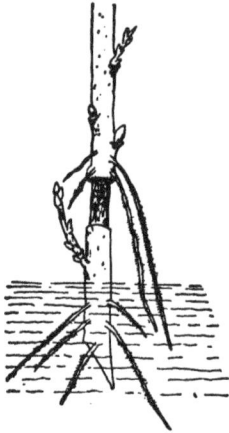

In what Form does Food pass through the Stem? — We have already seen that materials in solution (those substances which will dissolve in the water) will pass from cell to cell by the process of osmosis. This is shown in the experiment illustrated in the figure. Two thistle tubes are partly filled, one with starch and water, the other with sugar and water, and a piece of parchment paper is tied over the end of each. The lower ends of both tubes are placed in a glass dish under water. After twenty-four hours, the water in the dish is tested for starch, and then for sugar. We find that only the sugar, which has been dissolved by the water, can pass through the membrane.

Experiment to show that food material passes down in the inner bark.

Digestion. — Much of the food made in the leaves is stored in the form of starch. But starch, being insoluble, cannot be passed from cell to cell in a plant. It must be changed to a soluble form, for otherwise it could not pass through the delicate cell membranes. This is accomplished by the process of *digestion*. We have already seen that starch is changed to grape sugar in the corn by the action of a substance

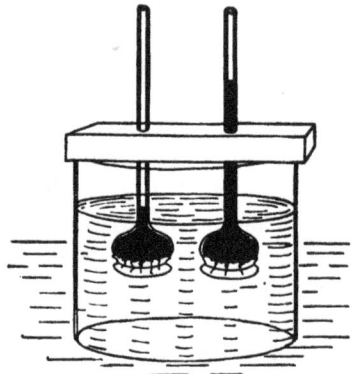

Experiment to show osmosis of sugar (right hand tube) and non-osmosis of starch (left hand tube).

(an enzyme) called *diastase*. This process of digestion seemingly may take place in all living parts of the plant, although most of it is done in the leaves. In the bodies of all animals, including man, starchy foods are changed in a similar manner, but by other enzymes, into soluble grape sugar.

The food material may be passed in a soluble form until it comes to a place where food storage is to take place, then it can be transformed to an insoluble form (starch, for example); later, when needed by the plant in growth, it may again be transformed and sent in a soluble form through the stem to the place where it will be used.

In a similar manner, protein seems to be changed and transferred to various parts of the plant. Some forms of protein substance are *soluble* and others *insoluble* in water. White of egg, for example, is slightly soluble, but can be rendered insoluble by heating it so that it coagulates. Insoluble proteins are digested within the plant; how and where is but slightly understood. In a plant, soluble proteins pass down the sieve tubes in the bast and then may be stored in the bast or medullary rays of the wood in an insoluble form, or they may pass into the fruit or seeds of a plant, and be stored there.

What forces Water up the Stem. — We have seen that the process of osmosis is responsible for taking in soil water, and that the enormous absorbing surface exposed by the root hairs makes possible the absorption of a large amount of water. Frequently this is more than the weight of the plant in every twenty-four hours.

Experiments have been made which show that at certain times in the year this water is in some way forced up the tiny tubes of the stem. During the

Diagram to show the areas in a plant through which the raw food materials pass up the stem and food materials pass down.

spring season, in young and rapidly growing trees, water has been proved to rise to a height of nearly ninety feet. The force that causes this rise of water in stems is known as *root pressure*.

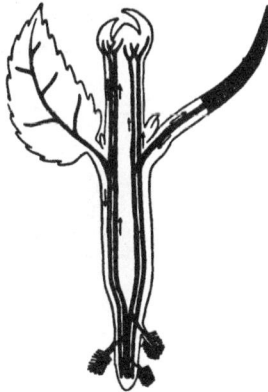

The greatest factor, however, is transpiration of water from leaves. This evaporation of water in the form of vapor seems to result in a kind of suction on the column of water in the stem. In the fall, after the leaves have gone, much less water is taken in by roots, showing that an intimate relation exists between the leaves and the root.

Summary of the Functions of Green Plants. — The processes which we have just described (with the exception of food making) are those which occur in the lives of any plant or animal. All plants and animals breathe, they oxidize their foods to release energy, carbon dioxide being given off as the result of the union of the carbon in the foods with the oxygen of the air. Both plants and animals digest their food; plants may do this in the cells of the root, stem, and leaf. Digestion must always occur so that food can be moved in a soluble condition from cell to cell in the plant's body.

Plants with Special Digestive Organs. — Some plants have special organs of digestion. One of these, the sundew, has leaves which are covered on one side with tiny glandular hairs. These

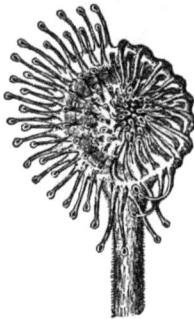

Leaf of sundew closing over
a captured insect.

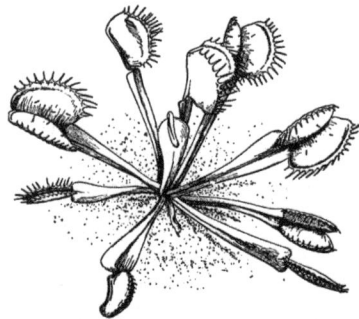

The Venus fly trap, showing open
and closed leaves.

attract insects and later serve to catch and digest the nitrogenous matter of these insects by means of enzymes poured out by the same hairs. Another plant, the Venus fly trap, catches insects in a sensitive leaf which folds up and holds the insect fast until enzymes poured out by the leaf slowly digest it. Still others, called pitcher plants, use as food the decayed bodies of insects which fall into their cuplike leaves and die there. In this respect plants are like those animals which have certain organs in the body set apart for the digestion of food.

Assimilation. — The assimilation of foods, or making of foods into living matter, is a process we know very little about. We know it takes place in the living cells of plants and animals. But how foods are changed into living matter is one of the mysteries of life which we have not yet solved.

Excretion. — The waste and repair of living matter seems to take place in both plants and animals. When living plants breathe, they give off carbon dioxide. In the process of starch-making, oxygen might be considered the waste product. Water is evaporated from leaves and stems. The leaves fall and carry away waste mineral substances which they contain.

Reproduction. — Finally, both plants and animals have organs of reproduction. We have seen that the flower gives rise, after pollination, to a fruit which holds the seeds. These seeds hold

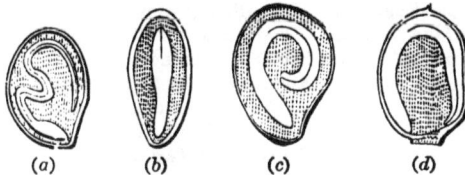

The embryos of (a) the morning glory, (b) the barberry, (c) the potato, (d) the four o'clock, showing the position of their food supply. (After Gray.)

the *embryo*. Thus the young plant is doubly protected for a time and is finally thrown off in the seed with enough food to give it a start in life. In much the same way we will find that animals reproduce, either by laying eggs which contain an *embryo* and food to start it in life or, as in the higher animals, by holding and protecting the embryo within the body of the mother until it is born, a helpless little creature, to be tenderly nourished by the mother until able to care for itself.

The Life Cycle. — Ultimately both plants and animals grow old and die. Some plants, for example the pea or bean, live but a season; others, such as the big trees of California, live for hundreds of years. Some insects exist as adults but a day, while the elephant is said to live almost two hundred years. The span of life from the time the plant or animal begins to grow until it dies is known as the *life cycle*.

REFERENCE BOOKS

ELEMENTARY

Hunter, *Laboratory Problems in Civic Biology*. American Book Company.
Andrews, *A Practical Course in Botany*, pages 112–127. American Book Company.
Atkinson, *First Studies of Plant Life*, Chaps. IV, V, VI, VIII, XXI. Ginn.
Coulter, *Plant Life and Plant Uses*, Chap. V. American Book Company.
Dana, *Plants and their Children*, pages 99–129. American Book Company.
Mayne and Hatch, *High School Agriculture*. American Book Company.
Hodge, *Nature Study and Life*, Chaps. IX, X, XI. Ginn and Company.
MacDougal, *The Nature and Work of Plants*. The Macmillan Company.

ADVANCED

Apgar, *Trees of the United States*, Chaps. II, V, VI. American Book Company.
Coulter, Barnes, and Cowles, *A Textbook of Botany*, Vol. I. American Book Company.
Duggar, *Plant Physiology*. The Macmillan Company.
Ganong, *The Teaching Botanist*. The Macmillan Company.
Goebel, *Organography of Plants*, Part V. Clarendon Press.
Goodale, *Physiological Botany*. American Book Company.
Gray, *Structural Botany*, Chap. V. American Book Company.
Kerner-Oliver, *Natural History of Plants*. Henry Holt and Company.
Strasburger, Noll, Schenck, and Karston, *A Textbook of Botany*. The Macmillan Company.
Ward, *The Oak*. D. Appleton and Company.
Yearbook, U.S. Department of Agriculture, 1894, 1895, 1898–1910.

IX. OUR FORESTS, THEIR USES AND THE NECESSITY FOR THEIR PROTECTION

Problem. — Man's relations to forests.
(a) What is the value of forests to man?
(b) What can man do to prevent forest destruction?

LABORATORY SUGGESTIONS

Demonstration of some uses of wood. Optional exercise on structure of wood. Method of cutting determined by examination. Home work on study of furniture trim, etc.

Visit to Museum to study some economic uses of wood.

Visit to Museum or field trip to learn some common trees.

The Economic Value of Trees. Protection and Regulation of Water Supply. — Trees form a protective covering for parts of

A forest in North Carolina. (U. S. G. S.)

the earth's surface.   They prevent soil from being washed away, and they hold moisture in the ground.   The devastation of immense areas in China and considerable damage by floods in parts of Switzerland, France, and in Pennsylvania has resulted where the forest covering has been removed.   No one who has tramped through our Adirondack forest can escape noticing the differences in the condition of streams surrounded by forest and those which flow through areas from

Working to prevent erosion after the removal of the forest in the French Alps.

which trees have been cut.   The latter streams often dry up entirely in hot weather, while the forest-shaded stream has a never failing supply of crystal water.

The city of New York owes much of its importance to its position at the mouth of a great river with a harbor large enough to float the navies of the world.   This river is supplied with water largely from the Adirondack and Catskill forests.   Should these forests be destroyed, it is not impossible that the frequent freshets which would follow would so fill the Hudson River with silt and débris that the ship channels in the bay,

Erosion at Sayre, Pennsylvania, by the Chemung River.   (Photograph by W. C. Barbour.)

already costing the government hundreds of thousands of dollars a year to keep dredged, would become too shallow for ships.   If this *should* occur, the greatest city in this country would soon lose its place and become of second-rate importance.

The story of how this very thing happened to the old Greek city of Poseidonia is graphically told in the following lines : —

" It was such a strange, tremendous story, that of the Greek Poseidonia, later the Roman Pæstum. Long ago those adventuring mariners from Greece had seized the fertile plain, which at that time was covered with forests of great oak and watered by two clear and shining rivers. They drove the Italian natives back into the distant hills, for the white man's burden even then included the taking of all the desirable things that were being wasted by incompetent natives, and they brought over colonists — whom the philosophers and moralists at home maligned, no doubt, in the same pleasant fashion of our own day. And the colonists cut down the oaks, and plowed the land, and built cities, and made harbors, and finally dusted their busy hands and busy souls of the grime of labor and wrought splendid temples in honor of the benign gods who had given them the possessions of the Italians and filled them with power and fatness.

" Every once in so often the natives looked lustfully down from the hills upon this fatness, made an armed snatch at it, were driven back with bloody contumely, and the heaping of riches upon riches went on. And more and more the oaks were cut down — mark that! for the stories of nations are so inextricably bound up with the stories of trees — until all the plain was cleared and tilled ; and then the foothills were denuded, and the wave of destruction crept up the mountain sides, and they, too, were left naked to the sun and the rains.

" At first these rains, sweeping down torrentially, unhindered by the lost forests, only enriched the plain with the long-hoarded sweetness of the trees ; but by and by the living rivers grew heavy and thick, vomiting mud into the ever shallowing harbors, and the land soured with the un-drained stagnant water. Commerce turned more and more to deeper ports, and mosquitoes began to breed in the brackish soil that was making fast between the city and the sea.

" Who of all those powerful landowners and rich merchants could ever have dreamed that little buzzing insects could sting a great city to death? But they did. Fevers grew more and more prevalent. The malaria haunted population went more and more languidly about their business. The natives, hardy and vigorous in the hills, were but feebly repulsed. Carthage demanded tribute, and Rome took it, and changed the city's name from Poseidonia to Pæstum. After Rome grew weak, Saracen corsairs came in by sea and grasped the slackly defended riches, and the little winged poisoners of the night struck again and again, until grass grew in the streets, and the wharves crumbled where they stood. Finally, the wretched remnant of a great people wandered away into the more wholesome hills, the marshes rotted in the heat and grew up in coarse reeds where corn and vine had flourished, and the city melted back into the wasted earth." [1]

Prevention of Erosion by Covering of Organic Soil. — We have shown how ungoverned streams might dig out soil and carry it

Result of deforestation in China.   This land has been ruined by erosion.
(Carnegie Institution Research in China.)

far from its original source.   Examples of what streams have done may be seen in the deltas formed at the mouths of great rivers. The forest prevents this by holding the water supply and letting it out gradually.   This it does by covering the inorganic soil with humus or decayed organic material.   In this way the forest floor

[1] Elizabeth Bisland and Anne Hoyt, *Seekers in Sicily*.   John Lane Company.

becomes like a sponge, holding water through long periods of drought. The roots of the trees, too, help hold the soil in place. The gradual evaporation of water through the stomata of the leaves cools the atmosphere, and this tends to precipitate the moisture in the air. Eventually the dead bodies of the trees themselves are added to the organic covering, and new trees take their place.

Other Uses of the Forest. — In some localities forests are used as windbreaks and to protect mountain towns against avalanches.

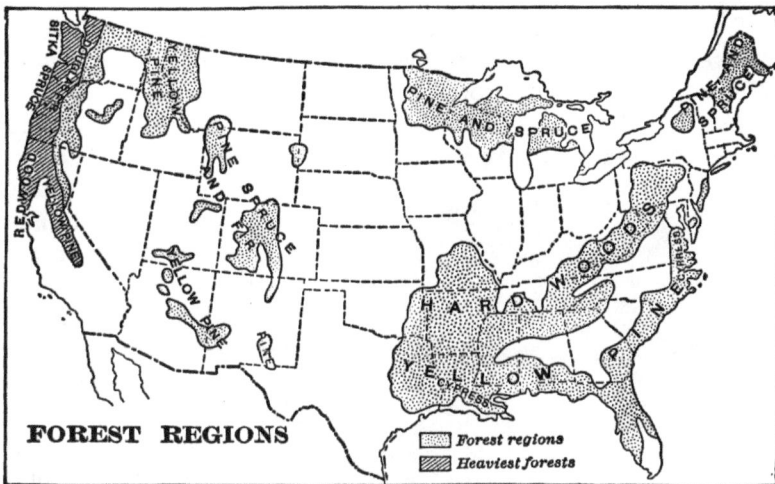

The forest regions of the United States.

In winter they moderate the cold, and in summer reduce the heat and lessen the danger from storms. Birds nesting in the woods protect many valuable plants which otherwise might be destroyed by insects.

Forests have great commercial importance. Pyrogallic and other acids are obtained from trees, as are tar, creosote, resin, turpentine, and many useful oils. The making of maple sirup and sugar forms a profitable industry in several states.

The Forest Regions of the United States. — The combined area of all the forests in the United States, exclusive of Alaska, is about 500,000,000 acres. This seemingly immense area is rapidly decreasing in acreage and in quality, thanks to the demands of an increasing population, a woeful ignorance on the part of the owners of the land, and wastefulness on the part of cutters and users alike.

A glance at the map on page 109 shows the distribution of our principal forests. Washington ranks first in the production of lumber. Here the great Douglas fir, one of the "evergreens," forms the chief source of supply. In the Southern states, especially Louisiana and Mississippi, yellow pine and cypress are the trees most lumbered.

Which states produce the most hardwoods? From which states do we get most of our yellow pine, spruce, red fir, redwood? Where are the heaviest forests of the United States?

Uses of Wood. — Even in this day of coal, wood is still by far the most used fuel. It is useful in building. It outlasts iron under water, in addition to being durable and light. It is cheap and, with care of the forests, inexhaustible, while our mineral wealth may some day be used up. Distilled wood gives wood alcohol. Partially burned wood is charcoal. In our forests much of the soft wood (the cone-bearing trees, spruce, balsam, hemlock, and pine),

Transportation of lumber in the West.
A logging train.

and poplars, aspens, basswood, with some other species, make paper pulp. The daily newspaper and cheap books are responsible for inroads on our forests which cannot well be repaired. It is not necessary to take the largest trees to make pulp wood. Hence many young trees of not more than six inches in diameter are sacrificed. Of the hundreds of species of trees in our forests, the conifers are probably most sought after for lumber. Pine, especially, is probably used more extensively than any other wood. It is used in all heavy construction work, frames of houses, bridges, masts, spars and timber of ships, floors, railway ties, and many other purposes. Cedar is used for shingles, cabinetwork, lead pencils, etc.; hemlock and spruce for heavy timbers and, as we have seen,

Transportation of lumber in the East. Logs are mostly floated down rivers to the mills.

for paper pulp. Another use for our lumber, especially odds and ends of all kinds, is in the packing-box industry. It is estimated that nearly 50 per cent of all lumber cut ultimately finds its way into the construction of boxes. Hemlock bark is used for tanning.

The hard woods — ash, basswood, beech, birch, cherry, chestnut, elm, maple, oak, and walnut —are used largely for the "trim" of our houses, for manufacture of furniture, wagon or car work, and endless other purposes.

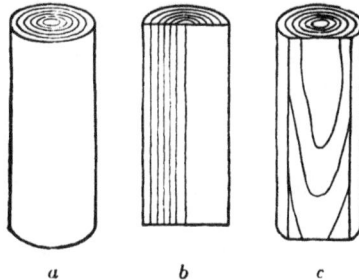

a　　　b　　　c

Diagrams of sections of timber. a, cross section; b, radial; c, tangential. (From Pinchot, U. S. Dept. of Agriculture.)

Methods of cutting Timber. — A glance at the diagram of the sections of timber shows us that a tree may be cut radially through the middle of the trunk or tangentially to the middle portion. Most lumber is cut tangentially. In wood cut in this manner the yearly rings take a more or less irregular course. The grain in wood is caused by the fibers not taking straight lines in their course in the tree trunk. In many cases the fibers of the wood take a spiral course up the trunk, or they may wave outward to form little projections. Boards cut out of such a piece of wood will show the effect seen in many of the school desks, where the annual rings appear to form elliptical markings. Quite a difference in color and structure is often seen between the heartwood, composed of the dead walls of cells occupying the central part of the tree trunk, and the sapwood, the living part of the stem.

Knots. — Knots, as can be seen from the diagram, are branches which at one time started in their outward growth and were for some reason

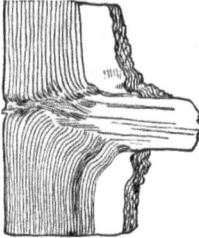

Section of a tree trunk
showing knot.

killed. Later, the tree, continuing in its outward growth, surrounded them and covered them up. A dead limb should be pruned before such growth occurs. The markings in bird's-eye maple are caused by buds which have not developed, and have been overgrown with the wood of the tree.

Destruction of the Forest. — *By Waste in Cutting.* — Man is responsible for the destruction of one of this nation's most valuable assets. This is primarily due to wrong and wasteful lumbering. Hundreds of thousands of dollars' worth of lumber is left to rot annually because the lumbermen do not cut the trees close enough to the ground, or because through careless felling of trees many other smaller trees are injured. There is great waste in the mills. In fact, man wastes in every step from the forest to the finished product.

By Fire. — Indirectly, man is responsible for fire, one of the greatest enemies of the forest. Most of the great forest fires of recent years, the losses from which total in the hundreds of millions, have been due either to railroads or to carelessness in making fires in the woods. It is estimated that in forest lands traversed by railroads from 25 per cent to 90 per cent of the fires are caused by coal-burning locomotives. For this reason laws have been made in New York State requiring locomotives passing through the Adirondack forest preserve to burn oil instead of coal. This has resulted in a considerable reduction in the number of fires. In addition to the loss in timber, the fires often burn out the organic matter in the soil (the " duff ") forming the forest floor, thus preventing the growth of forest there for many years to come. In New York and other states fires are fought by an organized corps

A forest in the far west totally destroyed by fire and wasteful lumbering.

of fire wardens, whose duty it is to watch the forest and to fight forest fires.

Other Enemies. — Other enemies of the forest are numerous fungus plants, insect parasites which bore into the wood or destroy the leaves, and grazing animals, particularly sheep. Wind and snow also annually kill many trees.

Forestry. — In some parts of central Europe, the value of the forests was seen as early as the year 1300 A.D., and many towns consequently bought up the surrounding forests. The city of Zurich has owned forests in its vicinity for at least 600 years and has found them a profitable investment. In this country only recently has the importance of preserving and caring for our forests been noted by our government. Now, however, we have a Forest Survey of the Department of Agriculture and numerous state and university schools of forestry which are rapidly teach-

The forest primeval. Trees are killing each other in the struggle for light and air.

ing the people of this country the best methods for the preservation of our forests. The Federal government has set aside a number of tracts of mountain forest in some of the Western states, making a total area of over 167,000,000 acres. New York has established for the same purpose the Adirondack Park, with nearly 1,500,000 acres of timberland. Pennsylvania has one of 700,000 acres, and many other states have followed their example.

Methods for Keeping and Protecting the Forests. — Forests should be kept thinned. Too many trees are as bad as too few. They struggle with one another for foothold and light, which only a few can enjoy. In cutting the forest, it should be considered as a harvest. The oldest trees are the "ripe grain," the younger trees being left to grow to maturity. Several methods of renewing the forest are in use in this country. (1) Trees may be cut down and young ones allowed to sprout from cut stumps. This is called coppice growth. This growth is well seen in parts of New Jersey. (2) Areas or strips may be cut out so that seeds from neighboring trees are carried there to start new

A German beech forest. The trees are kept thinned out so as to allow the young trees to get a start. Contrast this with the picture above.

growth. (3) Forests may be artificially planted. Two seedlings planted for every tree cut is a rule followed in Europe. (4) The most economical method is that shown in the lower picture on page 114, where the largest trees are thinned out over a large area so as to make room for the younger ones to grow up. The greatest dangers to the forests are from fire and from careless cutting, and these dangers may be kept in check by the efficient work of our national and state foresters.

A City's Need for Trees. — The city of Paris, well known as one of the most beautiful of European capitals, spends over $100,000 annually in caring for and replacing some of the 90,000 trees owned by the city. All over the United States the city governments are beginning to realize what European cities have long known, that trees are of great value to a

We must protect our city trees. This tree was badly wounded by being gnawed by a horse.

city. They are now following the example of European cities by planting trees and by protecting the trees after they are planted. Thousands of city trees are annually killed by horses which gnaw the bark. This may be prevented by proper protection of the trunk by means of screens or wire guards. Chicago has appointed a city forester, who has given the following excellent reasons why trees should be planted in the city : —

(1) Trees are beautiful in form and color, inspiring a constant appreciation of nature.

(2) Trees enhance the beauty of architecture.

(3) Trees create sentiment, love of country, state, city, and home.

(4) Trees have an educational influence upon citizens of all ages, especially children.

(5) Trees encourage outdoor life.

(6) Trees purify the air.

(7) Trees cool the air in summer and radiate warmth in winter.

(8) Trees improve climate and conserve soil and moisture.

(9) Trees furnish resting places and shelter for birds.

(10) Trees increase the value of real estate.

(11) Trees protect the pavement from the heat of the sun.

(12) Trees counteract adverse conditions of city life.

Let us all try to make Arbor Day what it should be, a day for caring for and planting trees, for thus we may preserve this most important heritage of our nation.

REFERENCE BOOKS

ELEMENTARY

Hunter, *Laboratory Problems in Civic Biology.* American Book Company.

Mayne and Hatch, *High School Agriculture.* American Book Company.

Murrill, *Shade Trees*, Bul. 205, Cornell University Agricultural Experiment Station.

Pinchot, *A Primer of Forestry*, Division of Forestry, U.S. Department of Agriculture.

ADVANCED

Apgar, *Trees of the United States*, Chaps. II, V, VI. American Book Company.

Coulter, Barnes, and Cowles, *A Textbook of Botany*, Part I and Vol. II. American Book Company.

Goebel, *Organography of Plants*, Part V. Clarendon Press.

Strasburger, Noll, Schenck, and Karston, *A Textbook of Botany.* The Macmillan Company.

Ward, *Timber and Some of its Diseases.* The Macmillan Company.

Yearbook, U.S. Department of Agriculture, Division of Forestry, Buls. 7, 10, 13, 16, 17, 18, 20, 26, 27.

X. THE ECONOMIC RELATION OF GREEN PLANTS TO MAN

Problems.—*How green plants are useful to man.*

(a) *As food.*

(b) *For clothing.*

(c) *Other uses.*

How green plants are harmful to man.

SUGGESTED LABORATORY WORK

If a commercial museum is available, a trip should be planned to work over the topics in this chapter. The school collection may well include most of the examples mentioned, both of useful and harmful plants.

A study of weeds and poisonous plants should be taken up in actual laboratory work, either by collection and identification or by demonstration.

Green Plants have a " Dollar and Cents " Value. — To the girl or boy living in the city green plants seem to have little direct value. Although we see vegetables for sale in stores and we know that fruits have a money value, we are apt to forget that the wealth of our nation depends more upon its crops than it does on its manufactories and business houses. The economic or " dollars and cents " value of plants is enormous and far too great for us to comprehend in terms of figures.

We have already seen some of the uses to mankind of the products of the forest; let us now consider some other plant products.

Leaves as Food. — Grazing animals feed almost entirely on tender shoots or leaves, blades of grass, and other herbage. Certain leaves and buds are used by man as food. Lettuce, beet tops, kale, spinach, broccoli, are examples. A cabbage head is nothing but a big bud which has been cultivated by

117

man. An onion is a compact budlike mass of thickened leaves which contain stored food.

Cabbage   Onions   Lettuce

Leaves used as food.

Stems as Food. — A city child would, if asked to name some stem used as food, probably mention asparagus. We sometimes forget that one of our greatest necessities, cane sugar, comes from the stem of sugar cane. Over seventy pounds of sugar is used each year by every person in the United States. To supply the growing demand beets are now being raised for their sugar in many parts of the world, so that nearly half the total supply of sugar comes from this source. Maple sugar is a well-known commodity which is obtained by boiling the sap of sugar maple until it crystallizes. Over 16,000 tons of maple sugar is obtained every spring, Vermont producing about 40 per cent of the total output. The sago palm is another stem

Celery   Kohl-rabi   Potato   Sugar cane

Stems used as food.

which supports the life of many natives in Africa. Another stem, living underground, forms one of man's staple articles of diet. This is the potato.

Roots as Food. — Roots which store food for plants form important parts of man's vegetable diet. Beets, radishes, carrots, parsnips, sweet potatoes, and many others might be mentioned.

The following table shows the proportion of foods in some of the commoner roots and stems : —

	WATER	PROTEINS	CARBO-HYDRATES	FAT	MINERAL MATTER
Potato	75	1.2	18	0.3	1.0
Carrot	89	0.5	5	0.2	1.0
Parsnip	81	1.2	8.7	1.5	1.0
Turnip	92.8	0.5	4.	0.1	0.8
Onion	91	1.5	4.8	0.2	0.5
Sweet potato	74	1.5	20.2	0.1	1.5
Beet	82.2	0.4	13.4	0.1	0.9

Fruits and Seeds as Foods. — Our cereal crops, corn, wheat, etc., have played a very great part in the civilization of man and are now of so much importance to him as food products that bread

Wheat   Nuts   Pear   Melon

Seeds and fruits used for food.

made from flour from the wheat has been called the " staff of life." Our grains are the cultivated progeny of wild grasses. Domestication of plants and animals marks epochs in the advance of civilization. The man of the stone age hunted wild beasts for food, and lived like one of them in a cave or wherever he happened to be; he was a nomad, a wanderer, with no fixed home. He may have discovered that wild roots or grains were good to eat; perhaps he stored some away for future use. Then came the idea of growing things at home instead of digging or gathering the wild fruits from the forest and plain. The tribes which first cultivated the soil made a great step in advance, for they had as a result a fixed place for habitation. The cultivation of grains and cereals gave them a store of food which could be used at times when other food was scarce. The word " cereal " (derived from Ceres, the Roman Goddess of Agriculture) shows the importance of this crop to Roman civilization. From earliest times the growing of grain and the progress of civilization have gone hand in hand. As nations have advanced in power, their dependence upon the cereal crops has been greater and greater.

" Indian corn," says John Fiske, in *The Discovery of America,* " has played a most important part in the history of the New World. It could be planted without clearing or plowing the soil. There was no need of threshing or winnowing. Sown in tilled land, it yields more than twice as much food per acre as any other kind of grain. This was of incalculable advantage to the English settlers in New England, who would have found it much harder to gain a secure foothold upon the soil if they had had to begin by preparing it for wheat or rye."

To-day, in spite of the great wealth which comes from our mineral resources, live stock, and manufactured products, the surest index of our country's prosperity is the size of the corn and wheat crop. According to the last census, the amount of capital invested in agriculture was over $20,000,000,000, while that invested in manufacture was less than one half that amount.

Corn. — About three billion bushels of corn were raised in the United States during the year 1910. This figure is so enormous that it has but little meaning to us. In the past half century our corn crop has increased over 350 per cent. Illinois and Iowa are the greatest corn-producing states, each having a yearly record of over four hundred million bushels.   The figure on this page shows the principal corn-producing areas in the United States.

Indian corn is put to many uses.   It is a valuable food.   It contains a large proportion of starch, from which glucose (grape sugar) and alcohol are made.   Machine oil and soap are made from it. The leaves and stalk are an excellent fodder; they can be made into paper and packing material.   Mattresses can be stuffed with

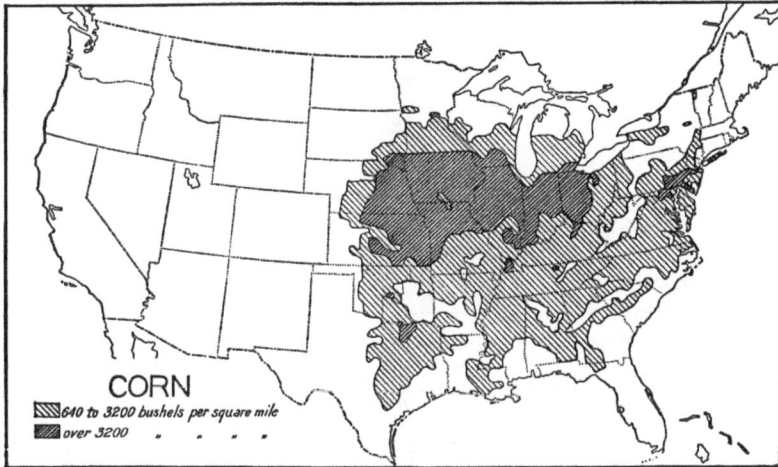

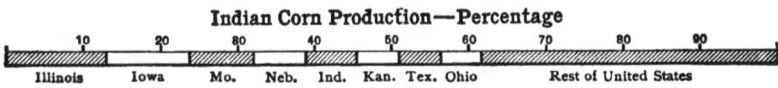

CORN
§§§640 to 3200 bushels per square mile
▨▨ over 3200   „   „   „   „

Indian Corn Production—Percentage

| Illinois | Iowa | Mo. | Neb. | Ind. | Kan. | Tex. | Ohio | Rest of United States |

the husks.   The pith is used as a protective belt placed below the water line of our huge battleships.   Corn cobs are used for fuel, one hundred bushels having the fuel value of a ton of coal.

Wheat. — Wheat is the crop of next greatest importance in size. Nearly seven hundred millions of bushels were raised in this country in 1910, representing a total money value of over $700,-000,000.   Seventy-two per cent of all the wheat raised comes from the North Central states and California.   About three fourths of the wheat crop is exported, nearly one half of it to Great Britain, thus indirectly giving employment to thousands of people on railways and steamships.   Wheat has its chief use in its manufacture into flour. The germ, or young wheat plant, is sifted out during this process and made into breakfast foods. Flour making forms

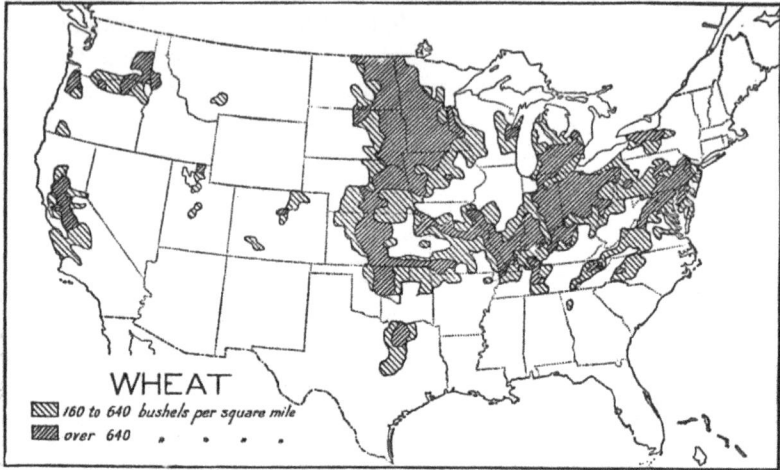

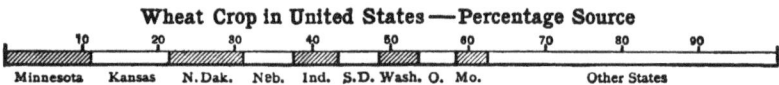

WHEAT

▨ 160 to 640 bushels per square mile
▨ over 640 , . . .

Wheat Crop in United States — Percentage Source

10 20 30 40 50 60 70 80 90

Minnesota Kansas N. Dak. Neb. Ind. S.D. Wash. O. Mo.   Other States

the chief industry of Minneapolis, Minnesota, and of several other large and wealthy cities in this country.

Other Grains. — Of the other grain and cereals raised in this country, oats are the most important crop, over one billion bushels having been produced in 1910. Barley is another grain, a staple of some of the northern countries of Europe and Asia. In this country, it is largely used in making malt for the manufacture of beer. Rye is the most important cereal crop of northern Europe, Russia, Germany, and Austro-Hungary producing over 50 per cent of the world's supply. One of the most important grain crops for the world (although relatively unimportant in the United States) is rice. The fruit of this grasslike plant, after thrashing, screening, and milling, forms the principal food of one third of the human race. Moreover, its stems furnish straw, its husks make a bran used as food for cattle, and the grain, when fermented and distilled, yields alcohol.

A field of rice, showing the conditions of culture.

Garden Fruits. — Green plants and especially vegetables have come to play an important part in the dietary of man. The diseases known as scurvy and beri-beri, the latter the curse of the far Eastern navies, have been largely prevented by adding vegetables and fruit juices to the dietary of the sailors. People in this country are beginning to find that more vegetables and less meat are better than the meat diet so often used. Market gardening forms the lucrative business of many thousands of people near our great cities. Some of the more important fruits are squash, cucumbers, pumpkins, melons, tomatoes, peppers, strawberries, raspberries, and blackberries. The latter fruits bring in an annual income of $25,000,000 to our market gardeners. Beans and peas are important as foods because of their relatively large amount of protein. Canning green corn, peas, beans, and tomatoes has become an important industry.

Orchard and Other Fruits. — In the United States over one hundred and seventy-five million bushels of apples are grown every year. Pears, plums, apricots, peaches, and nectarines also form large orchards, especially in California. Nuts form one of our important articles of food, largely because of the large amount of protein contained in them.

Picking apples, an important crop in some parts of the United States.

The grape crop of the world is commercially valuable, because of the raisins and wine produced. The culture of lemons, oranges, and grapefruit has come in recent years to give a living to many people in this country as well as in other parts of the world. Figs, olives, and dates are staple foods in the Mediterranean countries and are sources of wealth to the people there, as are coconuts, bananas, and many other fruits in tropical countries.

Beverages and Condiments. — The coffee and cacao beans, and leaves of the tea plant, products of tropical regions, form the basis of very important beverages of civilized man. Pepper, black and red, mustard, allspice, nutmegs, cloves, and vanilla are all products manufactured from various fruits or seeds of tropical plants.

Alcoholic liquors are produced from various plants in different parts of the world, the dried fruit of the hop vine being an important product of New York State used in the making of beer.

Raw Materials. — Besides use as food, green plants have many other uses. Many of our city industries would not be in existence, were it not for certain plant products which furnish the raw materials for many manufacturing industries. Many cities of the east and south, for example, depend upon cotton to give employment to thousands of factory hands.

Cotton. — Of our native plant products cotton is probably of the most importance to the outside world. Over eleven million bales of five hundred pounds each are raised annually.

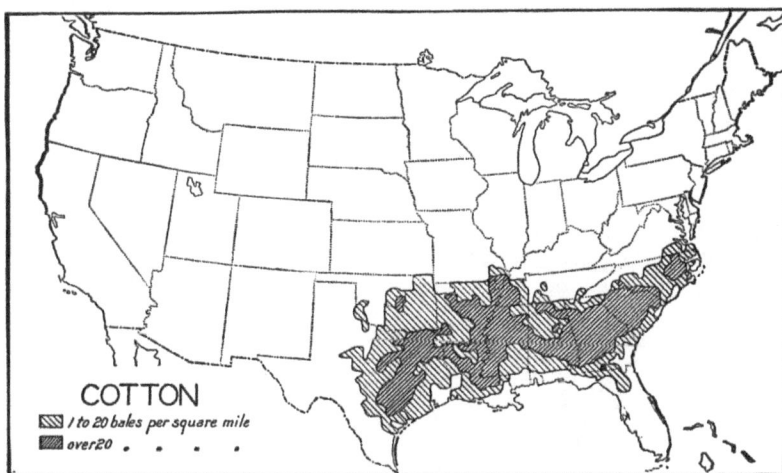

COTTON
/ to 20 bales per square mile
over 20

Cotton Crop in United States — Percentage Source

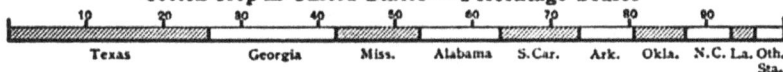

10	20	30	40	50	60	70	80	90

Texas   Georgia   Miss.   Alabama   S. Car.   Ark.   Okla.   N. C. La. Oth. Sta.

Cotton Crop in United States — Percentage Consumption

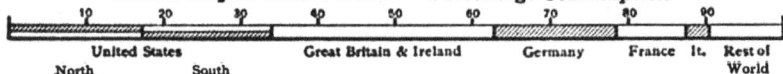

10	20	30	40	50	60	70	80	90

United States   Great Britain & Ireland   Germany   France It. Rest of
North   South   World

The cotton plant thrives in warm regions. Its commercial importance is gained because the seeds of the fruit have long filaments attached to them. Bunches of these filaments, after treatment, are easily twisted into threads from which are manufactured cotton cloth, muslin, calico, and cambric. In addition to the fiber, cottonseed oil, a substitute for olive oil, is made from the seeds, and the refuse remaining makes an excellent cattle fodder.

Cotton Boll Weevil. — The cotton crop of the United States has rather recently been threatened with destruction by a beetle called the cotton boll weevil. This insect, which bores into the young

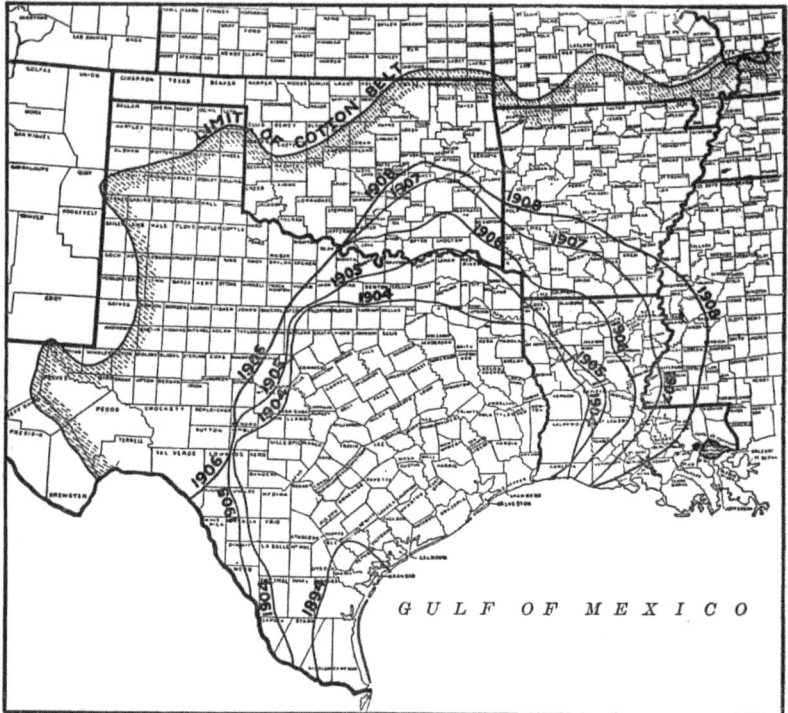

Map showing the spread of the cotton boll weevil. It was introduced from Mexico about 1894. What proportion of the cotton raising belt was infected in 1908?

pod of the cotton, develops there, stunting the growth of the fruit to such an extent that seeds are not produced. The loss in Texas alone is estimated at over $10,000,000 a year. The boll weevil, because of the protection offered by the cotton boll, is very difficult to exterminate. The weevils are destroyed by birds, the infected bolls and stalks are burnt, millions are killed each winter by cold, other insects prey on them, but at the present time they are one of the greatest pests the south knows.

The control of this pest seems to depend upon early planting so that the crop has an opportunity to ripen before the insects in the boll grow large enough to do harm. Ultimately the boll weevil

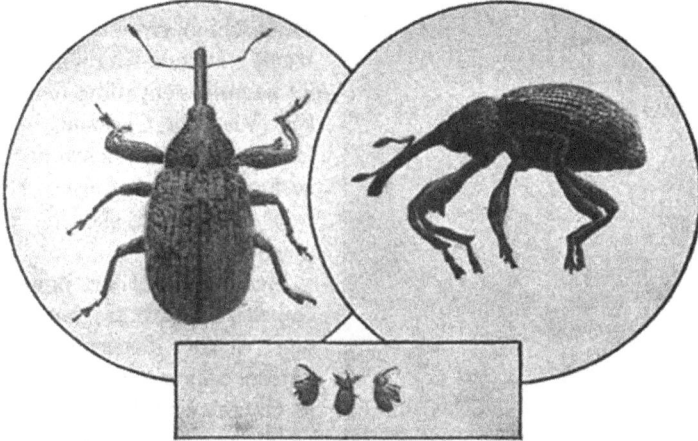

Mexican cotton boll weevil. Much enlarged, above; natural size, below. (Herrick.)

may do more good than harm by bringing into the market a type of cotton plant that ripens very early.

Vegetable Fibers. — Among the most important are Manila hemp, which comes from the leaf-stalks of a plant of the banana family and true hemp, which is the bast or woody fiber of a plant cultivated in most warm parts of the earth. Flax is also an important fiber plant, grown largely in Russia and other parts of Europe (see picture on next page). From the bast fibers of the stem of this herb linen cloth is made.

Vegetable Oils. — Some of the same plants which give fiber also produce oil. Cotton seed oil pressed from the seeds, linseed oil from the seeds of the flax plant, and coconut oil (the covering of the nut here producing the fiber) are examples.

Some Harmful Green Plants. — We have seen that on the whole green plants are useful to man. There are, however, some that

Flax grown for fiber.

are harmful. For example, the poison ivy is extremely poisonous to touch. The poison ivy is a climbing plant which attaches itself to the trees or walls by means of tiny air roots which grow out from the stem. It is distinguished from its harmless climbing neighbor, the Virginia Creeper, by the fact that its leaves are notched in *threes* instead of *fives*. Every boy and girl should know poison ivy.

Numerous other poisonous common plants are found, but one other deserves special notice because of its presence in vacant city lots. The Jim-

son Weed (*Datura*) is a bushy plant, from two to five feet high, bearing large leaves. It has white or purplish flowers, and later bears a four-valved seed pod containing several hundred seeds. These plants contain a powerful poison, and people are often made seriously ill by eating the roots or other parts by mistake.

Weeds. — From the economic standpoint the green plants which

Poison ivy, a climbing plant which is poisonous to touch. Notice the leaves in threes.

do the greatest damage are weeds. Those plants which provide best for their young are usually the most successful in life's race. Plants which combine with the ability to scatter many seeds over a wide territory the additional characteristics of rapid growth, resistance to dangers of extreme cold or heat, attacks of enemies, inedibility, and peculiar adaptations to cross-pollination or self-pollination, are usually spoken of as weeds. They flourish in the sterile soil of the roadside and in the fertile soil of the garden. By means of rapid growth they kill other plants of slower growth by usurping their territory. Slow-growing plants are thus actually exterminated. Many of our common weeds have been introduced from other countries and have, through their numerous adaptations, driven out other plants which stood in their way. Such is the Russian Thistle. A single plant of this kind will give rise to over 20,000 seeds. First introduced from Russia in 1873, it spread so rapidly that in twenty years it had appeared as a common weed over an area of some twenty-five thousand square miles. It is now one of the greatest pests in our Northwest.

REFERENCE BOOKS

ELEMENTARY

Hunter, *Laboratory Problems in Civic Biology*. American Book Company.
Gannet, *Commercial Geography*. American Book Company.
Sargent, *Plants and their Uses*. Henry Holt and Company.
Toothaker, *Commercial Raw Materials*. Ginn and Company.
U. S. Dept. of Agriculture, Farmers' Bulletin 86, *Thirty Poisonous Plants of the United States*, V. K. Chestnut. Bulletin 17. *Two Hundred Weeds, How to Know Them and How to Kill Them*, L. H. Dewey.

ADVANCED

Bailey, *Cyclopedia of American Agriculture*. The Macmillan Company.

XI. PLANTS WITHOUT CHLOROPHYLL IN THEIR RELATION TO MAN

Problems. — (*a*) *How molds and other saprophytic fungi do harm to man.*

(*b*) *What yeasts do for mankind.*

(*c*) *A study of bacteria with reference to*

(*1*) *Conditions favorable and unfavorable to growth.*

(*2*) *Their relations to mankind.*

(*3*) *Some methods of fighting harmful bacteria and diseases caused by them.*

LABORATORY SUGGESTIONS

Field work. — Presence of bracket fungi and chestnut canker.

Home experiment. — Conditions favorable to growth of mold.

Laboratory demonstration. — Growth of mold, structure, drawing.

Home experiment or laboratory demonstration. — Conditions unfavorable for growth of molds.

Demonstration. — Process of fermentation.

Microscopic demonstration. — Growing yeast cells. Drawing.

Home experiment. — Conditions favorable for growth of yeast.

Home experiment. — Conditions favorable for growth of yeast in bread.

Demonstration and experiment. — Where bacteria may be found.

Demonstration. — Methods of growth of bacteria, pure cultures and colonies shown.

Demonstration. — Foods preferred by bacteria.

Demonstration. — Conditions favorable for growth of bacteria.

Demonstration. — Conditions unfavorable for growth of bacteria.

Demonstration by charts, diagrams, etc. — The relation of bacteria to disease in a large city.

COLORLESS PLANTS ARE USEFUL AND HARMFUL TO MAN

The Fungi. — We have found that green plants on the whole are useful to mankind. But not all plants are green. Most of us are familiar with the edible mushroom sold in the markets or the so-called "toadstools" found in parks or lawns. These plants contain no chlorophyll and hence do not make their own food. They are members of the plant group called *fungi*. Such plants are almost as much dependent upon the green plants for food as are animals. But the fungi require for the most part dead organic matter for their food. This may be obtained from decayed vegetable or animal material in soil, from the bodies of dead plants and animals, or even from foods prepared for man. Fungi which feed upon *dead* organic material are known as *saprophytes*. Examples are the mushrooms, the yeasts, molds, and some bacteria, of which more will be learned later.

Some Parasitic Fungi. — Other fungi (and we will find this applies to some animals as well) prefer *living* plants or animals for their food. Thus a tiny plant, recently introduced into this country, known as the chestnut canker, is killing our chestnut trees by the thousands in the eastern part of the United States. It produces millions of tiny reproductive cells known as *spores;* these spores, blown about by the wind, light on the trees, sprout, and send in under the bark a thread-like structure which sucks in the food circulating in the living cells, eventually causing the death of the tree. *A plant or animal which lives at the expense of another living plant or animal is called a parasite.*

Chestnut trees in a New York City park; killed by a parasite, the chestnut canker.

The chestnut canker is a dangerous parasite. Later we shall see that animal and plant parasites destroy yearly crops and trees valued at hundreds of millions of dollars and cause untold misery and suffering to humanity.

Another fungus which does much harm to the few trees found in large towns and cities is the shelf or bracket fungus. The part of the body visible on the tree looks like a shelf or bracket, hence the name. This bracket is in reality the reproductive part of the plant; on its lower surface are formed millions of little bodies called *spores*. These spores are capable, under favorable conditions, of reproducing new plants. The true body of the plant, a network of threads, is found under the bark. This fungus begins its life as a spore in some part of the tree which has become *diseased* or *broken*. Once established, it spreads rapidly. There is no remedy except to kill the tree and burn it, so as to destroy the spores. Many fine trees, sound except for a slight bruise or other injury, are annually infected and eventually killed. In cities thousands of trees become infected through careless hitching of horses so that the horse may gnaw the tree, thus exposing a fresh surface on which spores may obtain lodgment and grow (see page 115).

Shelf fungi.
(Photographed by W. C. Barbour.)

Suggestions for Field Work. — A field trip to a park or grove near home may show the great destruction of timber by this means. Count the number of perfect trees in a given area. Compare it with the number of trees attacked by the fungus. Does the fungus appear to be transmitted from one tree to another near at hand? In how many instances can you discover the point where the fungus first attacked the tree?

Fungi of our Homes. — But not all fungi are wild. Some have become introduced into our homes and these live on food or other materials. *These plants are very important because of their relation to life in a town or crowded city.*[1]

[1] Experiments on conditions favorable to growth of mold should be introduced here.

The Growth of Bread Mold. — If a piece of moist bread is exposed to the air of the schoolroom, or in your own kitchen for a few minutes and then covered with a glass tumbler and kept in a warm place, in a day or two a fuzzy whitish growth will appear on the surface of the bread.   This growth shortly turns black.   If we now examine a little piece of the bread with a lens or low-powered microscope, we find a tangled mass of threads (the *mycelium*) covering the surface of the bread. From this mass of threads project tiny upright stalks bearing round black bodies, the fruit. Little rootlike structures known as *rhizoids* dip down into the bread, and absorb food for its threadlike body.   The upright

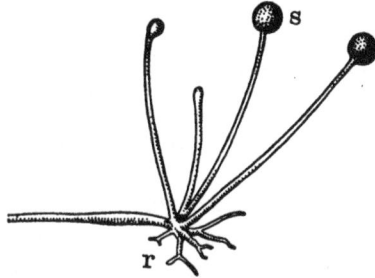

Bread mold; *r*, rhizoids; *s*, fruiting bodies containing spores.

threads with the balls at the end contain many tiny bodies called *spores*.   These spores have been formed by the division of the protoplasm making up the fruiting bodies into many separate cells.   When grown under favorable conditions, the spores will produce more mycelia, which in turn bear fruiting bodies.

Physiology of the Growth of Mold. — Molds, in order to grow rapidly, need oxygen, moisture, and moderate heat.   They seem to prefer dark, damp places where there is not a free circulation of air, for if the bell jar is removed from growing mold for even a short time, the mold wilts.   Too great or very little heat will prevent growth and kill everything except the spores. They obtain their food from the material on which they live. This they are able to do by means of digestive enzymes given out by the rootlike parts, by means of which the molds cling to the bread.   These digestive enzymes change the starch of the bread to sugar and the protein to a soluble form which will pass by osmosis into cells of the mold.   Thus the mold is able to absorb food material.   These foods are then used to supply energy and make protoplasm.   This seems to be the usual method by which saprophytes make use of the materials on which they live.

What can Molds live On? — We have seen that black mold lives upon bread. We would find that it or some other mold (*e.g.* green or blue mold) live upon decaying or overripe fruit, — apples, peaches, and plums being especially susceptible to their growth. Molds feed upon all cakes or breads, upon meat, cheese, and many raw vegetables. They are almost sure to grow upon flour if it is allowed to get damp. Moisture seems necessary for their growth. Jelly is a substance particularly favorable to molds for this reason. Shoes, leather, cloth, paper, or. even moist wood will give food enough to support their growth. At least one troublesome disease, *ringworm*, is due to the growth of molds in the skin.

What Mold does to Foods. — Mold usually changes the taste of the material it grows upon, rendering it " musty " and sometimes unfit to eat. Eventually it will spoil food completely because decay sets in. Decay, as we will see later, is not entirely due to mold growth, but is usually caused by another group of organisms, the *bacteria*. Molds, however, in feeding *do* cause chemical changes which result in decay or putrefaction. Some molds are useful. They give the flavor to Roquefort, Gorgonzola, Camembert, and Brie cheeses. But on the whole molds are pests which the housekeeper wishes to get rid of.

How to prevent Molds.[1] — As we have seen, moisture is favorable for mold growth ; conversely, dryness is unfavorable. Inasmuch as the spores of mold abound in the air, materials which cannot be kept dry should be covered. Jelly after it is made should at once be tightly covered with a thin layer of paraffin, which excludes the air and possible mold spores. Or waxed paper may be fastened over the surface of the jelly so as to exclude the spores. To prevent molds from attacking fresh fruit, the surface of the fruit should be kept dry and, if possible, each piece of fruit should be wrapped in paper. Why? Heating with dry heat to 212° for a few moments will kill any mold spores that happen to be in food. Moldy food, if heated after removing surface on which the mold grew, is perfectly good to eat.

[1] An experiment to show conditions unfavorable for growth of molds should be shown at this point.

Dry dusting or sweeping will raise dust, which usually contains mold spores. Use a dampened broom or dust cloth frequently in the kitchen if you wish to preserve foods from molds.

Other Moldlike Fungi. — Mildews are near relatives of the molds found in our homes. They may attack leather, cloth, etc., in a damp house. Other allied forms may do damage to living plants. Some of these live upon the lilac, rose, or willow. These fungi do not penetrate the host plant to any depth, for they obtain their food from the outer layer of cells in the leaf of their host and cover the leaves with the whitish threads of the mycelium. Hence they may be killed by means of applications of some fungus-killing fluid, as Bordeaux mixture.[1] Among the useful plants preyed upon by mildews are the plum, cherry, and peach trees. (The diseases known as black knot and peach curl are thus caused.) Another important member of this group is the tiny parasite found on rye and other grains, which gives us the drug ergot.

Among other parasitic fungi are rusts and smuts. Wheat rust is probably the most destructive parasitic fungus. Indirectly this parasite is of considerable importance to the citizen of a great city because of its effect upon the price of wheat.

Yeasts in their Relation to Man

Fermentation. — It is of common knowledge to country boys or girls that the juice of fresh apples, grapes, and some other fruits, if allowed to stand exposed to the air for a short time will *ferment*. That is, the sweet juice will begin to taste sour and to have a peculiar odor, which we recognize as that of alcohol. The fermenting juice appears to be full of bubbles which rise to the surface. If we collect enough of these bubbles of gas to make a test, we find it to be carbon dioxide.

Evidently something changed some part of the apple or grape, the sugar, ($C_6H_{12}O_6$), into alcohol, $2\,(C_2H_6O)$, and carbon dioxide, $2\,(CO_2)$. This chemical process is known as *fermentation*.

[1] See Goff and Mayne, *First Principles of Agriculture*, page 59, for formula of Bordeaux mixture.

Yeast causes Fermentation. — Let us now take a compressed yeast cake, shake up a small portion of it in a solution of molasses and water, and fill a fermentation tube with the mixture. Leave the tube in a warm place overnight. In the morning a gas will be found to have been collected in the closed end of the tube (see Figure on page **138**). The taste and odor of the liquid shows alcohol to be present, and the gas, if tested, is proven carbon dioxide. Evidently yeast causes fermentation.

Apparatus to show effect of fermentation. *N*, molasses, water and yeast plants; *C*, bubbles of carbon dioxide.

What are Yeasts? — If now part of the liquid from the fermentation tube which contains the settlings be drawn off, a drop placed on a slide and a little weak iodine added and the mixture examined under the compound microscope, two kinds of structures will be found (see Figure below), starch grains which are stained deep blue, and other smaller ovoid structures of a brownish yellow color. The latter are yeast plants.

Size and Shape, Manner of Growth, etc. — The common compressed yeast cake contains millions of these tiny plants. In its simplest form a yeast plant is a single cell. The shape of such a plant is ovoid, each cell showing under the microscope the granular appearance of the protoplasm of which it is formed. Look for tiny clear areas in the cells;

Yeast and starch grains. Notice that the starch grains around which are clustered yeast cells have been rounded off by the yeast plants. How do you account for this?

these are vacuoles, or spaces filled with fluid. The nucleus is hard to find in a yeast cell. Many of the cells seem to have others

attached to them, sometimes there being several in a row. Yeast cells reproduce very rapidly by a process of budding, a part of the parent cell forming one or more smaller daughter cells which eventually become free from the parent.

Conditions favorable to growth of Yeast. — *Experiment.* — Label three pint fruit jars A, B, and C. Add one fourth of a compressed yeast cake to two cups of water containing two tablespoonfuls of molasses or sugar. Stir the mixture well and divide it into three equal parts and pour them into the jars. Place covers on the jars. Put jar A in the ice box on the ice, and jar B over the kitchen stove or near a radiator; pour the contents of jar C into a small pan and boil for a few minutes. Pour back into C, cover and place it next to B. After forty-eight hours, look to see if any bubbles have made their appearance in any of the jars. If the experiment has been successful, only jar B will show bubbles. After bubbles have begun to appear at the surface, the fluid in jar B will be found to have a sour taste and will smell unpleasantly. The gas which rises to the surface, if collected and tested, will be found to be carbon dioxide. The contents of jar B have fermented. Evidently, the growth of yeast will take place only under conditions of moderate warmth and moisture.

Carbohydrates necessary to Fermentation. — Sugar must be present in order for fermentation to take place. The wild yeasts cause fermentation of the apple or grape juice because they live on the skin of the apple or grape. Various peoples recognize this when they collect the juice of certain fruits and, exposing it to the air, allow it to ferment. Such is the *saki* or rice wine of the Japanese, the *tuba* or sap of the coconut palm of the Filipinos and the *pulqué* of the Mexicans.

Beer and Wine Making. — Brewers' yeasts are cultivated with the greatest care; for the different flavors of beer seem to depend largely upon the condition of the yeast plants. Beer is made in the following manner. Sprouted barley, called malt, in which the starch of the grain has been changed to grape sugar by digestion, is killed by drying in a hot kiln. The malt is dissolved in water, and hops are added to give the mixture a bitter taste. Now comes the addition of the yeast plants, which multiply rapidly under the favorable conditions of food and heat. Fermentation results on a large scale from the breaking down of the grape sugar, the alcohol remaining in the fluid, and the carbon dioxide passing off into the air. At the right time the beer is stored either in bottles or casks, but fermentation slowly continues, forming carbon dioxide in the bottles. This gives the sparkle to beer when it is poured from the bottle.

In wine making the wild yeasts growing on the skin of the grapes set up a slow fermentation. It takes several weeks before the wine is ready to bottle. In sparkling wines a second fermentation in the bottles gives rise to carbon dioxide in such quantity as to cause a decided frothing when the bottle is opened.

Commercial Yeast. — Cultivated yeasts are now supplied in the home as compressed or dried yeast cakes. In both cases the yeast plants are mixed with starch and other substances and pressed into a cake. But the compressed yeast cake must be used fresh, as the yeast plants begin to die rapidly after two or three days. The dried yeast cake, while it contains a much smaller number of yeast plants, is nevertheless probably more reliable if the yeast cannot be obtained fresh.

The cut illustrates an experiment that shows how yeast plants depend upon food in order to grow. In each of three fermentation tubes were placed an equal amount of a compressed yeast cake. Then tube *a* was filled with distilled water, tube *b* with a solution of glucose and water, and tube *c* with a nutrient solution containing nitrogenous matter as well as glucose. The quantity of gas (CO_2) in each tube is an index of the amount of growth of the yeast cells. In which tube did the greatest growth take place ?

Bread Making. — Most of us are familiar with the process of bread making. The materials used are flour, milk or water or both, salt, a little sugar to hasten the process of fermentation, or " *rising*," as it is called, some butter or lard, and yeast.

After mixing the materials thoroughly by a process called "kneading," the bread is put aside in a warm place (about 75° Fahrenheit) to " rise." If we examine the dough at this time, we find it filled with holes, which give the mass a spongy appearance. The yeast plants, owing to favorable conditions, have grown rapidly and filled the cavities with carbon dioxide. Alcohol is present, too, but this is evaporated when the dough is baked. The baking cooks the starch of the bread, drives off the carbon dioxide and alcohol, and kills the yeast plants, besides forming a protective crust on the loaf.

Sour Bread. — If yeast cakes are not fresh, sour bread may result from their use. In such yeast cakes there are apt to be present other tiny one-celled plants, known as *bacteria*. Certain of these plants form acids after fermentation takes place. The sour taste of the bread is usually due to this cause. The remedy would be to have fresh yeast, to have good and fresh flour, and to have clean vessels with which to work.

Importance of Yeasts. — Yeasts in their relation to man are thus seen to be for the most part useful. They may get into canned substances put up in sugar and cause them to "work," giving them a peculiar flavor. But they can be easily killed by heating to the temperature of boiling. On the other hand, yeast plants are necessary for the existence of all the great industries which depend upon fermentation. And best of all they give us leavened bread, which has become a necessity to most of mankind.

BACTERIA IN THEIR RELATION TO MAN

What Bacteria do and Where They May be Found. — A walk through a crowded city street on any warm day makes one fully alive to odors which pervade the atmosphere. Some of these unpleasant odors, if traced, are found to come from garbage pails, from piles of decaying fruit or vegetables, or from some butcher shop in which decayed meat is allowed to stand. This characteristic phenomena of decay is one of the numerous ways in which we can detect the presence of bacteria. These tiny plants, "man's invisible friends and foes," are to be found "anywhere, but not everywhere," in nature. They swarm in stale milk, in impure water, in soil, in the living bodies of plants and animals and in their dead bodies as well. Most "catching" diseases we know to be caused directly by them; the processes of decay, souring of milk, acid fermentation, the manufacture of nitrogen for plants are directly or indirectly due to their presence. It will be the purpose of the next paragraphs to find some of the places where bacteria may be found and how we may know of their presence.

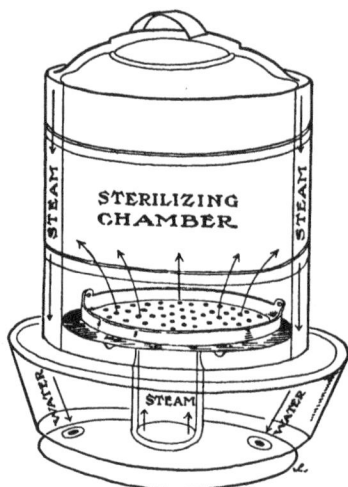

A steam sterilizer.

How we catch Bacteria to Study Them. — To study bacteria it is first necessary to find some material in which they will grow, then kill all living matter in this food material by heating to boiling point (212°) for half an hour or more (this is called *sterilization*), and finally protect the *culture medium*, as this food is called, from other living things that might grow upon it.

One material in which bacteria seem to thrive is a mixture of beef extract, digested protein and gelatine or agar-agar, the latter a preparation derived from seaweed. This mixture, after sterilization, is poured into flat dishes with loose-fitting covers. These *petri* dishes, so called after their inventor, are the traps in which we collect and study bacteria.

Where Bacteria might Grow. — Expose a number of these sterilized dishes, each for the same length of time, to some of the following conditions:

(a) exposed to the air of the schoolroom.

(b) exposed in the halls of the school while pupils are passing.

(c) exposed in the halls of the school when pupils are not moving.

(d) exposed at the level of a dirty and much-used city street.

(e) exposed at the level of a well-swept and little-used city street.

(f) exposed in a city park.

(g) exposed in a factory building.

(h) dirt from hands placed in dish.

(i) rub interior of mouth with finger and touch surface of dish.

(j) touch surface of dish with decayed vegetable or meat.

(k) touch surface of dish with dirty coin or bill.

(l) place in dish two or three hairs from boy's head.

This list might be prolonged indefinitely.

Now let us place all of the dishes together in a moderately warm place (a closet in the schoolroom will do) and watch for results. After a day or two little spots, brown, yellow, white, or red, will begin to appear. These spots, which grow larger day by day, are *colonies* made up of millions of bacteria. But probably each colony arose from a single bacterium which got into the dish when it was exposed to the air.

How we may isolate Bacteria of Certain Kinds from Others. — In order to get a number of bacteria of a given kind to study, it becomes necessary to grow them in what is

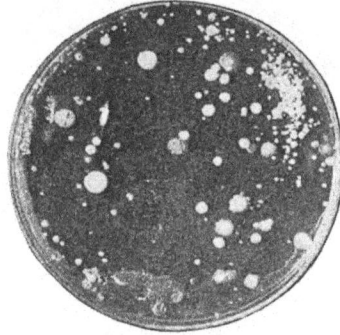

Colonies of bacteria growing in a petri dish.

known as a pure culture. This is done by first growing the bacteria in some medium such as beef broth, gelatin, or on potato.[1] Then as growth follows the colonies of bacteria appear in the culture media or the beef broth becomes cloudy. If now we wish to study one given form, it becomes necessary to isolate them from the others. This is done by the following process: a platinum needle is first passed through a flame to *sterilize* it; that is, to kill all living things that may be on the needle point.

---

[1] For directions for making a culture medium, see Hunter, *Laboratory Problems in Civic Biology*. Culture tubes may be obtained, already prepared, from Parke, Davis, and Company or other good chemists.

Then the needle, which cools very quickly, is dipped in a colony containing the bacteria we wish to study. This mass of bacteria is quickly transferred to another sterilized plate, and this plate is immediately covered to prevent any other forms of bacteria from entering. When we have succeeded in isolating a certain kind of bacterium in a given dish, we are said to have a *pure culture*. Having obtained a pure culture of bacteria, they may easily be studied under the compound microscope.

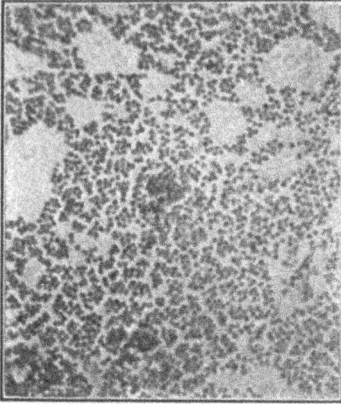

A pure culture of bacteria. Notice that the bacteria are all the same size and shape.

Size and Form. — In size, bacteria are the most minute plants known. A bacterium of average size is about $\frac{1}{10000}$ of an inch in length, and perhaps $\frac{1}{50000}$ of an inch in diameter. Some species are much larger, others smaller. A common spherical form is $\frac{1}{50000}$ of an inch in diameter. They are so small that several million are often found in a single drop of impure water or sour milk. Three well-defined forms of bacteria are recognized: a spherical form called a *coccus*, a rod-shaped bacterium, the *bacillus*, and a spiral form, the *spirillum*. Some bacteria are capable of movement when living in a fluid. Such movement is caused by tiny lashlike threads of protoplasm called *flagella*. The flagella project from the body, and by a rapid movement cause locomotion to take place. Bacteria reproduce with almost incredible rapidity. It is estimated that a single bacterium, by a process of division called *fission*, will give rise to over 16,700,000 others in twenty-four hours. Under unfavorable conditions they stop dividing and form rounded bodies called spores. This spore is usually protected by a wall and may withstand very unfavorable conditions of dryness or heat; even boiling for several minutes will not kill some forms.

Where Bacteria are most Numerous. — As the result of our experiments, we can make some generalizations concerning the presence of bacteria in our own environment. They are evidently present in the air, and in greater quantity in air that is moving than quiet air. Why? That they stick to particles of dust can be proven by placing a little dust from the schoolroom in a culture dish. Bacteria are present in greater numbers where crowds of people live and move, the air from dusty streets of a populous city contains many more bacteria than does the air of a village street. The air of a city park contains relatively few bacteria as compared with the near-by street. The air of the woods or high mountains fewer still. Why? Our previous experiment has shown that dirt on our hands, the mouth and teeth, decayed meat and vegetables, dirty money, the very hairs of our head are all carriers of bacteria.

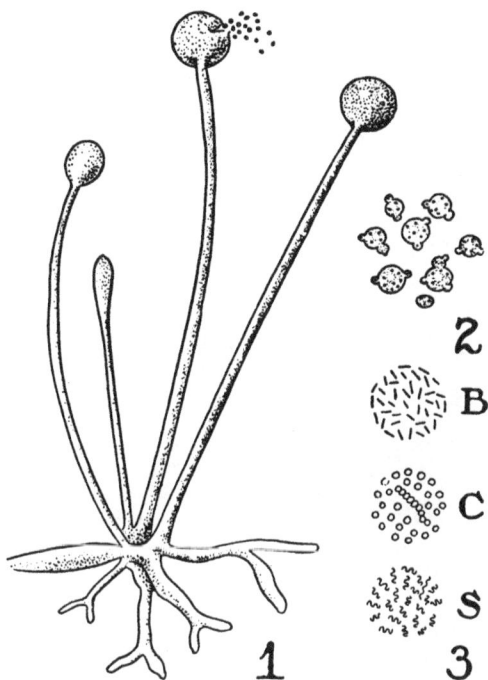

A figure to show the relative size and shape of (1) a black mold, (2) yeast cells, and (3) different forms of bacteria; B, bacillus; C, coccus; S, spirillum forms. The yeast and bacteria are drawn to scale, they are much enlarged in proportion to the black mold, being actually much smaller than the mold spores seen at the top of the picture.

Fluids the Favorite Home of Bacteria. — Tap water, standing water, milk, vinegar, wine, cider all can be proven to contain bacteria by experiments similar to those quoted above. Spring or artesian well water would have very few, if any, bacteria, while the same quantity of river water, if it held any sewage, might contain untold millions of these little organisms.

Foods preferred by Bacteria. — If bacteria are living and contain no chlorophyll, we should expect them to obtain protein food in order to grow. Such is not always the case, for some bacteria seem to be able to build up protein out of simple inorganic nitrogenous substances. If, however, we take several food substances, some containing much protein and others not so much, we will find that the bacteria cause decay in the proteins almost at once, while other food substances are not always attacked by them.

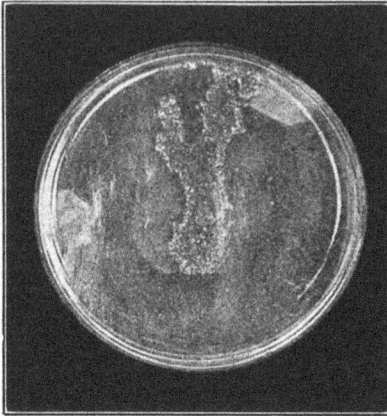

Growth of bacteria in a drop of impure water allowed to run down a sterilized culture in a dish.

What Bacteria do to Foods. — When bacteria feed upon a protein they use part of the materials in the food so that it falls to pieces and eventually rots. The material left behind after the bacteria have finished their meal is quite different from its original form. It is broken down by the action of the bacteria into gases, fluids, and some solids. It has a characteristic "rotten" odor and it has in it poisons which come as a result of the work of the bacteria. These poisonous wastes, called *ptomaines*, we shall learn more about later.

Conditions Favorable and Unfavorable to the Growth of Bacteria. — **Moisture and Dryness.** — *Experiment.* — Take two beans, remove the skins crush one, soak the second bean overnight and then crush it. Place in test tubes, one dry, the second with water. Leave in a warm place two or three days, then smell each tube. In which is decay taking place? In which tube are bacteria at work? How do you know?

Moisture. — Moisture is an absolute need for bacterial growth, consequently keeping material dry will prevent the growth of germs upon its surface. Foods, in order to decay, must contain enough water to make them moist. Bacteria grow most freely in fluids.

Light. — If we cover one half of a petri dish in which bacteria are growing with black paper and then place the dish in a light warm place for a few days, the growth of bacteria in the light part of the dish will be found to be checked, while growth continues in the covered part. It is a matter of common knowledge that disease germs thrive where dirt and darkness exist and are killed by any long exposure to sunlight. This shows us the need of light in our homes, especially in our bedrooms.

Air. — We have seen that plants need oxygen in order to perform the work that they do. This is equally true of all animals. But not all bacteria need *air* to live; in fact, some are killed by the presence of air. Just how these organisms get the oxygen necessary to oxidize their food is not well understood. The fact that some bacteria grow without air makes it necessary for us to use the one sure weapon we have for their extermination, and that is heat.

Heat. — *Experiment.* — Take four cultures containing bouillon, inoculate each tube with bacteria and plug each tube with absorbent cotton. Place one tube in the ice box, a second tube in a dark closet at a moderate temperature, a third in a warm place (about 100° Fahrenheit), and boil the contents of the fourth tube for ten minutes, then place it with tube number two. In which tubes does growth take place most rapidly? Why?

Bacteria grow very slowly if at all in the temperature of an ice box, very rapidly at the room temperature of from 70° to 90° and much less rapidly at a higher temperature. All bacteria except those which have formed spores can be instantly killed as soon as boiling point is reached, and most spores are killed by a few minutes boiling.

Sterilization. — The practical lessons drawn from *sterilization* are many. We know enough now to boil our drinking water if we are uncertain of its purity; we sterilize any foods that we believe might harbor bacteria, and thus keep them from spoiling. The industry of canning is built upon the principle of sterilization.

Canning. — Canning is simply a method by which first the bacteria in a substance are killed by heating and then the substance is put into vessels into which no more bacteria may gain entrance. This is usually done at home by boiling the fruit

HUNTER, CIV. BI. — 10

or vegetable to be canned either in salt and water or with sugar and water, either of which substances aids in preventing the growth of bacteria. The time of boiling will be long or short, depending upon the materials to be canned. Some vegetables, as peas, beans, and corn, are very difficult to can, probably because of spores of bacteria which may be attached to them. Fruits, on the other hand, are usually much easier to preserve. After boiling for the proper time, the food, now free from all bacteria, must be put into jars or cans that are themselves absolutely *sterile* or free from germs. This is done by first boiling the jars, then pouring the boiling hot material into the hot jars and sealing them so as to prevent the entrance of bacteria later.

Uses of Canning. — Canning as an industry is of immense importance to mankind. Not only does it provide him with fruits and vegetables at times when he could not otherwise get them, but it also cheapens the cost of such things. It prevents the waste of nature's products at a time when she is most lavish with them, enabling man to store them and utilize them later. Canning has completely changed the life of the sailor and the soldier, who in former times used to suffer from various diseases caused by lack of a proper balance of food.

Pasteurizing milk. Why should this be done?

Pasteurization. — Milk is one of the most important food supplies of a great city. It is also one of the most difficult supplies to get in good condition. This is in part due to the fact that milk is produced at long distances from the city and must be brought first from farms to the railroads, then shipped by train, again taken to the milk supply depot by wagon, there bottled, and again shipped by delivery wagons to the consumers. When we remember that much of the milk used in New York City is forty-eight hours old and when we realize that bacteria grow *very* rapidly in milk, we see the need of finding some way to protect the supply so as to make it safe, particularly for babies and young children.

This is done by *pasteurization*, a method named after the French bacteriologist Louis Pasteur. To pasteurize milk we heat it to a temperature of not over 170° Fahrenheit for from ten minutes to half an hour. By such a process all harmful germs will be killed and the keeping qualities of the milk greatly lengthened. Most large milk companies pasteurize their city supply by a rapid pasteurization at a much higher temperature, but this method slightly changes the flavor of the milk.

Cold Storage. — Man has also come to use cold to keep bacteria from growing in foods. The ice box at home and cold storage on a larger scale enables one to keep foods for a more or less lengthy period. If food is frozen, as in cold storage, it might keep without growth of bacteria for years. But fruits and vegetables cannot be frozen without spoiling their flavor. And all foods after freezing seem particularly susceptible to the bacteria of decay. For that reason products taken from cold storage must be used at once.

Ptomaines. — Many foods get their flavor from the growth of molds or bacteria in them. Cheese, butter, the gamey taste of certain meats, the flavor of sauerkraut, are all due to the work of bacteria. But if bacteria are allowed to grow so as to become very numerous, the ptomaines which result from their growth in foods may poison the person eating such foods. Frequently ptomaine poisoning occurs in the summer time because of the rapid growth of bacteria. Much of the indigestion and diarrhœa which attack people during the summer is doubtless due to this kind of poisoning.

Preservatives.[1] — This leads us to ask if we may not preserve food in ways other than those mentioned so as to protect ourselves from danger of ptomaine poisoning. Many substances check the development of bacteria and in this way they *preserve*

[1] Perform experiment here to determine the value of different preservatives. Use sugar, salt, vinegar, boracic acid, benzoic acid, formaldehyde, and alcohol.

the food. Preservatives are of two kinds, those harmless to man and those that are poisonous. Of the former, salt and sugar are examples; of the latter, formaldehyde and possibly benzoic acid.

Sugar. — We have noted the use of sugar in canning. Small amounts of sugar will be readily attacked by yeasts, molds, and bacteria, but a 40 to 50 per cent solution will effectually keep out bacteria. Preserves are fruits boiled in about their own weight of sugar. Condensed milk is preserved by the sugar added to it; so are candied and, in part, dried fruits.

Salt. — Salt has been used for centuries to keep foods. Meats are smoked, dried, and salted; some are put down in strong salt solutions. Fish, especially cod and herring, are dried and salted. The keeping of butter is also due to the salt mixed with it. Vinegar is another preservative. It, like salt, changes the flavor of materials kept in it and so cannot come into wide use. Spices are also used as preservatives.

Harmful Preservatives. — Certain chemicals and drugs, used as preservatives, seem to be on the border line of harmfulness. Such are benzoic acid, borax, or boracic acid. Such drugs *may* be harmless in small quantities, but unfortunately in canned goods we do not always know the amount used. The national government in 1906 passed what is known as the Pure Food Law, which makes it illegal to use any of these preservatives (excepting benzoic acid in very small amounts). Food which contains this preservative will be so labeled and should not be given to children or people with weak digestion. Unfortunately people do not always read the labels and thus the pure food law is ineffective in its working. Infrequently formaldehyde or other preservatives are used in milk. Such treatment renders milk unfit for ordinary use and is an illegal process.

Disinfectants.[1] — Frequently it becomes necessary to destroy bacteria which cause diseases of various kinds. This process is called *disinfecting*. The substances commonly used are carbolic acid, formalin or formaldehyde, lysol, and bichloride of mercury.

---

[1] Experiment to determine the most effective disinfectants. Use tubes of bouillon containing different strength solutions of formaldehyde, lysol, iodine, carbolic acid, and bichloride of mercury. Results. Conclusions.

Of these, the last named is the most powerful as well as the most dangerous to use. As it attacks metal, it should not be used in a metal pail or dish. It is commonly put up in tablets which are mixed to form a 1 to 1000 solution. Such tablets should be carefully safeguarded because of possible accidental poisoning.

Formaldehyde used in liquid form is an excellent disinfectant. When burned in a formalin candle, it sets free an intensely pungent gas which is often used for disinfecting sick rooms after the patient has been removed.

Carbolic acid is perhaps the best disinfectant of all. If used in a solution of about 1 part to 25 of water, it will not burn the skin. It is of particular value to disinfect skin wounds, as it heals as well as cleanses when used in a weak solution. Its rather pleasant odor makes it useful to cover up unpleasant smells of the sick room.

The fumes of burning sulphur, which are so often used for disinfecting, are of little real value.

Bacteria cause Decay. — Let us next see in what ways the bacteria directly influence man upon the earth. Have you ever stopped to consider what life would be

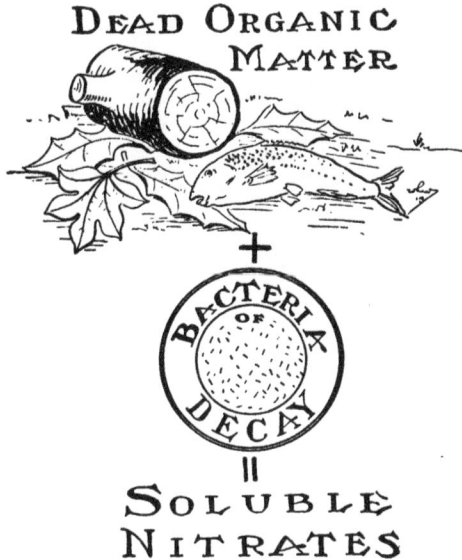

DEAD ORGANIC MATTER

+

BACTERIA OF DECAY

=

SOLUBLE NITRATES

This shows how organic matter is broken down by bacteria so it may be used again by green plants.

like on the earth if things did not decay? The sea would soon be filled and the land covered with dead bodies of plants and animals. Conditions of life would become impossible and living things on the earth would cease to exist.

Fortunately, bacteria cause decay. All organic matter, in

whatever form, is sooner or later decomposed by the action of untold millions of bacteria which live in the air, water, and soil. These soil bacteria are most numerous in rich damp soils containing large amounts of organic material. They are very numerous around and in the dead bodies of plants and animals. To a considerable degree, then, these bacteria are useful in feeding upon these dead bodies, which otherwise would soon cover the surface of the earth to the exclusion of everything else. Bacteria may thus be scavengers. They oxidize organic materials, changing them to compounds that can be absorbed by plants and used in building protoplasm. Without bacteria and fungi it would be impossible for life to exist on the earth, for green plants would be unable to get the raw food materials in forms that could be used in making food and living matter. In this respect bacteria are of the greatest service to mankind.

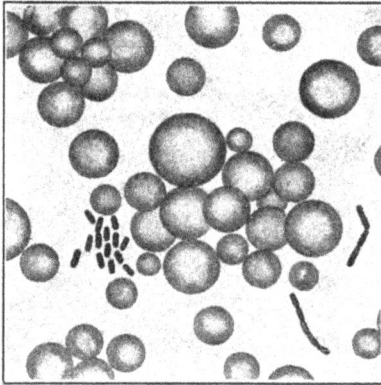

Microscopic appearance of ordinary milk, showing fat globules and bacteria which cause the souring of milk.

Relation to Fermentation. — They may incidentally, as a result of this process of decay, continue the process of fermentation begun by the yeasts. In making vinegar the yeasts first make alcohol (see page 135) which the bacteria change to acetic acid. The lactic acid bacteria, which sour milk, changing the milk sugar to an acid, grow very rapidly in a warm temperature; hence milk which is cooled immediately and kept cool or which is pasteurized and kept in a cool place will not sour readily. Why? These same lactic acid bacteria may be useful when they sour the milk for the cheese maker.

Other Useful Bacteria. — Certain bacteria give flavor to cheese and butter, while still other bacteria aid in the " curing " of tobacco, in the production of the dye indigo, in the preparation of certain fibers of plants for the market, as hemp, flax, etc., in the rotting of animal matter from the skeletons of sponges, and in the process of tanning hides to make leather.

Nitrogen-fixing Bacteria. — Still other bacteria, as we have seen before, "change over" nitrogen in organic material in the soil and even the free nitrogen of the air so that it can be used by plants in the form of a compound of nitrogen. The bacteria living in tubercles on the roots of clover, beans, peas, etc., have the power of thus "fixing" the free nitrogen in the air found between particles of soil. This fact is made use of by farmers who rotate their crops, growing first a crop of clover or other plants having root tubercles, which produce the bacteria, then plowing these in and planting another crop, as wheat or corn, on the same area. The latter plants, making use of the nitrogen compounds there, produce a larger crop than when grown in ground containing less nitrogenous material.

A field of alfalfa, a plant which harbors the nitrogen-fixing bacteria.

Bacteria cause Disease. — The most harmful bacteria are those which cause diseases of plants and animals. Certain diseases of plants — blights, rots, and wilts — are of bacterial nature. These do much annual damage to fruits and other parts of growing plants useful to man as food. But by far the most important are the bacteria which cause disease in man. They accomplish this by becoming parasites in the human body. Millions upon millions of bacteria exist in the human body at all times — in the mouth, on the teeth, in the blood, and especially in the lower part of the food tube. Some in the food tube are believed to be useful, some harmless, and some harmful; others in the mouth cause decay of the teeth, while a few kinds, if present in the body, may cause disease.

It is known that bacteria, like other living things, feed and give off organic waste from *their own* bodies. This waste, called a *toxin,*

Tubercles on the roots of the soy bean. They contain the nitrogen-fixing bacteria. (Fletcher's Soils.) Copyright by Doubleday, Page and Company.

is poison to the host on which the bacteria live, and it is usually the production of this toxin that causes the symptoms of disease. Some forms, however, break down tissues and plug up the small blood vessels, thus causing disease.

Diseases caused by Bacteria. — It is estimated that bacteria cause annually over 50 per cent of the deaths of the human race. As we will later see, a very large proportion of these diseases might be prevented if people were educated sufficiently to take the proper precautions to prevent their spread. These precautions might save the lives of some 3,000,000 of people yearly in Europe and America. Tuberculosis, typhoid fever, diphtheria, pneumonia, blood poisoning, diarrhea, and a score of other germ diseases ought not to exist. A good deal more than half of the present misery of this world might be prevented and this earth made cleaner and better by the coöperation of the young people now growing up to be our future home makers.

How we take Germ Diseases. — Germ or contagious diseases cither enter the body by way of the mouth, nose, or other body openings, or through a break in the skin. They may be carried by means of air, food, or water, but are usually *transmitted directly* from the person who has the disease to a well person. This may be done through personal contact or by handling articles used by the sick person or by drinking or eating foods which have received some of the germs. From this it follows that if we know the methods by which a given disease is communicated, we may protect ourselves from it and aid the civic authorities in preventing its spread.

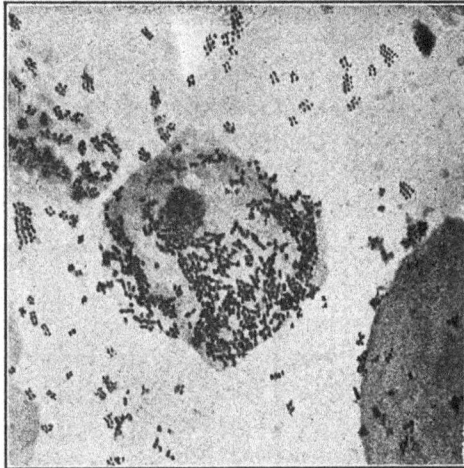

A single cell scraped from the roof of the mouth and highly magnified. The little dots are bacteria, most of which are harmless. Notice the comparative size of bacteria and cell.

Tuberculosis. — The one disease responsible for the greatest number of deaths — perhaps one seventh of the total on the globe — is tuberculosis. It is estimated that of all people alive in the United States to-day, 5,000,000 will die of this disease. But this disease is slowly but surely being overcome. It is believed that within perhaps one hundred years, with the aid of good laws and sanitary living, it will be almost extinct.

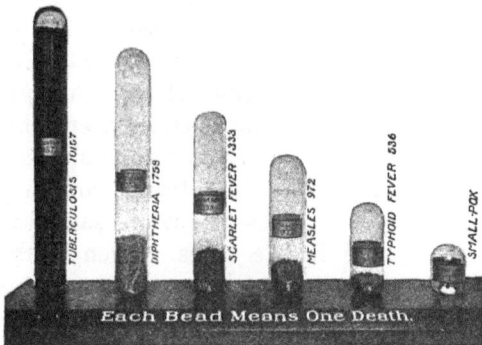

Deaths from tuberculosis compared with other contagious diseases in the city of New York in 1908.

Tuberculosis is caused by the growth of bacteria, called the *tubercle bacilli*, within the lungs or other tissues of the human body. Here they form little tubers full of germs, which close up the delicate air passages in the lungs, while in other tissues they give rise to hip-joint disease, scrofula, lupus, and other diseases, depending on the part of the body they attack. Tuberculosis may be contracted by taking the bacteria into the throat or lungs or possibly by eating meat or drinking milk from tubercular cattle. Especially is it communicated from a consumptive to a well person by kissing, by drinking or eating from the same cup or plate, using the same towels, or in coming in direct contact with the person having the germs in his body. Although there are always some of the germs in the air of an ordinary city street, and though we may take some of these germs into our bodies at any time, yet the bacteria seem able to gain a foothold only

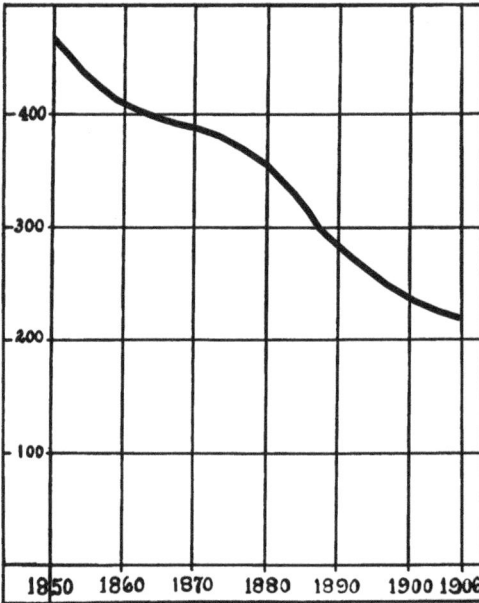

This curve shows a decreasing death rate from tuberculosis. Explain.

under certain conditions. It is only when the tissues are in a worn-out condition, when we are "run down," as we say, that the parasite may obtain a foothold in the lungs. Even if the disease gets a foothold, it is quite possible to cure it if it is taken in time. The germ of tuberculosis is killed by exposure to bright sunlight and fresh air. Thus the course of the disease may be arrested, and a permanent cure brought about, by a life in the open air, the patient sleeping out of doors, taking

plenty of nourishing food and very little exercise. **See also**
Chapter XXIV.

Typhoid Fever. — One of the most common germ diseases in
this country and Europe is typhoid fever.   This is a disease which
is conveyed by means of water and food, especially milk, oysters,
and uncooked vegetables.   Typhoid fever germs live in the intestine and from there get into the blood and are carried to all parts
of the body.   A poison which they give off causes the fever so
characteristic of the disease.   The germs multiply very rapidly

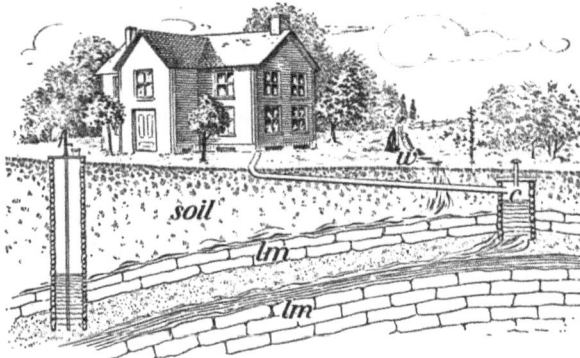

This figure shows how sewage from a cesspool (*c*) might get into the
water supply: *lm*, layer of rock; *w*, wash water.

in the intestine and are passed off from the body with the excreta
from the food tube.   If these germs get into the water supply
of a town, an epidemic of typhoid will result.   Among the recent
epidemics caused by the use of water containing typhoid germs
have been those in Butler, Pa., where 1364 persons were made ill;
Ithaca, N. Y., with 1350 cases; and Watertown, N. Y., where
over 5000 cases occurred.   Another source of infection is milk.
Frequently epidemics have occurred which were confined to users
of milk from a certain dairy.   Upon investigation it was found
that a case of typhoid had occurred on the farm where the milk
came from, that the germs had washed into the well, and that this
water was used to wash the milk cans.   Once in the milk, the bacteria multiplied rapidly, so that the milkman gave out cultures of

typhoid in his milk bottles. Proper safeguarding of our water and milk supply is necessary if we are to keep typhoid away.

Blood Poisoning. — The bacterium causing blood poisoning is another toxin-forming germ. It lives in dust and dirt and is often found on the skin. It enters the body through cuts or bruises. It seems to thrive best in less oxygen than is found in the air. It is therefore important not to close up with court-plaster wounds which such germs may have entered. It, with typhoid, is responsible for four times as many deaths as bullets and shells in time of battle. The wonderfully small death rate of the Japanese army in their war with Russia was due to the fact that the Japanese soldiers always boiled their drinking water before using it, and their surgeons always dressed all wounds on the battlefield, using powerful antiseptics in order to kill any bacteria that might have lodged in the exposed wounds.

Other Diseases. — Many other diseases have been traced to bacteria. Diphtheria is one of the best known. As it is a throat disease, it may easily be conveyed from one person to another by kissing, putting into the mouth objects which have come in contact with the mouth of the patient, or by food into which the germs have been carried. Grippe, pneumonia, whooping cough, and certain kinds of colds, all undoubtedly germ

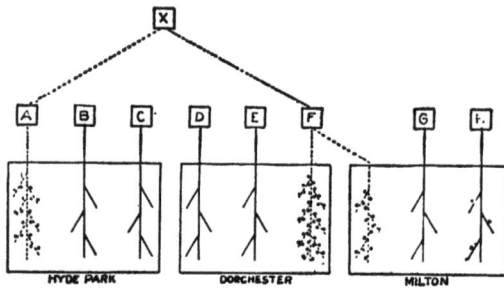

This figure shows how a milk route might be instrumental in spreading diphtheria. *X* is a farm on which a case of diphtheria occurred that was responsible for all the cases along milk routes *A* and *F* in Hyde Park, Dorchester, and Milton. How would you explain this?

diseases, are contracted in a similar manner. *Contact* with the bacteria causing the disease must occur in order that a person take the disease. This may mean actual contact with the sick person or an indirect transfer of the germs by the means mentioned above. The germs which cause diarrhea of babies, a disease which takes such a toll of death each summer, may be prevented by pasteurizing the milk before using, so as to kill the harmful bacteria. Other diseases, as malaria, yellow fever, sleeping sickness, and probably smallpox, scarlet fever, and measles, are due to the attack of one-celled animal parasites. Of these we shall learn later in Chapter XV.

Immunity. — It has been found that after an attack of a germ disease the body will not soon be again attacked by the same disease. This immunity, of which we will learn more later, seems to be due to a manufacture in the blood of substances which fight the bacteria or their poisons. If a person keeps his body in good physical condition and lives carefully, he will do much toward acquiring this natural immunity.

Acquired Immunity. — Modern medicine has discovered means of protecting the body from some contagious diseases. Vaccination as protection against smallpox, the use of antitoxins (of which more later) against diphtheria, and inoculation against typhoid are all ways in which we may be protected against diseases.

Methods of fighting Germ Diseases. — As we have seen, diseases produced by bacteria may be caused by the bacteria being *directly* transferred from one person to another, or the disease may obtain a foothold in the body from food, water, or by taking them into the blood through a cut or a wound or a body opening.

It is evident that as individuals we may each do something to prevent the spread of germ diseases, especially in our homes. We may keep our bodies, especially our hands and faces, clean. Sweeping and dusting may be done with damp cloths so as not to raise a dust; our milk and water, when from a suspicious supply, may be *sterilized* or pasteurized. Wounds through which bacteria might obtain foothold in the body should be washed with some *antiseptic* such as carbolic acid (1 part to 25 water), which kills the germs. In a later chapter we shall learn more of how we may coöperate with the authorities to combat disease and make our city or·town a better place in which to live.[1]

[1] Teachers may take up parts or all of Chapter XXIV at this point. I have found it advisable to repeat much of the work on bacteria *after* the students have taken up the study of the human organism.

REFERENCE BOOKS

ELEMENTARY

Hunter, *Laboratory Problems in Civic Biology.* American Book Company.
Bigelow, *Introduction to Biology.* The Macmillan Company.
Conn, *Bacteria, Yeasts, and Molds in the Home.* Ginn and Company.
Conn, *Story of Germ Life.* D. Appleton and Company.
Davison, *The Human Body and Health.* American Book Company.
Frankland, *Bacteria in Daily Life.* Longmans, Green, and Company.
Overton, *General Hygiene.* American Book Company.
Prudden, *Dust and its Dangers.* G. P. Putnam's Sons.
Prudden, *The Story of the Bacteria.* G. P. Putnam's Sons.
Ritchie, *Primer of Sanitation.* World Book Company.
Sharpe, *Laboratory Manual in Biology*, pages 123–132. American Book Company.

ADVANCED

Conn, *Agricultural Bacteriology.* P. Blakiston's Sons and Company.
Coulter, Barnes, and Cowles, *A Textbook of Botany*, Vol. I. American Book Company
De Bary, *Comparative Morphology and Biology of the Fungi, Mycetozoa, and Bacteria.* Clarendon Press.
Duggar, *Fungous Diseases of Plants.* Ginn and Company.
Hough and Sedgwick, *The Human Mechanism.* Ginn and Company.
Hutchinson, *Preventable Diseases.* Houghton, Mifflin and Company.
Lee, *Scientific Features of Modern Medicine.* Columbia University Press.
Muir and Ritchie, *Manual of Bacteriology.* The Macmillan Company.
Newman, *The Bacteria.* G. P. Putnam's Sons.
Sedgwick, *Principles of Sanitary Science and Public Health.* The Macmillan Company.

XII. THE RELATIONS OF PLANTS TO ANIMALS

Problems. — To determine the general biological relations existing between plants and animals.

(a) As shown in a balanced aquarium.

(b) As shown in hay infusion.

Demonstration of life in a "balanced" and "unbalanced" aquarium. — Determination of factors causing balance.

Demonstration of hay infusion. — Examination to show forms of animal and plant life.

Tabular comparison between balanced aquarium and hay infusion.

Some Ways in which Plants affect Animals. — We have been studying the life of plants in order better to understand the life of animals and men. We have seen first that green plants play indirectly a tremendous part in man's welfare by supplying him with food. We have found that the colorless plants directly affected his welfare by causing disease, and by causing decay, thus making usable the nitrogen locked up in dead bodies of plants and animals, and by some even supplying nitrogen from the atmosphere. The dependence of animals upon plants has been shown and the interdependence of plants on animals has also been seen in cross-pollination and in the supply of raw food materials to plants by animals.

Study of a Balanced Aquarium. — Perhaps the best way for us to understand the interrelation between plants and animals is to study an aquarium in which plants and animals live and in which a balance has been established between the plant life on one side and animal life on the other. Aquaria containing green pond weeds, either floating or rooted, a few snails, some tiny animals known as water fleas, and a fish or two will, if kept near a light window, show this relation.

159

We have seen that green plants under favorable conditions of sunlight, heat, moisture, and with a supply of raw food materials, give off oxygen as a by-product while manufacturing food in their green cells. We know the necessary raw materials for starch manufacture are carbon dioxide and water, while nitrogenous material is necessary for the making of proteins within the plant.

A balanced aquarium.   Explain the term "balanced."

In previous experiments we have proved that carbon dioxide is given off by any living thing when oxidation occurs in the body. The crawling snails and the swimming fish give off carbon dioxide, which is dissolved in the water; the plants themselves, at all times, oxidize food within their bodies, and so must *pass off* some carbon dioxide. The green plants in the daytime *use up* the carbon dioxide obtained from the various sources and, with the water taken in, manufacture starch. While this process is going on, oxygen is given off to the water of the aquarium, and this free oxygen is used by the animals there.

But the plants are continually growing larger. The snails and fish, too, eat parts of the plants. Thus the plant life gives food to the animals within the aquarium. The animals give off certain nitrogenous wastes of which we shall learn more later. These materials, with other nitrogenous matter from the dead parts of the plants or animals, form part of the raw material used for protein manufacture in the plant. This nitrogenous matter is prepared for use by several different kinds of bacteria which first break the dead bodies down and then give it to the plants in the form of soluble nitrates. The green plants manufacture food, the animals eat the plants and give off organic waste, from which the plants in turn make their food and living matter.

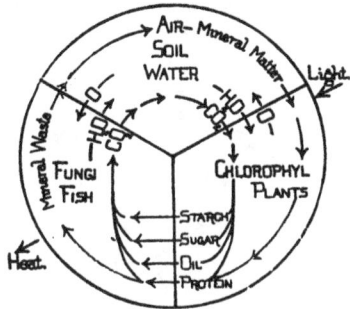

This diagram shows that plants and animals on the earth hold the same relation to each other as plants and animals in a balanced aquarium. Explain the diagram in your notebook.

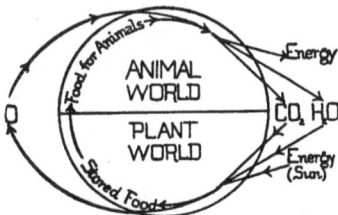

The plants give off oxygen to the animals, and the animals give carbon dioxide to the plants. Thus a balance exists between the plants and animals in the aquarium. Make a table to show this balance.

The carbon and oxygen cycle in the balanced aquarium. Trace by means of the arrows the carbon from the time plants take it in as CO_2 until animals give it off. Show what happens to the oxygen.

Relations between Green Plants and Animals. — What goes on in the aquarium is an example of the relation existing between all green plants and all animals. Everywhere in the world green plants are making food which becomes, sooner or later, the food of animals. Man does not feed to a great extent upon leaves, but he eats roots, stems, fruits, and seeds. When he does not feed directly upon plants, he eats the flesh of plant eating animals, which in turn feed directly upon plants. And so it is the world over; the plants are the food makers and supply the animals.

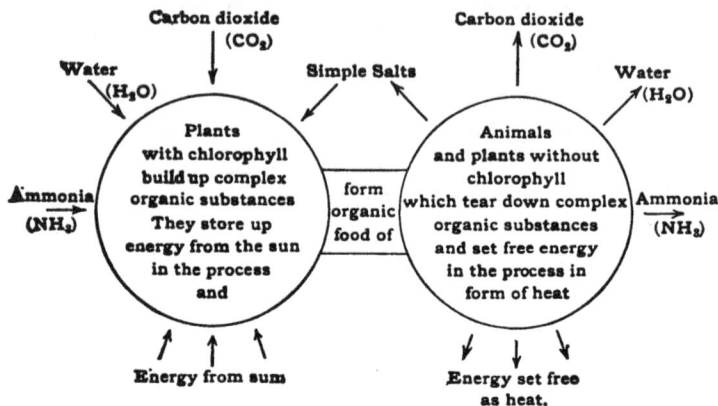

Carbon dioxide (CO_2)

Water (H_2O)

Simple Salts

Carbon dioxide (CO_2)

Water (H_2O)

Ammonia (NH_3)

Plants with chlorophyll build up complex organic substances They store up energy from the sun in the process and

form organic food of

Animals and plants without chlorophyll which tear down complex organic substances and set free energy in the process in form of heat

Ammonia (NH_3)

Energy from sun

Energy set free as heat.

The relations between green plants and animals.

Green plants also give a very considerable amount of oxygen to the atmosphere every day, which the animals may use.

The Nitrogen Cycle. — The animals in their turn supply much of the carbon dioxide that the plant uses in starch making. They

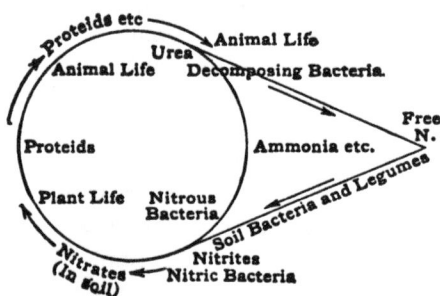

The nitrogen cycle. Trace the nitrogen from its source in the air until it gets back again into the air.

also supply some of the nitrogenous matter used by the plants, part being given the plants from the dead bodies of their own relatives and part being prepared from the nitrogen of the air through the agency of bacteria, which live upon the roots of certain plants. These bacteria are the only organisms that can take nitrogen from the air. Thus, in spite of all the nitrogen of the atmosphere, plants and animals are limited in the amount available. And the available supply is used over and over again, perhaps in nitrogenous food by an animal, then it may be given off as organic waste, get into the soil, and be taken up by a plant through the roots. Eventually the nitrogen forms part of the food supply in the body of the plant, and then may become part of its living matter. When the plant dies, the nitrogen is returned to the soil. Thus the usable nitrogen is kept in circulation.[1]

Symbiosis. — We have seen that in the balanced aquarium the animals and plants, in a wide sense, form a sort of unconscious partnership. *This process of living together for mutual advantage is called symbiosis.* Some animals thus combine with plants; for example, the tiny animal known as the hydra with certain of the one-celled algæ, and, if we accept the term in a wide sense, all green plants and animals live in this relation of mutual give and take. Animals also frequently live in this relation to each other, as the crab, which lives within the shell of the oyster; the sea anemones, which are carried around on the backs of some hermit crabs, aiding the crab in protecting it from its enemies, and being carried about by the crab to places where food is plentiful.

A Hay Infusion. — Still another example of the close relation between plants and animals may be seen in the study of a hay infusion. If we place a wisp of hay or straw in a small glass jar nearly full of water, and leave it for a few days in a warm room, certain changes are seen to take place in the contents of the jar; after a little while the water gets cloudy and darker in color, and a scum appears on the surface. If some of this scum is examined under the compound microscope, it will be found to consist almost entirely of bacteria. These bacteria evidently aid in the decay which (as the unpleasant odor from the jar testifies) is beginning to take place. As we have learned, bacteria flourish wherever the food supply is abundant. The water within the jar has come to contain much of the food material which was once within the leaves of the grass, — organic nutrients, starch, sugar, and proteins, formed in the leaf by the action of the sun on the chlorophyll

[1] A small amount of nitrogen gas is returned to the atmosphere by the action of the decomposing bacteria on the ammonia compounds in the soil. (See figure of nitrogen cycle.)

of the leaf, and now released into the water by the breaking down
of the walls of the cells of the leaves. The bacteria themselves
release this food from the hay by causing it to decay. After a
few days small one-celled animals appear; these multiply with
wonderful rapidity, so that in some cases the surface of the water
seems to be almost white with active one-celled forms of life. If
we ask ourselves where these animals come from, we are forced

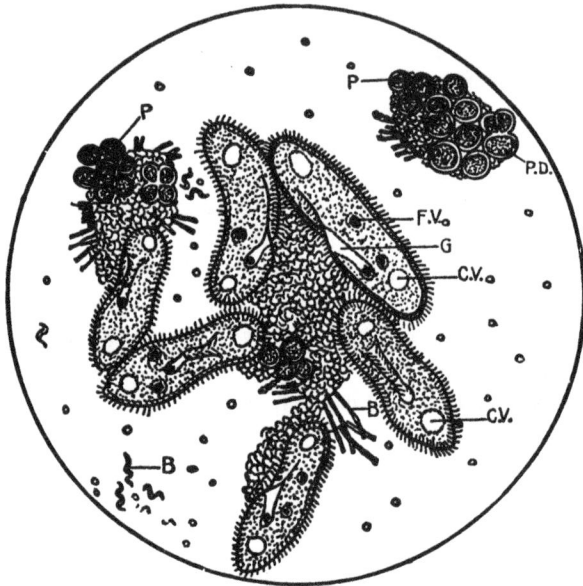

Life in the late stage of a hay infusion. *B*, bacteria, swimming or forming masses
of food upon which the one-celled animals, the paramœcia, are feeding;
G, gullet; *F.V.*, food vacuole; *C.V.*, contractile vacuole; *P*, pleurococcus;
P.D., pleurococcus dividing. (Drawn from nature by J. W. Teitz.)

to the conclusion that they must have been in the water, in the
air, or on the hay. Hay is dried grass and may have been cut
in a field near a pool containing these creatures. When the
pool dried up, the wind may have scattered some of these little
organisms in the dried mud or dust. Some may have existed in
a dormant state on the hay and the water awakened them to active

life.   In the water, too, there may have been some living cells, plants and animals.

At first the multiplication of the tiny animals within the hay infusion is extremely rapid; there is food in abundance and near at hand.   After a few days more, however, several kinds of one-celled animals may appear, some of which prey upon others.   Consequently a struggle for life takes place, which becomes more and more intense as the food from the hay is used up.   Eventually the end comes for all the animals unless some green plants obtain a foothold within the jar.   If such a thing happens, food will be manufactured within their bodies, a new food supply arises for the animals within the jar, and a balance of life may result.

REFERENCE BOOKS

ELEMENTARY

Hunter, *Laboratory Problems in Civic Biology*.   American Book Company.
Sharpe, *A Laboratory Manual for the Solution of Problems in Biology*, pp. 133–138.   American Book Company.

ADVANCED

Eggerlin and Ehrenberg, *The Fresh Water Aquarium and its Inhabitants*.   Henry Holt and Company.
Furneaux, *Life in Ponds and Streams*.   Longmans, Green, and Company.
Parker, *Biology*.   The Macmillan Company.
Sedgwick and Wilson, *Biology*.   Henry Holt and Company.

XIII. SINGLE–CELLED ANIMALS CONSIDERED AS ORGANISMS

Problems. — *To determine:*

(a) *How a one-celled animal is influenced by its environment.*

(b) *How a single cell performs its functions.*

(c) *The structure of a single-celled animal.*

LABORATORY SUGGESTIONS

Laboratory study. — Study of paramœcium under compound microscope in its relation to food, oxygen, etc. Determination of method of movement, turning, avoiding obstructions, sensitiveness to stimuli. Drawings to illustrate above points.

Laboratory demonstration. — Living paramœcium to show structure of cell. Demonstration with carmine to show food vacuoles, and action of cilia. Use of charts and stained specimens to show other points of cell structure. Laboratory demonstration of fission.

The Simplest Plants. — We have seen that perhaps the simplest plant would be exemplified by one of the tiny bacteria we have just read about. A typical one-celled plant, however, would contain green coloring matter or chlorophyll, and would have the

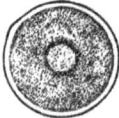

Pleurococcus. A very simple plant cell.

power to manufacture its own food under conditions giving it a moderate temperature, a supply of water, oxygen, carbon dioxide, and sunlight. Such a simple plant is the *pleurococcus*, the tiny green plants seen on the shady sides of trees, stones, or city houses. This plant would meet one definition of a cell, as it is a minute mass of protoplasm containing a nucleus. It is surrounded by a wall of a woody material formed by the activity of the living matter within the cell. It also contains a little mass of protoplasm colored green. Of the work of the chlorophyll in the manufacture of organic food we have already learned. Such is a simple plant cell. Let us now examine a simple animal cell in order to compare it with that of a plant.

Where to find Paramœcium. — If we examine very carefully the surface of a hay infusion, we are likely to notice in addition to the scum formed of bacteria, a mass of whitish tiny dots collected along the edge of the jar close to the surface of the water. More attentive observation shows us that these objects move, and that they are never found far from the surface.

The Life Habits of Paramœcium. — If we place on a slide a drop of water containing some of these moving objects and examine it under the compound microscope, we find each minute whitish dot is a cell, elongated, oval, or elliptical in outline and somewhat flattened. This is a one-celled animal known as the *paramœcium* or the slipper animalcule (because of its shape).

Seen under the low power of the microscope, it appears to be extremely active, rushing about now rapidly, now more slowly, but seemingly always taking a definite course. The narrower end of the body (the *anterior*) usually goes first. If it pushes its way past any dense substance in the water, the cell body is seen to change its shape temporarily as it squeezes through.

Response to Stimuli. — Many of these little creatures may be found collected around masses of food, showing that they are attracted by it. In another part of the slide we may find a number of the paramœcia lying close to the edge of an air bubble with the greatest possible amount of their surface exposed to its surface. These animals are evidently taking in oxygen by osmosis. They are breathing. A careful inspection of the jar containing paramœcia shows thousands of tiny whitish bodies collected near the surface of the jar. In the paramœcium, as in the one-celled plants, the protoplasm composing the cell responds to certain agencies acting upon it, coming from without; these agencies we call *stimuli*. Such stimuli may be light, differences of temperature, presence of food, electricity, or other factors of its surroundings. Plant and animal cells may react differently to the same stimulus. In general, however, we know that protoplasm is *irritable* to some of these factors. To severe stimuli, protoplasm usually responds by *contracting*, another power which it possesses. We know, too, that plant and animal cells take in food and change the food to protoplasm, that is, that they *assimilate* food ; and that they may waste away and repair themselves. Finally, we know that new plant and animal cells are *reproduced* from the original bit of protoplasm, a single cell.

The Structure of Paramœcium. — The cell body is almost transparent, and consists of semifluid protoplasm which has a granular grayish appearance under the microscope. This protoplasm appears to be bounded by a very delicate membrane through which project numerous delicate threads of protoplasm called *cilia*. (These are usually invisible under the microscope).

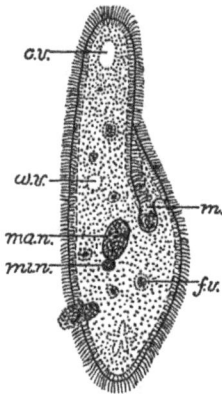

The locomotion of the paramœcium is caused by the movement of these cilia, which lash the water like a multitude of tiny oars. The cilia also send particles of food into a funnel-like opening, the *gullet*, on one side of the cell. Once inside the cell body, the particles of food materials are gathered into little balls within the almost transparent protoplasm. These masses of food seem to be inclosed within a little area containing fluid, called a *vacuole*. Other vacuoles appear to be clear; these are spaces in which food has been digested. One or two larger vacuoles may be found; these are the *contractile vacuoles;* their purpose seems to be to pass off waste material from the cell body. This is done by pulsation of the vacuole, which ultimately bursts, passing fluid waste to the outside. Solid wastes are passed out of the cell in somewhat the same manner. No breathing organs are seen, because osmosis of oxygen and carbon dioxide may take place anywhere through the cell membrane. The nucleus of the cell is not easily visible in living specimens. In a cell that has been stained it has been found to be a double

A paramœcium. *c.v.*, contractile vacuole; *f.v.*, food vacuole; *m*, mouth; *ma.n.*, macronucleus; *mi.n.*, micronucleus; *w.v.*, water vacuole.

structure, consisting of one large and one small portion, called, respectively, the *macronucleus* and the *micronucleus*.

Reproduction of Paramœcium. — Sometimes a paramœcium may be found in the act of dividing by the process known as *fission*, to form two new cells, each of which contains half of the original cell. This is a method of *asexual* reproduction. The original cell may thus form in succession many hundreds of cells in every respect like the original parent cell.

Amœba.[1] — In order to understand more fully the life of a simple bit of protoplasm, let us take up the study of the *amœba*, a type of the simplest form of animal life. Unlike the plant and animal cells we have examined, the amœba has no fixed form.

Paramœcium dividing by fission. *M*, mouth; *MAC.*, macronucleus; *MIC.*, micronucleus. (After Sedgwick and Wilson.)

Viewed under the compound microscope, it has the appearance of an irregular mass of granular protoplasm. Its form is constantly changing as it moves about. This is due to the pushing out of tiny projections of the protoplasm of the cell, called *pseudopodia* (false feet). The locomotion is accomplished by a streaming or flowing of the semifluid protoplasm. The pseudopodia are pushed forward in the direction which the animal is to go, the rest of the body following. In the central part of the cell is the nucleus. This im-

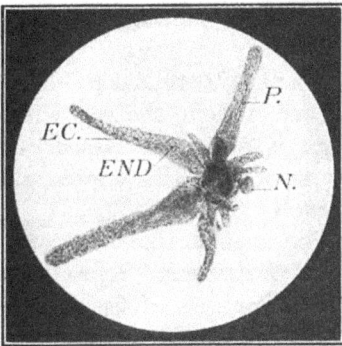

Amœba, with pseudopodia (*P.*) extended; *EC*, ectoplasm; *END*, endoplasm; the dark area (*N.*) is the nucleus. (From a photograph loaned by Professor G. N. Calkins.)

[1] Amœbæ *may* be obtained from the hay infusion, from the dead leaves in the bottom of small pools, from the same source in fresh-water aquaria, from the roots of duckweed or other small water plants, or from green algæ growing in quiet localities. No *sure* method of obtaining them can be given.

portant organ is difficult to see, except in cells that have been stained.

Although but a single cell, still the amœba appears to be aware of the existence of food when it is near at hand. Food may be taken into the body at any point, the semifluid protoplasm simply rolling over and engulfing the food material. Within the body, as in the paramœcium, the food becomes inclosed within a fluid space or vacuole. The protoplasm has the power to take out such material as it can use to form new protoplasm or give energy.

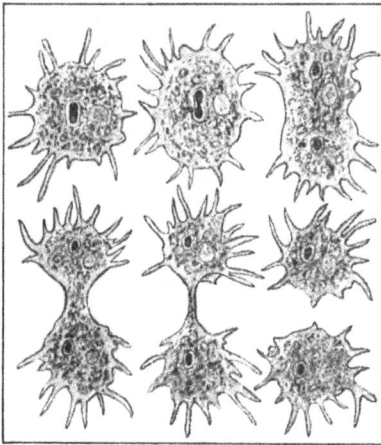

Amœba, showing the changes which take place during division of the cell. The dark body in each figure is the nucleus; the transparent circle, the contractile vacuole; the large granular masses, the food vacuoles. Much magnified.

Circulation of food material is accomplished by the constant streaming of the protoplasm within the cell.

The cell absorbs oxygen from the water by osmosis through its delicate membrane, giving up carbon dioxide in return. Thus the cell " breathes " through any part of its body covering.

Waste nitrogenous products formed within the cell when work is done are passed out by means of the contractile vacuole.

The amœba, like other one-celled organisms, reproduces by the process of fission. A single cell divides by splitting into two others, each of which resembles the parent cell, except that they are of less bulk. When these become the size of the parent amœba, they each in turn divide. This is a kind of asexual reproduction.

When conditions unfavorable for life come, the amœba, like some one-celled plants, encysts itself within a membranous wall. In this condition it may become dried and be blown through the air. Upon return to a favorable environment, it begins life again, as before. In this respect it resembles the spore of a plant.

The Cell as a Unit. — In the daily life of a one-celled animal we find the single cell performing all the general activities which we shall later find the many-celled animal is able to perform. In the amœba no definite parts of the cell appear to be set off to perform certain functions; but any part of the cell can take in food, can absorb oxygen, can change the food into protoplasm, and excrete the waste material. The single cell is, in fact, an organism able to carry on the business of living almost as effectually as a very complex animal.

Complex One-celled Animals. — In the paramœcium we find a single cell, but we find certain parts of the cell having certain definite functions: the cilia are used for locomotion; a definite part of the cell takes in food, while the waste passes out at another definite spot. In another one-celled animal called *vorticella*, part of the cell has become elongated and is contractile. By this stalk the little animal

Vorticella. *e*, gullet; *n*, nucleus; *cv*, contractile vacuole; *a*, axis; *s*, sheath; *fv*, food vacuole. (From Herrick's *General Zoölogy*.)

is fastened to a water plant or other object. The stalk may be said to act like a muscle fiber, as its sole function seems to be movement; the cilia are located at one end of the cell and serve to create a current of water which will bring food particles to the mouth. Here we have several parts of the cell, each doing a different kind of work. This is known as *physiological division of labor.*

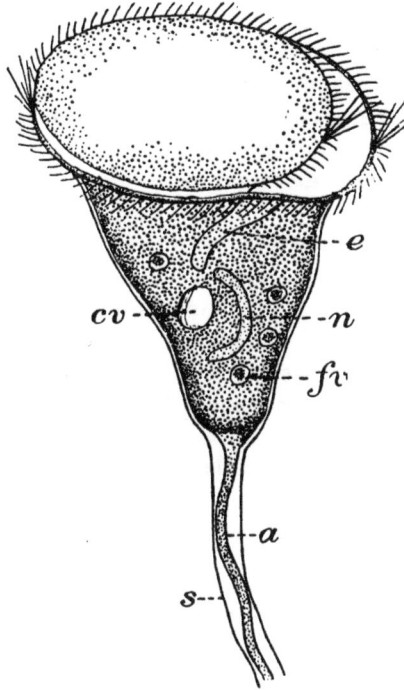

Habitat of Protozoa. — Protozoa are found almost everywhere in shallow water, especially close to the surface. They appear to be attracted near to the surface by the supply of oxygen. Every fresh-water lake swarms with them; the ocean contains countless myriads of many different forms.

Use as Food. — They are so numerous in lakes, rivers, and the ocean as to form the food for many animals higher in the scale of life. Almost all fish that do not take the hook and that travel in schools, or companies, migrating from one place to another, live partly on such food. Many feed on slightly larger animals, which in turn eat the Protozoa. Such fish have on each side of the mouth attached to the gills a series of small structures looking like tiny rakes. These are called the *gill rakers*, and aid in collecting tiny organisms from the water as it passes over the gills. The whale, the largest of all mammals, strains protozoans and other small animals and plants out of the water by means of hanging plates of whalebone or baleen, the slender filaments of which form a sieve from the top to the bottom of the mouth.

Protozoa cause Disease. — Protozoa of certain kinds play an important part in causing malaria, yellow fever, and other diseases, as we shall see later.[1] (See page 217.)

REFERENCE BOOKS

ELEMENTARY

Hunter, *Laboratory Problems in Civic Biology.* American Book Company.
Davison, *Human Body and Health.* American Book Company.
Jordan, Kellogg and Heath, *Animal Studies.* D. Appleton and Company.
Sharpe, *Laboratory Manual*, pp. 140–143. American Book Company.

ADVANCED

Calkins, *The Protozoa.* Macmillan Company.
Jennings, *Study of the Lower Organisms.* Carnegie Institution Report.
Parker, *Lessons in Elementary Biology.* The Macmillan Company.
Wilson, *The Cell in Development and Inheritance.* The Macmillan Company.

[1] Teachers may find it expedient to take up the study of protozoan diseases at this point.

XIV. DIVISION OF LABOR. THE VARIOUS FORMS OF PLANTS AND ANIMALS

Problems. — *The development and forms of plants.*
The development of a simple animal.
What is division of labor? In what does it result?
How to know the chief characters of some great animal groups.

LABORATORY SUGGESTIONS

A visit to a botanical garden or laboratory demonstration. — Some of the forms of plant life. Review of essential facts in development of bean or corn embryo.

Demonstration. — Charts or models showing the development of a many-celled animal from egg through gastrula stage.

Demonstration. — Types which illustrate increasing complexity of body form and division of labor.

Museum trip. — To afford pupil a means of identification of examples of principal phyla. This should be preceded by objective demonstration work in school laboratory.

Reproduction in Plants. — Although there are very many plants and animals so small and so simple as to be composed of but a single cell, by far the greater part of the animal and plant world is made up of individuals which are collections of cells living together.

In a simple plant like the pond scum, a string or filament of cells is formed by a single cell dividing crosswise, the two cells formed each dividing into two

A cell of pond scum. How might it divide to form a long thread made up of cells?

more. Eventually a long thread of cells is thus formed. At times, however, a cell is formed by the union of two cells, one from each of two adjoining filaments of the plant. At length a hard coat forms around this cell, which has now become a *spore*. The tough covering protects it from unfavorable changes in the surroundings. Later, when conditions become favorable for its germination, the spore may form a new filament of pond scum.

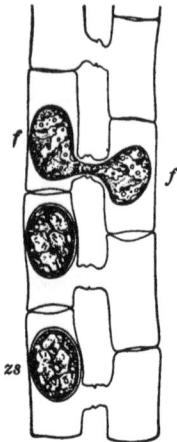

In molds, in yeasts, and in the bacteria we also found spores could be formed by the protoplasm of the plant cutting up into a number of tiny spores. These spores are called *asexual* (without sex) because they are not formed by the union of two cells, and may give rise to other tiny plants like themselves. Still other plants, mosses and ferns, give rise to two kinds of spores, sexual and asexual. All of these collectively are called *spore plants*.

Reproduction in Seed Plants. — Another great group of plants we have studied, plants of varied shapes and sizes, produce seeds. They bear flowers and fruits.

The formation of spores in pond scum. *zs*, zygospore; *f*, fusion in progress.

The embryo develops from a single fertilized "egg," growing by cell division into two, four, eight, and a constantly increasing number of cells until after a time a baby plant is formed, which as in the bean, either contains some stored food to give it a start in life, or, as in the corn, is surrounded with food which it can digest and absorb into its own tiny body. We have seen that these young plants in the seed are able to develop when conditions are favorable. Furthermore, the young of each kind of plant will eventually develop into the kind of plant its parent was and into no other kind. Thus the plant world is divided into many tribes or groups.

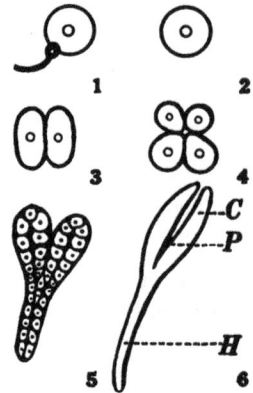

The formation and growth of a plant embryo. 1, the sperm and egg cell uniting; 2, a fertilized egg; 3, two cells formed by division; 4, four cells formed from two; 5, a many-celled embryo; 6, young plant; *H*, hypocotyl; *P*, plumule; *C*, cotyledons.

Plants are placed in Groups. — If we plant a number of peas so that they will all germinate under the same conditions of soil, tem-

perature, and sunlight, the
seedlings that develop will
each differ one from an-
other in a slight degree.[1]
But in a general way they
will have many characters
in common, as the shape
of the leaves, the possession of tendrils, form of
the flower and fruit. A
species of plants or animals
is a group of individuals so
much alike in their char-

A colony of trilliums, a flowering plant.
(Photograph by W. C. Barbour.)

acters that they might have had the same parents. Individuals of
such species differ slightly ; for no two individuals are exactly alike.

Rock fern, *polypody*. Notice the underground
stem giving off roots from its lower surface,
and leaves (C), (S), from its upper surface.

Species are grouped to-
gether in a larger group
called a genus. For ex-
ample, many kinds of peas
— the wild beach peas, the
sweet peas, and many
others — are all grouped in
one genus (called *Lathyrus*,
or vetchling) because they
have certain structural
characteristics in common.

Plant and animal genera
are brought together in still
larger groups, the classification based on general like-
nesses in structure. Such
groups are called, as they
become successively larger,

Family, Order, and *Class.* Thus both the plant and animal king-
doms are grouped into divisions, the smallest of which contains

[1] NOTE TO TEACHERS. — A trip to the Botanical Garden or to a Museum should
be taken at this time.

individuals very much alike; and the largest of which contains very many groups of individuals, the groups having some characters in common.   This is called a system of classification.

Classification of the Plant Kingdom. — The entire plant kingdom has been divided into four sub-kingdoms by botanists: —

1. *Spermatophytes.* $\begin{cases} Angiosperms, \text{ true flowering plants.} \\ Gymnosperms, \text{ the pines and their allies.} \end{cases}$
2. *Pteridophytes.*   The fern plants and their allies.
3. *Bryophytes.*   The moss plants and their allies.

Rockweed, a brown algæ, showing its distribution on rocks below highwater mark.

4. *Thallophytes.* The Thallophytes form two groups: the Algæ and the Fungi; the algæ being green, while the fungi have no chlorophyll.

The extent of the plant kingdom can only be hinted at; each year new species are added to the lists.   There are about 110,000 species of flowering plants and nearly as many flowerless plants. The latter consist of over 3500 species of fernlike plants, some 16,500 species of mosses, over 5600 lichens (plants consisting of a partnership between algæ and fungi), approximately 55,000 species of fungi, and about 16,000 species of algæ.

Development of a Simple Animal. — Many-celled animals are formed in much the same way as are many-celled seed plants. A common bath sponge, an earthworm, a fish, or a dog, — each and all of them begin life in the same manner. In a many-celled animal the life history begins with a single cell, the fertilized egg. As in the flowering plant, this cell has been formed by the union of two other cells, a tiny (usually motile) cell, the *sperm*, and a large cell, the *egg*. After the egg is fertilized by a sperm cell, it splits into two, four, eight, and sixteen cells; as the number of cells increases, a hollow ball of cells called the *blastula* is formed; later this ball sinks in on one side, and a double-walled cup of cells, now called a *gastrula*, results. Practically all animals pass through the above stages in their development from the egg, although these stages are often not plain to see because of the presence of food material (yolk) in the egg.

In animals the body consists of three layers of cells: those of the outside, developed from the outer layer of the gastrula, are called *ecto-*

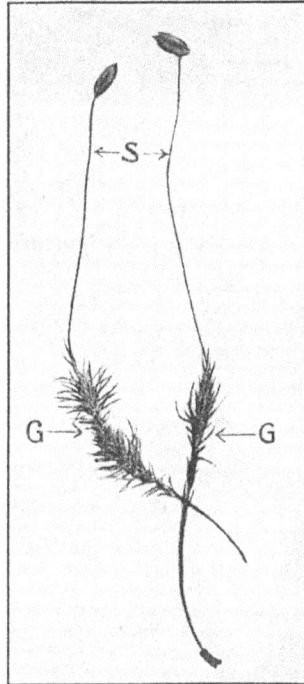

A moss plant. *G*, the moss body; *S*, the spore-bearing stalk (fruiting body).

derm, which later gives rise to the skin, nervous system, etc.; an inner layer, developed from the inner layer of the gastrula, the *endoderm*, which forms the lining of the digestive organs, etc.; a middle layer, called the *mesoderm*, lying between the ectoderm and the endoderm, is also found. In higher animals this layer gives rise to muscles, the skeleton, and parts of other internal structures.

Physiological Division of Labor. — If we compare the amœba and the paramœcium, we find the latter a more complex organism

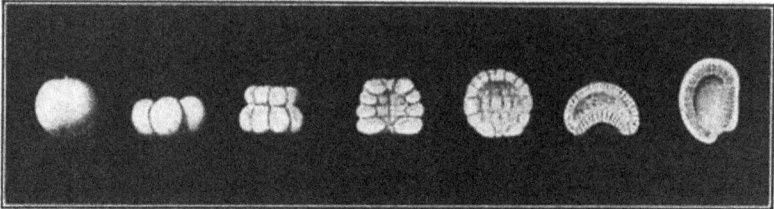

Stages in the development of a fertilized egg into the gastrula stage. Read your text, then draw these stages and name each stage.

than the former. An amœba may take in food through any part of the body; the paramœcium has a definite gullet; the amœba

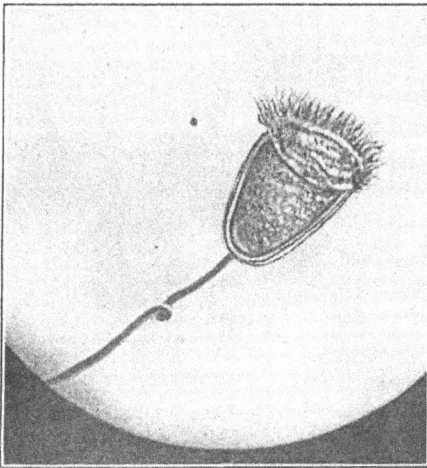

Photograph of a living *vorticella*, showing the contractile stalk and the cilia around the mouth. Compare this figure with that of the paramœcium. Which cell shows greater division of labor?

may use any part of the body for locomotion; the paramœcium has definite parts of the cell, the cilia, fitted for this work. Since the structure of the paramœcium is more complex, we say that it is a "higher" animal. In the vorticella, a still more complex cell, part of the cell has grown out like a stalk, has become contractile, and acts like muscle.

As we look higher in the scale of life, we invariably find that certain parts of a plant or animal are set apart to do certain work, and only that work. Just as in a community of people, there are some men who do rough manual work, others who are skilled workmen, some who are shopkeepers, and still others who are professional men, so among plants and animals, wherever *collections* of cells live together to form an organism, there is division of labor, some cells being fitted to do one kind of work, while others are fitted to do work of another sort. This

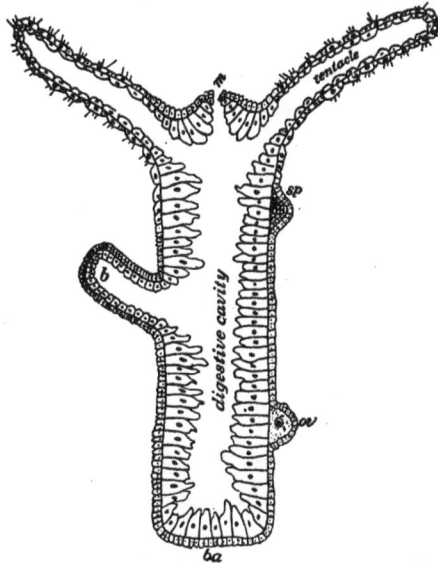

Enlarged lengthwise section of the hydra, a very simple animal which shows slight division of labor. *ba*, base; *b*, bud; *m*, mouth; *ov*, ovary; *sp*, spermary.

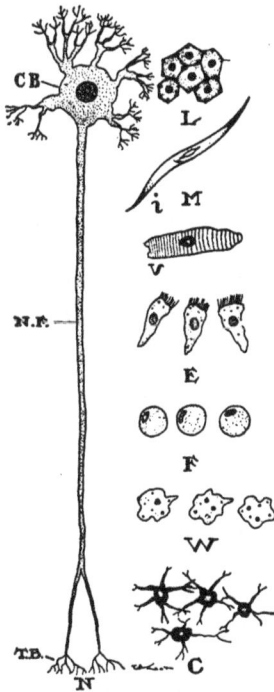

Different forms of tissue cells. *C*, bone making cells; *E*, epithelial cells; *F*, fat cells; *L*, liver cells; *M*, muscle cell; *i*, involuntary; *v*, voluntary; *N*, nerve cell; *CB*, cell body; *N.F.*, nerve fiber; *T.B.*, nerve endings; *W*, colorless blood cells.

is called physiological division of labor.

As we have seen, the higher plants are made up of a vast number of cells of many kinds. Collections of cells alike in structure and performing the same function we have called a *tissue*. Examples of animal tissues are the highly contractile cells set apart for movement, *muscles;* those which cover the body or line the inner parts of organs, the skin, or *epithelium;* the cells which form secretions or *glands* and the sensitive cells forming the *nervous* tissues.

Frequently several tissues have certain functions to perform in conjunction with one another. The arm of the human body performs movement. To do this, several tissues, as muscles, nerves, and bones, must act together. A collection of tissues performing certain work we call an *organ*.

In a simple animal like a sponge, division of labor occurs between the cells; some cells which line the pores leading inward create a current of water, and feed upon the minute organisms which come within reach, other cells build the skeleton of the sponge, and still others become eggs or sperms. In higher animals more complicated in structure and in which the tissues are found working together to form organs, division of labor is much more highly specialized. In the human arm, an organ fitted for certain movements, think of the number of tissues and the complicated actions which are possible. The most extreme division of labor is seen in the organism which has the most complex actions to perform and whose organs are fitted for such work, for there the cells or tissues which do the particular work do it quickly and very well.

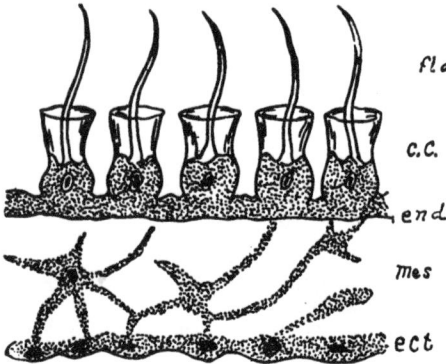

Part of a sponge, showing how cells perform division of labor. *ect*, ectoderm; *mes*, mesoderm; *end*, endoderm; *c.c.*, ciliated cells, which take in food by means of their flagellæ or large cilia (*fla*).

In our daily life in a town or city we see division of labor between individuals. Such division of labor may occur among other animals, as, for example, bees or ants. But it is seen at its highest in a great city or in a large business or industry. In the stockyards of Chicago, division of labor has resulted in certain men performing but a single movement during their entire day's work, but this movement repeated so many times in a day has resulted in wonderful accuracy and speed. Thus division of labor obtains its end.

Organs and Functions Common to All Animals. — The same general functions performed by a single cell are performed by a many-celled animal. But in the many-celled animals the various functions of the single cell are taken up by the organs. In a complex organism, like man, the organs and the functions they perform may be briefly given as follows: —

(1) The organs of *food taking*: food may be taken in by individual cells, as those lining the pores of the sponge, or definite parts of a food tube may be set apart for this purpose, as the mouth and parts which place food in the mouth.

(2) The organs of *digestion*: the food tube and collections of cells which form the glands connected with it. The enzymes in the fluids secreted by the latter change the foods from a solid form (usually insoluble) to that of a *fluid*. Such fluid may then pass by osmosis, through the walls of the food tube into the blood.

(3) The organs of *circulation*: the tubes through which the blood, bearing its organic foods and oxygen, reaches the tissues of the body. In simple animals, as the sponge and hydra, no such organs are needed, the fluid food passing from cell to cell by osmosis.

(4) The organs of *respiration*: the organs in which the blood receives oxygen and gives up carbon dioxide. The outer layer of the body serves this purpose in very simple animals; gills or lungs are developed in more complex animals.

(5) The organs of *excretion*: such as the kidneys and skin, which pass off nitrogenous and other waste matters from the body.

(6) The organs of *locomotion*: muscles and their attachments and connectives; namely, tendons, ligaments, and bones.

(7) The organs of *nervous control*: the central nervous system, which has control of coördinated movement. This consists of scattered cells in low forms of life; such cells are collected into groups and connected with each other in higher animals.

(8) The organs of *sense*: collections of cells having to do with the reception and transmission of sight, hearing, smell, taste, touch, pressure, and temperature sensations.

(9) The organs of *reproduction*: the sperm and egg-forming organs.

Almost all animals have the functions mentioned above. In most, the various organs mentioned are more or less developed, although in the simpler forms of animal life some of the organs mentioned above are either very poorly developed or entirely lacking. But in the so-called " higher " animals each of the above-named functions is assigned to a certain organ or group of organs. The work is done better and more quickly than in the " lower " animals. Division of labor is thus a guide in helping us to determine the place of animals in the groups that exist on the earth.

The Animal Series. — We have found that a one-celled animal can perform certain functions in a rather crude manner. Man can perform these same functions in an extremely efficient manner. Division of labor is well worked out, extreme complexity of structure is seen. Between these two extremes are a great many groups of animals which can be arranged more or less as a series, showing the gradual evolution or development of life on the earth. It will be the purpose of the following pages to show the chief characteristics of the great groups of the animal kingdom.

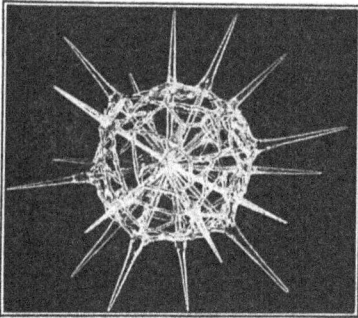

The glasslike skeleton of a *radiolarian*, a protozoan. (From model at American Museum of Natural History.)

I. **Protozoa.** — Animals composed of a single cell, reproducing by cell division.

The following are the principal classes of Protozoa, examples of which we may have seen or read about : —

Class I. *Rhizopoda* (Greek for *root-footed*). Having no fixed form, with pseudopodia. Either naked as *Amœba* or building limy (*Foraminifera*) or glasslike skeletons (*Radiolaria*).

Class II. *Infusoria* (*in infusions*). Usually active ciliated Protozoa. Examples, *Paramœcium*, *Vorticella*.

Class III. *Sporozoa* (*spore animals*). Parasitic and usually nonactive. Example, *Plasmodium malariœ*.

II. **Sponges.** — Because the body contains many pores through which water bearing food particles enters, these animals are called *Porifera*. They are classed according to the skeleton they possess into limy, glasslike, and horny fiber sponges. The latter are the sponges of commerce. With but few exceptions sponges live in salt water and are never free swimming.

III. **Cœlenterates.** — The hydra and its saltwater allies, the jellyfish, hydroids, and corals, belong to a group of animals known as the *Cœlenterata.* The word " cœlenterate " (*cœlom* = body cavity, *enteron* = food tube) explains

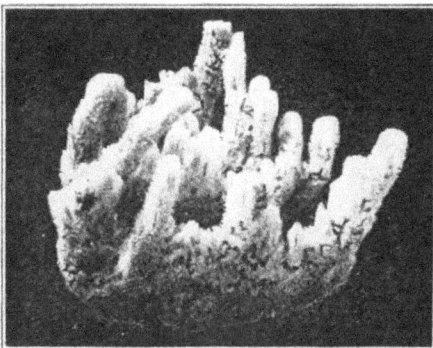

A horny fiber sponge. Notice that it is a colony. One fourth natural size.

the structure of the group. They are animals in which the real body cavity is lacking, the animal in its simplest form being little more than a bag. Some examples are the hydra, shown on page 179, salt-water forms known as hydroids, colonial forms which have part of their life free swimming as jellyfish ; sea anemones and coral polyps, tiny colonial hydra like forms which build a living or secreted covering.

Sea anemones. One half natural size. The right hand specimen is expanded and shows the mouth surrounded by the tentacles. The left hand specimen is contracted. (From model at the American Museum of Natural History.)

IV. **Worms.** — The wormlike animals are grouped into *flatworms, roundworms,* and segmented or *jointed* worms.

(*a*) Flatworms are sometimes parasitic, examples being the tapeworm and liver fluke. They are usually small, ribbon- or leaf-like and flat and live in water.

(*b*) Roundworms, minute threadlike creatures, are not often seen by the city girl or boy. Vinegar eels, the horsehair worm, the pork worm or trichina and the dread hookworm are examples.

(*c*) Segmented worms are long, jointed creatures composed of body rings or segments. Examples are the earth-worm, the sandworm (known to New York boys as the fishworm), and the leeches or bloodsuckers.

A jointed worm. The sandworm. Slightly reduced.

The common starfish seen from below to show the tube feet. About one half natural size.

V. **Echinoderms.** — These are spiny-skinned animals, which live in salt water. They are still more complicated in structure

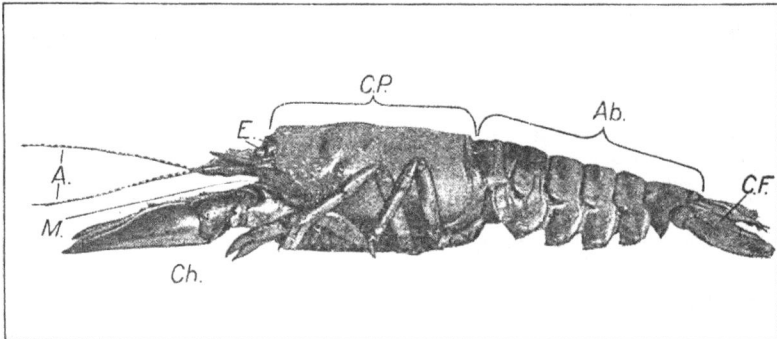

The crayfish, a crustacean. *A*, antenna; *M*, mouth; *E*, compound stalked eye; *Ch*, pincher claw; *C.P.*, cephalothorax; *Ab*, abdomen; *C.F.*, caudal fin. A little reduced.

than the worms and may be known by the spines in their skin. They show radial symmetry. Starfish or sea urchins are examples.

VI. **Arthropods.** — These animals are distinguished by having jointed body and legs. They form two great groups. The higher forms of the *Crustacea* have only two regions in the body, a fused head and thorax, called the *cephalothorax*, and an abdominal region. A second group is the *Insecta*, of which we know some-thing already. Crustacea breathe by means of *gills*, which are structures for taking oxygen out of the water, while adult insects breathe through air tubes called *trachea*.

Two smaller groups of arthropods also exist, the *Arachnida*, consisting of spiders, scorpions, ticks, and mites, and the *Myriapoda*, examples being the "thousand leggers" found in some city houses.

A common snail, a mollusk. (From a photograph by Davison.)

VII. **Mollusca.** — Another large group is the Mollusca. This phylum gets its name from the soft, unsegmented body (*mollis* = soft). Mollusks usually have a shell, which may be of one piece, as a snail, or two pieces or *valves*, as the clam or oyster.

VIII. **The Vertebrates.** — All of the animals we have studied thus far agree in having whatever skeleton or hard parts they possess on the outside of the body. Collectively, they are called *Invertebrates*. This exoskeleton differs from the main or axial

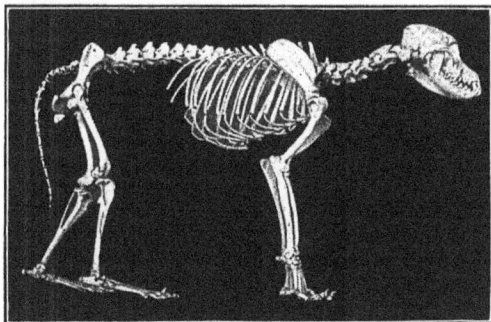

The skeleton of a dog; a typical vertebrate.

skeleton of the higher animals, the latter being inside of the body. The exoskeleton is dead, being secreted by the cells lining the body, while the endo-skeleton is, in part at least, alive and is capable of growth, *e.g.* a broken arm or leg bone will grow together. But a man has certain parts of the skeleton, as nails or hair, formed by the skin and in addition possesses inside bones to which the muscles are attached. Some of the bones are arranged in a flexible column in the *dorsal* (the back) side of the body. This *vertebral column*, as it is called, is distinctive of all *vertebrates*. Within its bony protection lies the delicate central nervous system, and to this column are attached the big bones of the legs and arms. The vertebrate animals deserve more of our attention than other forms of life because man himself i a vertebrate.

The sand shark, an elasmobranch. Note the slits leading from the gills. (From a photograph loaned by the American Museum of Natural History.)

Five groups or classes of vertebrates exist. *Fishes, Amphibians, Reptiles, Birds,* and *Mammals.* Let us see how to distinguish one class from another.

Fishes. — Fishes are familiar animals to most of us. We know that they live in the water, have a backbone, and that they have fins. They breathe by means of gills, delicate organs fitted for taking oxygen out of the water. The heart has two chambers, an auricle and a ventricle. They have a skin in which are glands

The sturgeon, a ganoid fish.

secreting mucus, a slimy substance which helps them go through the water easily. They usually lay very many eggs.

CLASSIFICATION OF FISHES

ORDER I. *The Elasmobranchs.* Fishes which have a soft skeleton made of cartilage and exposed gill slits. Examples: sharks, skates, and rays.

ORDER II. *The Ganoids.* Fishes which once were very numerous on the earth, but which are now almost extinct. They are protected by platelike scales. Examples: gars, sturgeon, and bowfin.

ORDER III. *The Teleosts, or Bony Fishes.* They compose 95 per cent of all living fishes. In this group the skeleton is bony, the gills are protected by an operculum, and the eggs are numerous. Most of our common food fishes belong to this class.

A bony fish.

ORDER IV. *The Dipnoi, or Lung Fishes.* This is a very small group. In many respects they are more like amphibians than fishes, the swim bladder being used as a lung. They live in tropical Africa, South America, and Australia, inhabiting the rivers and lakes there.

Characteristics of Amphibia. — The frog belongs to the class of vertebrates known as Amphibia. As the name indicates (*amphi*, both, and *bia*, life), members of this group live both in water and on land. In the earlier stages of their development they take oxygen into the blood by means of gills. When adult, however, they breathe by means of lungs. At all times, but especially during the winter, the skin serves as a breathing organ. The

Newt. (From a photograph loaned by the American Museum of Natural History.) About natural size.

skin is soft and unprotected by bony plates or scales. The heart has three chambers, two auricles and one ventricle. Most amphibians undergo a complete metamorphosis, or change of form, the young being unlike the adults.

CLASSIFICATION OF AMPHIBIA

ORDER I. *Urodela.* Amphibia having usually poorly developed appendages. Tail persistent through life. Examples: mud puppy, newt, salamander.

ORDER II. *Anura.* Tailless Amphibia, which undergo a metamorphosis, breathing by gills in larval state, by lungs in adult state. Examples: toad and frog.

The leopard frog, an amphibian.

Characteristics of Reptilia. — These animals are characterized by having scales developed from the skin. In the turtle they have become bony and are connected with the internal skeleton. Reptiles always breathe by means of lungs, differing in this respect from the amphibians. They show their distant relationship to birds in that their large eggs are incased in a leathery, limy shell.

CLASSIFICATION OF REPTILES

ORDER I. *Chelonia* (turtles and tortoises). Flattened reptiles with body inclosed in bony case. No teeth or sternum (breastbone). Examples: snapping turtle, box tortoise.

ORDER II. *Lacertilia* (lizards). Body covered with scales, usually having two-paired appendages. Breathe by lungs. Examples: fence lizard, horned toad.

Box tortoise, a land reptile. (From photograph loaned by the American Museum of Natural History.) About one fourth natural size.

The gila monster, a poisonous lizard. About one twelfth natural size.

ORDER III. *Ophidia* (snakes). Body elongated, covered with scales. No limbs present. Examples : garter snake, rattlesnake.

ORDER IV. *Crocodilia.* Fresh-water reptiles with elongated body and bony scales on skin. Two-paired limbs. Examples : alligator, crocodile.

The common garter snake. Reduced to about one tenth natural size.

Birds. — Birds among all other animals are known by their covering of feathers and the presence of wings. The feathers are developed from the skin. These aid in flight, and protect the body from the cold.

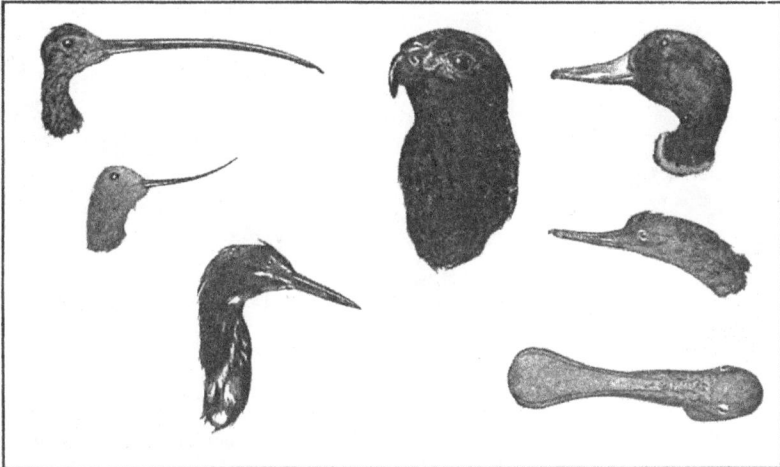

Adaptations in the bills of birds. Could we tell anything about the food of a bird from its bill? Do these birds all get their food in the same manner? Do they all eat the same kind of food?

The form of the bill in particular shows adaptation to a wonderful degree. A duck has a flat bill for pushing through the mud and straining out the food ; a bird of prey has a curved or hooked beak for tearing; the woodpecker has a sharp, straight bill for piercing the bark of trees in search of the insect larvæ which are hidden underneath. Birds do not have teeth.

The rate of respiration, of heartbeat, and the body temperature are all higher in the bird than in man. Man breathes from twelve to fourteen times per minute. Birds breathe from twenty to sixty times a minute. Because of the increased activity of a bird, there comes a necessity for a greater and more rapid supply of oxygen, an increased blood supply to carry the material to be used up in the release of energy, and a means of rapid excretion of the wastes resulting from the process of oxidation. Birds are

Common tern and young, showing nesting and feeding habits. (From group at American Museum of Natural History.)

large eaters, and the digestive tract is fitted to digest the food quickly, by having a large crop in which food may be stored in a much softened condition. As soon as the food is part of the blood, it may be sent rapidly to the places where it is needed, by means of the large four-chambered heart and large blood vessels.

The high temperature of the bird is a direct result of this rapid oxidation; furthermore, the feathers and the oily skin form an insulation which does not readily permit of the escape of heat. This insulating cover is of much use to the bird in its flights at high altitudes, where the temperature is often very low. Birds lay eggs and usually care for their young.

CLASSIFICATION OF BIRDS

ORDER I. *Cursores.* Running birds with no keeled breastbone. Examples: ostrich, cassowary.

ORDER II. *Passeres.* Perching birds; three toes in front, one behind. Over one half of all species of birds are included in this order. Examples: sparrow, thrush, swallow.

ORDER III. *Gallinæ.* Strong legs; feet adapted to scratching. Beak stout. Examples: jungle fowl, grouse, quail, domestic fowl.

ORDER IV. *Raptores.* Birds of prey. Hooked beak. Strong claws. Examples: eagle, hawk, owl.

ORDER V. *Grallatores.* Waders. Long neck, beak, and legs. Examples: snipe, crane, heron.

ORDER VI. *Natatores.* Divers and swimmers. Legs short, toes webbed. Examples: gull, duck, albatross.

ORDER VII. *Columbinæ.* Like Gallinæ, but with weaker legs. Examples: dove, pigeon.

ORDER VIII. *Pici.* Woodpeckers. Two toes point forward, two backward, and adaptation for climbing. Long, strong bill.

African ostrich, one of the largest living birds.

ORDER IX. *Psittaci.* Parrots, hooked beak and fleshy tongue.

ORDER X. *Coccyges.* Climbing birds, with powerful beak. Examples: kingfisher, toucan, and cuckoo.

ORDER XI. *Macrochires.* Birds having long-pointed wings, without scales on metatarsus. Examples: swift, humming bird, and goatsucker.

Mammals. — Dogs and cats, sheep and pigs, horses and cows, all of our domestic animals (and man himself) have characters of structure which cause them to be classed as mammals. They, like some other vertebrates, have lungs and warm blood. They also have a hairy covering and bear young developed to a form similar to their own,[1] and nurse them with milk secreted by glands known as the *mammary glands;* hence the term " mammal."

[1] With the exception of the monotremes.

Adaptations in Mammalia. — Of the thirty-five hundred species, most inhabit continents; a few species are found on different islands, and some, as the whale, inhabit the ocean. They vary in size from the whale and the elephant to tiny shrew mice and moles. Adaptations to different habitat and methods of life abound; the seal and whale have the limbs modified into flippers, the sloth and squirrel have limbs peculiarly adapted to climbing, while the bats have the fore limbs modeled for flight.

The bison, an almost extinct mammal.

Lowest Mammals. — The lowest are the monotremes, animals which lay eggs like the birds, although they are provided with hairy covering like other mammals. Such are the Australian spiny anteater and the duck mole.

All other mammals bring forth their young developed to a form similar to their own. The kangaroo and opossum, however, are provided with a pouch on the under side of the body in which the very immature, blind, and helpless young are nourished until they are able to care for themselves. These pouched animals are called *marsupials*.

The other mammals may be briefly classified as follows: —

CLASSIFICATION OF HIGHER MAMMALS

ORDER I. *Edentata.* Toothless or with very simple teeth. Examples: anteater, sloth, armadillo.

ORDER II. *Rodentia.* Incisor teeth chisel-shaped, usually two above and two below. Examples: beaver, rat, porcupine, rabbit, squirrel.

ORDER III. *Cetacea.* Adapted to marine life. Examples: whale, porpoise

ORDER IV. *Ungulata.* Hoofs, teeth adapted for grinding. Examples: (a) odd-toed, horse, rhinoceros, tapir; (b) even-toed, ox, pig, sheep, deer.

ORDER V. *Carnivora.* Long canine teeth, sharp and long claws. Examples: dog, cat, lion, bear, seal, and sea lion.

ORDER VI. *Insectivora.* Example: mole.

ORDER VII. *Cheiroptera.* Fore limbs adapted to flight, teeth pointed. Example: bat.

ORDER VIII. *Primates.* Erect or nearly so, fore appendage provided with hand. Examples: monkey, ape, man.

Increasing Complexity of Structure and of Habits in Plants and Animals. — In our study of biology so far we have attempted to get some notion of the various factors which act upon living things. We have seen how plants and animals interact upon each other. We have learned something about the various physiological processes of plants and animals, and have found them to be in many respects identical. We have found grades of complexity in plants from the one-celled plant, bacterium or pleurococcus, to the complicated flowering plants of considerable size and with many

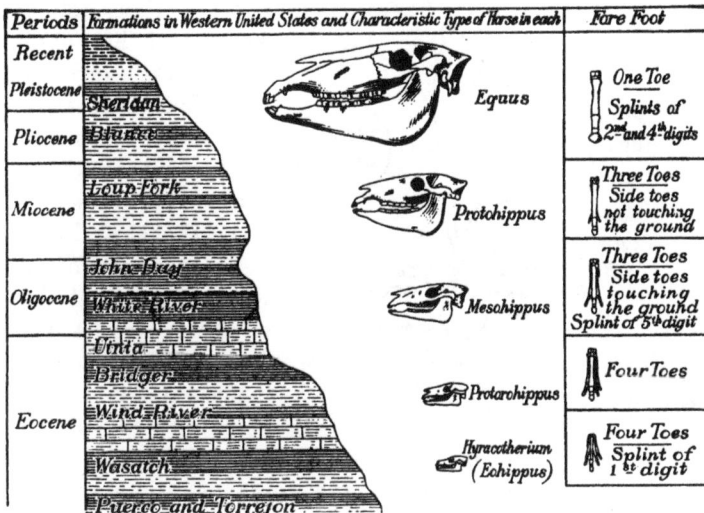

Periods	Formations in Western United States and Characteristic Type of Horse in each	Fore Foot
Recent		One Toe
Pleistocene	Sheridan	Equus, Splints of 2nd and 4th digits
Pliocene	Blanco	
Miocene	Loup Fork, Protohippus	Three Toes, Side toes not touching the ground
Oligocene	John Day, White River, Mesohippus	Three Toes, Side toes touching the ground, Splint of 5th digit
Eocene	Uinta, Bridger, Protorohippus	Four Toes
	Wind River, Wasatch, Hyracotherium (Echippus)	Four Toes, Splint of 1st digit
	Puerco and Torrejon	

The geological history of the horse. (After Mathews, in the American Museum of Natural History.)   Ask your teacher to explain this diagram.

organs. So in animal life, from the Protozoa upward, there is constant change, and the change is toward greater complexity of structure and functions. An insect is a higher type of life than a protozoan, because its structure is more complex and it can perform its work with more ease and accuracy. A fish is a higher type of animal than the insect for these same reasons, and also for another. The fish has an internal skeleton which forms a pointed column of bones on the *dorsal* side (the back) of the animal. It is a vertebrate animal.

The Doctrine of Evolution. — We have now learned that animal forms may be arranged so as to begin with very simple one-celled forms and culminate with a group which contains man himself. This arrangement is called the *evolutionary series*. Evolution means change, and these groups are believed by scientists to represent stages in complexity of development of life on the earth. Geology teaches that millions of years ago, life upon the earth was very simple, and that gradually more and more complex forms of life appeared, as the rocks formed latest in time show the most highly developed forms of animal life. The great English scientist, Charles Darwin, from this and other evidence, explained the theory of evolution. This is the belief that simple forms of life on the earth slowly and gradually gave rise to those more complex and that thus ultimately the most complex forms came into existence.

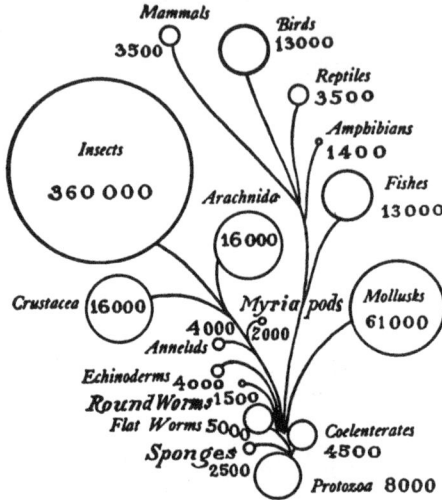

The evolutionary tree. Modified from Galloway. Copy this diagram in your notebook. Explain it as well as you can.

The Number of Animal Species. — Over 500,000 species of animals are known to exist to-day, as the following table shows.

Protozoa	8,000	Arachnids	16,000
Sponges	2,500	Crustaceans . . .	16,000
Cœlenterates . . .	4,500	Mollusks	61,000
Echinoderms . . .	4,000	Fishes	13,000
Flatworms . . .	5,000	Amphibians . . .	1,400
Roundworms . . .	1,500	Reptiles	3,500
Annelids	4,000	Birds	13,000
Insects	360,000	Mammals	3,500
Myriapods	2,000	Total	518,900

Man's Place in Nature. — Although we know that man is separated mentally by a wide gap from all other animals, in our study of physiology we must ask where we are to place man. If we attempt to classify man, we see at once he must be placed with the vertebrate animals because of his possession of a vertebral column. Evidently, too, he is a mammal, because the young are nourished by milk secreted by the mother and because his body has at least a partial covering of hair. Anatomically we find that we must place man with the apelike mammals, because of these numerous points of structural likeness. The group of mammals which includes the monkeys, apes, and man we call the *primates.*

Although anatomically there is a greater difference between the lowest type of monkey and the highest type of ape than there is between the highest type of ape and the lowest savage, yet there is an immense mental gap between monkey and man.

Instincts. — Mammals are considered the highest of vertebrate animals, not only because of their complicated structure, but because their instincts are so well developed. Monkeys certainly seem to have many of the mental attributes of man.

Professor Thorndike of Columbia University sums up their habits of learning as follows : —

" In their method of learning, although monkeys do not reach the human stage of a rich life of ideas, yet they carry the animal method of learning, by the selection of impulses and association of them with different sense-impressions, to a point beyond that reached by any other of the lower animals. In this, too, they resemble man ; for he differs from the lower animals not only in the possession of a new sort of intelligence, but also in the tremendous extension of that sort which he has in common with them. A fish learns slowly a few simple habits. Man learns quickly an infinitude of habits that may be highly complex. Dogs and cats learn more than the fish, while monkeys learn more than they. In the number of things he learns, the complex habits he can form, the variety of lines along which he can learn them, and in their permanence when once formed, the monkey justifies his inclusion with man in a separate mental genus."

Evolution of Man. — Undoubtedly there once lived upon the earth races of men who were much lower in their mental organization than the present inhabitants. If we follow the early history of man upon the earth, we find that at first he must have been little better than one of the lower animals. He was a nomad, wandering from place to place, feeding upon whatever living things he could kill with his hands. Gradually he must have learned to use weapons, and thus kill his prey, first using rough stone implements for this purpose. As man became more civilized, implements of bronze and of iron were used. About this time the subjugation and domestication of animals began to take place. Man then began to cultivate the fields, and to have a fixed place of abode other than a cave. The beginnings of civilization were long ago, but even to-day the earth is not entirely civilized.

The Races of Man. — At the present time there exist upon the earth five races or varieties of man, each very different from the other in instincts, social customs, and, to an extent, in structure. These are the Ethiopian or negro type, originating in Africa; the Malay or brown race, from the islands of the Pacific; the American Indian; the Mongolian or yellow race, including the natives of China, Japan, and the Eskimos; and finally, the highest type of all, the Caucasians, represented by the civilized white inhabitants of Europe and America.

REFERENCE BOOKS

ELEMENTARY

Hunter, *Laboratory Problems in Civic Biology*, American Book Company.
Bulletin of U.S. Department of Agriculture, *Division of Biological Survey*, Nos. 1, 6, 13, 17.
Davison, *Practical Zoölogy.* American Book Company.
Ditmars, *The Reptiles of New York.* Guide Leaflet 20. Amer. Mus. of Nat. History.
Sharpe, *A Laboratory Manual in Biology*, pp. 140–150, American Book Company.
Walker, *Our Birds and Their Nestlings.* American Book Company.
Walter, H. E. and H. A., *Wild Birds in City Parks.* Published by authors.

ADVANCED

Apgar, *Birds of the United States.* American Book Company.
Beebe, *The Bird.* Henry Holt and Company.
Ditmars, *The Reptile Book.* Doubleday, Page and Company.
Hegner, *Zoölogy.* The Macmillan Company.
Hornaday, *American Natural History.*
Jordan and Evermann, *Food and Game Fishes.* Doubleday, Page and Company
Parker and Haswell, *Textbook of Zoölogy.* The Macmillan Company.
Riverside Natural History. Houghton, Mifflin and Company.
Weed and Dearborn, *Relation of Birds to Man.* Lippincott.

XV. THE ECONOMIC IMPORTANCE OF ANIMALS

Problems. — *I. To determine the uses of animals.*

(a) Indirectly as food.

(b) Directly as food.

(c) As domesticated animals.

(d) For clothing.

(e) Other direct economic uses.

(f) Destruction of harmful plants and animals.

II. To determine the harm done by animals.

(a) Animals destructive to those used for food.

(b) Animals harmful to crops and gardens.

(c) Animals harmful to fruit and forest trees.

(d) Animals destructive to stored food or clothing.

(e) Animals indirectly or directly responsible for disease.

LABORATORY SUGGESTIONS

Inasmuch as this work is planned for the winter months the laboratory side must be largely museum and reference work. It is to be expected that the teacher will wish to refer to much of this work at the time work is done on a given group. But it is pedagogically desirable that the work as planned should be *varied*. Interest is thus held. Outlines prepared by the teacher to be filled in by the student are desirable because they lead the pupil to individual selection of what seems to *him* as important material. Opportunity should be given for laboratory exercises based on original sources. The pupils should be made to use reports of the U. S. Department of Agriculture, the Biological Survey, various States Reports, and others.

Special home laboratory reports may be well made at this time, for example: determination at a local fish market of the fish that are cheap and fresh at a given time. Have the students give reasons for this. Study conditions in the meat market in a similar manner. Other local food conditions may also be studied first hand.

197

USES OF ANIMALS

Indirect Use as Food. — Just as plants form the food of animals, so some animals are food for others. Man may make use of such food directly or indirectly. Many mollusks, as the barnacle and mussel, are eaten by fishes. Other fish live upon tiny organisms, water fleas and other small crustaceans. These in turn feed upon still smaller animals, and we may go back and back until finally we come to the Protozoa and one-celled water plants as an ultimate source of food.

Direct Use as Food. Lower Forms. — The forms of life lower than the Crustacea are of little use directly as food, although the Chinese are very fond of one of the Echinoderms, a holothurian.

Crustacea as Food. — Crustaceans, however, are of considerable value for food, the lobster fisheries in particular being of importance. The lobster is highly esteemed as food, and is rapidly disappearing from our coasts as the result of overfishing. Between twenty and thirty million are yearly taken on the North Atlantic coast. This means a value at present prices of about $15,000,000. Laws have been enacted in New York and other states against overfishing. Egg-carrying lobsters must be returned to the water; all smaller than six to nine inches in length (the law varies in different states) must be put back; other restrictions are placed upon the taking of the animals, in hope of saving the race from extinction. Some states now hatch and care for the young for a period of time; the United States Bureau of Fisheries is also doing much good work, in the hope of restocking to some extent the now almost depleted waters.

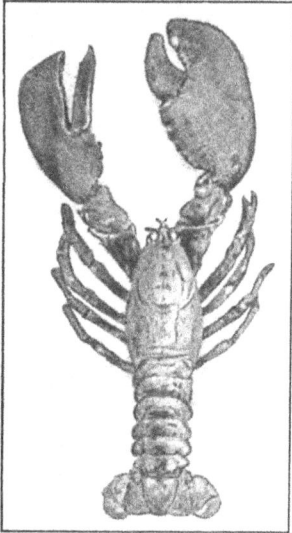

North American lobster. This specimen, preserved at the U. S. Fish Commission at Woods Hole, was of unusual size and weighed over twenty pounds.

Several other common crustaceans are near relatives of the crayfish. Among them are the shrimp and prawn, thin-shelled, active crustaceans common along our eastern coast. In spite of the fact that they form a large part of the food supply of many marine animals, especially fish, they do not appear to be decreasing in numbers. They are also used as food by man, the shrimp fisheries in this country aggregating over $1,000,000 yearly.

Another edible crustacean of considerable economic importance is the blue crab. Crabs are found inhabiting muddy bottoms; in such localities they are caught in great numbers in nets or traps baited with decaying meat. They are, indeed, among our most valuable sea scavengers, although they are carnivorous hunters as well.

The edible blue crab. (From a photograph loaned by the American Museum of Natural History.)

The young crabs differ considerably in form from the adult. They undergo a complete *metamorphosis* (change of form). Immediately after molting or shedding of the outer shell in order to grow larger, crabs are greatly desired by man as an article of food. They are then known as " shedders," or soft-shelled crabs.

Mollusks as Food. — Oysters are never found in muddy localities, for in such places they would be quickly smothered by the sediment in the water. They are found in nature clinging to stones or on shells or other objects which project a little above the bottom. Here food is abundant and oxygen is obtained from the water surrounding them. Hence oyster raisers throw oyster shells into the water and the young oysters attach themselves.

The oyster.

In some parts of Europe and this country where oysters are raised artificially, stakes or brush are sunk in shallow water so that the young oyster, which is at first free-swimming, may escape the danger of smothering on the bottom. After the oysters are a year or two old, they are taken up and put down in deeper water as seed oysters. At the age of three and four years they are ready for the market.

The oyster industry is one of the most profitable of our fisheries. Nearly

$15,000,000 a year has been derived during the last decade from such sources. Hundreds of boats and thousands of men are engaged in dredging for oysters. Three of the most important of our oyster grounds are Long Island Sound, Narragansett Bay, and Chesapeake Bay.

Sometimes oysters are artificially " fattened " by placing them on beds near the mouths of fresh-water streams. Too often these streams are the bearers of much sewage, and the oyster, which lives on microscopic organisms, takes in a number of bacteria with other food. Thus a person might become infected with the typhoid bacillus by eating raw oysters. State and city supervision of the oyster industry makes this possibility very much less than it was a few years ago, as careful bacteriological analysis of the surrounding water is constantly made by competent experts.

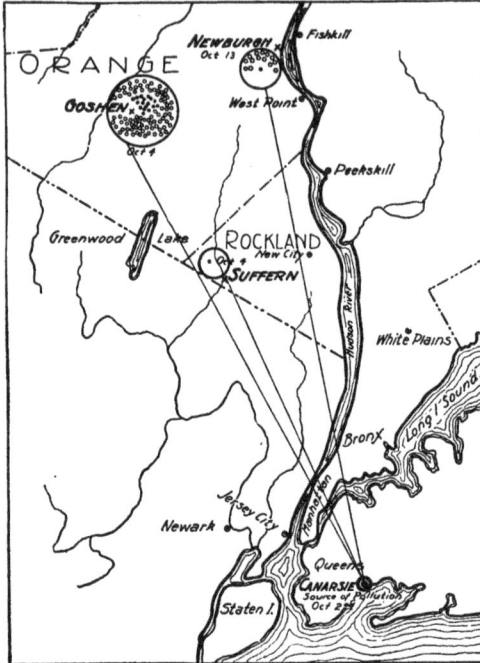

This diagram shows how cases of intestinal disease (typhoid and diarrhea) have been traced to oysters from a locality where they were "fattened" in water contaminated with sewage. (Loaned by American Museum of Natural History.)

Clams. — Other bivalve mollusks used for food are clams and scallops. Two species of the former are known to New Yorkers, one as the "round," another as the "long" or "soft-shelled" clams. The former (*Venus mercenaria*) was called by the Indians "quahog," and is still so called in the Eastern states. The blue area of its shell was used by the Indians to make wampum, or money. The quahog is now extensively used as food. The "long" clam (*Mya arenaria*) is considered better eating by the inhabitants of Massachusetts and Rhode Island. This clam was highly prized as food by the Indians. The clam industries of the eastern coast aggregate nearly $1,000,000 a year. The dredging for scallops, another molluscan delicacy, forms an important industry along certain parts of the eastern coast.

Fish as Food. — Fish are used as food the world over. From very early times the herring were pursued by the Norsemen. Fresh-water fish, such as whitefish, perch, pickerel, pike, and the various members of the trout family, are esteemed food and, especially in the Great Lake region, form important fisheries. But by far the most important food fishes are those which are taken in salt water. Here we have two types of fisheries, those where the fish comes up a river to spawn, as the salmon, sturgeon, or shad,

Salmon leaping a fall on their way to their spawning beds. (Photographed by Dr. John A. Sampson.)

and those in which fishes are taken on their feeding grounds in the open ocean. Herring are the world's most important catch, though not in this country. Here the salmon of the western

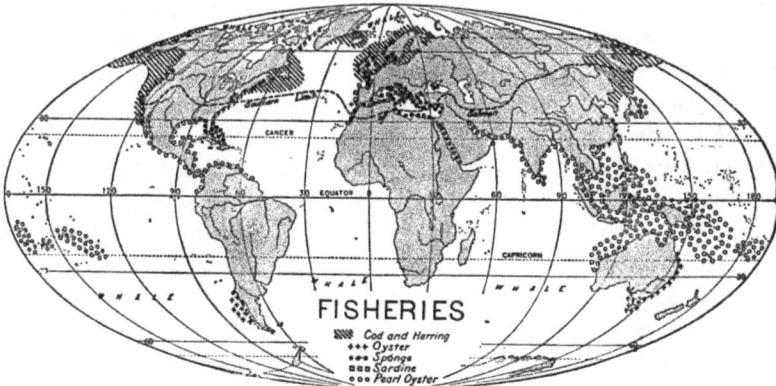

FISHERIES

Cod and Herring
Oyster
Sponge
Sardine
Pearl Oyster

Globe Fisheries.

coast is taken to the value of over $13,000,000 a year. Cod fishing also forms an important industry; over 7000 men being employed and over $2,000,000 of codfish being taken each year in this country.

Hundreds of other species of fish are used as food, the fish that is nearest at hand being often the cheapest and best. Why, for example, is the flounder so cheap in the New York markets? In what waters are the cod and herring fisheries, sardine, oyster, sponge, pearl oyster? (See chart on page 201.)

Amphibia and Reptiles as Food. — Frogs' legs are esteemed a delicacy. Certain reptiles are used as food by people of other nationalities, the Iguana, a Mexican lizard, being an example. Many of the sea-water turtles are of large size, the leatherback and the green turtle often weighing six hundred to seven hundred pounds each. The flesh of the green turtle and especially of the diamond-back terrapin, an animal found in the salt marshes along our southeastern coast, is highly esteemed as food. Unfortunately for the preservation of the species, these animals are usually taken during the breeding season when they go to sandy beaches to lay their eggs.

Birds as Food. — Birds, both wild and domesticated, form part of our food supply. Unfortunately our wild game birds are disappearing so fast that we should not consider them as a source of food. Our domestic fowls, turkey, ducks, etc., form an important food supply and poultry farms give lucrative employment to many people. Eggs of domesticated birds are of great importance as food, and egg albumin is used for other purposes, — clarifying sugars, coating photographic papers, etc.

Mammals as Food. — When we consider the amount of wealth invested in cattle and other domesticated animals bred and used for food in the United States, we see the great economic importance of mammals. The United States, Argentina, and Australia are the greatest producers of cattle. In this country hogs are largely raised for food. They are used fresh, salted, smoked as ham and bacon, and pickled. Sheep, which are raised in great quantities in Australia, Argentina, Russia, Uruguay, and this country, are one of the world's greatest meat supplies.

Goats, deer, many larger game animals, seals, walruses, etc., give food to people who live in parts of the earth that are less densely populated.

Domesticated Animals. — When man emerged from his savage state on the earth, one of the first signs of the beginning of civilization was the domestication of animals. The dog, the cow, sheep, and especially the horse, mark epochs in the advance of civilization. Beasts of burden are used the world over, horses almost all over the world, certain cattle, as the water buffalo, in tropical Malaysia; camels, goats, and the llama are also used as draft animals in some other countries.

Feeding silkworms. The caterpillars are the white objects in the trays.

Man's wealth in many parts of the world is estimated in terms of his cattle or herds of sheep. So many products come from these sources that a long list might be given, such as meats, milk, butter, cheese, wool, or other body coverings, leather, skins, and hides used for other purposes. Great industries are directly dependent upon our domesticated animals, as the making of shoes, the manufacture of woolen cloth, the tanning industry, and many others.

Uses for Clothing. — The manufacture of silk is due to the production of raw silk by the silkworm, the caterpillar of a moth. It lives upon the mulberry and makes a cocoon from which the silk is wound. The Chinese silkworm is now raised to a slight extent in southern California. China, Japan, Italy, and France, because of cheaper labor, are the most successful silk-raising countries.

The use of wool gives rise to many great industries. After the wool is cut from the sheep, it has to be washed and scoured to get out the dirt and grease. This wool fat or lanoline is used in making soap and ointments. The wool is next " carded," the fibers being interwoven by the fine teeth of the carding machine or " combed," the fibers here being pulled out parallel to each other. Carded wool becomes woolen goods; combed wool, worsted goods. The wastes are also utilized, being mixed with

Polar bear, a fur-bearing mammal which is rapidly being exterminated. Why?

" shoddy " (wool from cloth cuttings or rags) to make woolen goods of a cheap grade.

Goat hair, especially that of the Angora and the Cashmere goat, has much use in the clothing industries. Camel's hair and alpaca are also used.

Fur. — The furs of many domesticated and wild animals are of importance. The Carnivora as a group are of much economic importance as the source of most of our fur. The fur seal fisheries alone amount to many millions of dollars annually. Otters, skunks, sables, weasels, foxes, and minks are of considerable importance as fur producers. Even cats are now used for fur, usually masquerading under some other name. The fur of the beaver, one of the largest of the rodents or gnawing mammals, is of considerable value, as are the coats of the chinchilla, muskrats, squirrels, and other rodents. The fur of the rabbit and nutria are used in the manufacture of felt hats. The quills of the porcupines (greatly developed and stiffened hairs) have a slight commercial value.

Conservation of Fur-bearing Animals Needed. — As time goes on and the furs of wild animals become scarcer and scarcer through overkilling, we find the need for protection and conservation of many of these fast-vanishing wild forms more and more imperative. Already breeding of some fur-bearing animals has been tried with success, and cheap substitutes for wild animal skins are coming more and more into the markets. Black-fox breeding has been tried successfully in Prince Edward Island, Canada, $2500 to $3000 being given for a single skin. Skunk, marten, and mink are also being bred for the market. Game preserves in this country and Canada are also helping to preserve our wild fur-bearing animals.

Animal Oils. — Whale oil, obtained from the fat or " blubber " of whales, is used extensively for lubricating. Neat's-foot oil comes from the feet of cattle and is also used in lubrication. Tallow and lard, two fats from cattle, sheep, and pigs, have so many well-known uses that comment is unnecessary. Cod-liver oil is used medically and is well known. But it is not so widely known that a fish called the menhaden or " moss bunkers " of the Atlantic coast produces over 3,000,000 gallons of oil every year and is being rapidly exterminated in consequence.

Hides, Horns, Hoofs, etc. — Leathers, from cattle, horses, sheep, and goats, are used everywhere. Leather manufacture is one of the great industries of the Eastern states, hundreds of millions of dollars being invested in its manufacturing plants. Horns and bones are utilized for making combs, buttons, handles for brushes, etc. Glue is made from the animal matter in bones. Ivory, obtained from elephant, walrus, and other tusks, forms a valuable commercial product. It is largely used for knife handles, piano keys, combs, etc.

Perfumes. — The musk deer, musk ox, and muskrat furnish a valuable perfume called musk. Civet cats also give us a somewhat similar perfume. Ambergris, a basis for delicate perfumes, comes from the intestines of the sperm whale.

Protozoa. — The Protozoa have played an important part in rock buildings. The chalk beds of Kansas and other chalk formations are made up to a large extent of the tiny skeletons of *Protozoa*, called *Foraminifera*. Some limestone rocks are also composed in large part of such skeletons. The skeletons of some species are used to make a polishing powder.

Sponges. — The sponges of commerce have the skeleton composed of tough fibers of material somewhat like that of cow's horn. This fiber is elastic and has the power of absorbing water. In a living state, the horny fiber sponge is a dark-colored fleshy mass, usually found attached to rocks. The warm waters of the Mediterranean Sea and the West Indies furnish most of our sponges. The sponges are pulled up from their resting place on the bottom, by means of long-handled rakes operated by men in boats or are secured by divers. They are then spread out on the shore in the sun, and the living tissues allowed to decay; then after treatment consisting of beating, bleaching, and trimming, the bath sponge is ready for the market. Some forms of coral are of commercial value. The red coral of the Mediterranean Sea is the best example.

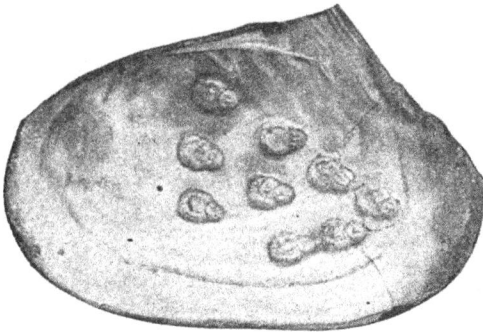

In some countries little metal images of Buddha are placed within the shells of living pearl oysters or clams. Over these the mantle of the animal secretes a layer of mother of pearl as is shown in the picture.

Pearls and Mother of Pearl. — Pearls are prized the world over. It is a well-known fact that even in this country pearls of some value are sometimes found within the shells of the fresh-water mussel and the oyster. Most of the finest, however, come from the waters around Ceylon. If a pearl is cut open and examined carefully, it is found to be a deposit of the mother-of-pearl layer of the shell around some central structure. It has been believed that any foreign substance, as a grain of sand, might irritate the mantle at a given point, thus stimulating it to secrete around the substance. It now seems likely that most perfect pearls are due to the growth within the mantle of the clam or oyster of certain parasites, stages in the development of a flukeworm. The irritation thus set up in the tissue causes mother of pearl to be deposited around the source of irritation, with the subsequent formation of a pearl.

The pearl-button industry in this country is largely dependent upon the fresh-water mussel, the shells of which are used. This mussel is being so rapidly depleted that the national government is working out a means of artificial propagation of these animals.

Honey and Wax. — Honeybees[1] are kept in hives. A colony consists of a queen, a female who lays the eggs for the colony, the drones, whose duty it is to fertilize the eggs, and the workers.

The cells of the comb are built by the workers out of wax secreted from the under surface of their bodies. The wax is cut off in thin plates by means of the wax shears between the two last joints of the hind legs. These cells are used to place the eggs of the queen in, one egg to each cell, and the young are hatched after three days, to begin life as footless white grubs.

The young are fed for several days, then shut up in the cells

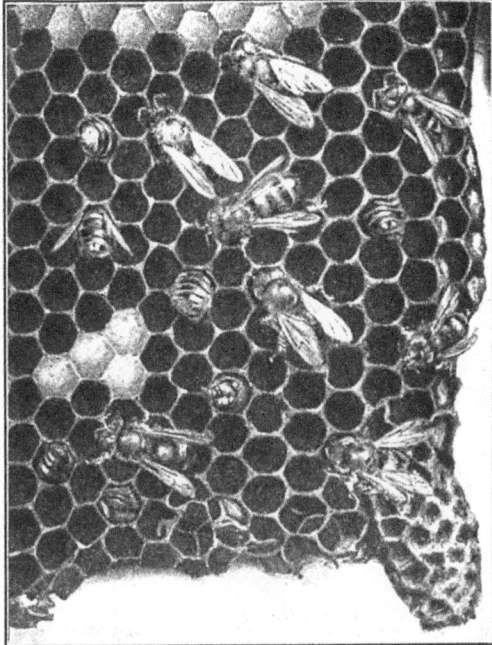

Cells of honeycomb, queen cell on right at bottom.

and allowed to form pupæ. Eventually they break their cells and take their place as workers in the hive, first as nurses for the young and later as pollen gatherers and honey makers.

We have already seen (pages 37 to 39) that the honeybee gathers nectar, which she swallows, keeping the fluid in her crop until her return to the hive. Here it is forced out into cells of

[1] Their daily life may be easily watched in the schoolroom, by means of one of the many good and cheap observation hives now made to be placed in a window frame. Directions for making a small observation hive for school work can be found in Hodge, *Nature Study and Life*, Chap. XIV. Bulletin No. 1, U. S. Department of Agriculture, entitled *The Honey Bee*, by Frank Benton, is valuable for the amateur beekeeper. It may be obtained for twenty-five cents from the Superintendent of Documents, Union Building, Washington, D.C.

the comb. It is now thinner than what we call honey. To thicken it, the bees swarm over the open cells, moving their wings very rapidly, thus evaporating some of the water. A hive of bees have been known to make over thirty-one pounds of honey in a single day, although the average is very much less than this. It is estimated from twenty to thirty millions of dollars' worth of honey and wax are produced each year in this country.

Cochineal and Lac. — Among other products of insect origin is cochineal, a red coloring matter, which consists of the dried bodies of a tiny insect, one of the plant lice which lives on the cactus plants in Mexico and Central America. The lac insect, another one of the plant lice, feeds on the juices of certain trees in India and pours out a substance from its body which after treatment forms shellac. Shellac is of much use as a basis for varnish.

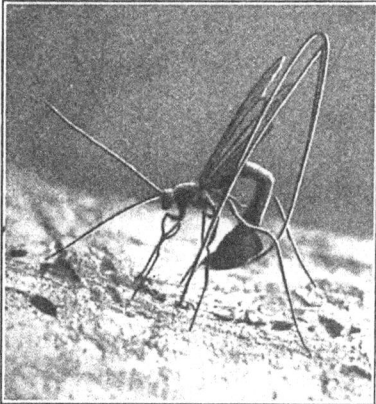

An insect friend of man. An ichneumon fly boring in a tree to lay its eggs in the burrow of a boring insect harmful to that tree.

Gall Insects. — Oak galls, growths caused by the sting of wasp-like insects, give us products used in ink making, in tanning, and in making pyrogaliic acid which is much used in developing photographs.

Insects destroy Harmful Plants or Animals. — Some forms of animal life are of great importance because of their destruction of harmful plants or animals.

A near relative of the bee, called the ichneumon fly, does man indirectly considerable good because of its habit of laying its eggs and rearing the young in the bodies of caterpillars which are harmful to vegetation. Some of the ichneumons even bore into trees in order to deposit their eggs in the larvæ of wood-boring insects. It is safe to say that the ichneumons save millions of dollars yearly to this country.

Several beetles are of value to man. Most important of these is the natural enemy of the orange-tree scale, the ladybug, or ladybird beetle. In New York state it may often be found feeding upon the plant lice, or aphids, which live on rosebushes. The carrion beetles and many water beetles act as scavengers. The sexton beetles bury dead carcasses of animals. Ants in tropical countries are particularly useful as scavengers.

Insects, besides pollinating flowers, often do a service by eating harmful weeds. Thus many harmful plants are kept in check. We have noted that they spin silk, thus forming clothing; that in many cases they are preyed upon, and that they supply an enormous multitude of birds, fishes, and other animals with food.

Use of the Toad. — The toad is of great economic importance to man because of its diet. No less than eighty-three species of insects, mostly injurious, have been proved to enter into the dietary. A toad has been observed to snap up one hundred and twenty-eight flies in half an hour. Thus at a low estimate it could easily destroy one hundred insects during a day and do an immense service to the garden during the summer. It has been estimated by Kirkland that a single toad may, on account of the cutworms which it kills, be worth $19.88 each season it lives, if the damage done by each cutworm be estimated at only one cent. Toads also feed upon slugs and other garden pests.

The common toad, an insect eater.

Birds eat Insects. — The food of birds makes them of the greatest economic importance to our country. This is because of the relation of insects to agriculture. A large part of the diet of most of our native birds includes insects harmful to vegetation. Investigations undertaken by the United States Department of

Agriculture (Division of Biological Survey) show that a surprisingly large number of birds once believed to harm crops really perform a service by killing injurious insects. Even the much maligned crow lives to some extent upon insects. Swallows in the Southern states kill the cotton-boll weevil, one of our worst insect pests. Our earliest visitor, the bluebird, subsists largely on injurious insects, as do woodpeckers, cuckoos, kingbirds, and many others. The robin, whose presence in the cherry tree we resent, during the rest of the summer does much good by feeding upon noxious insects. Birds use the food substances which are most abundant around them at the time.[1]

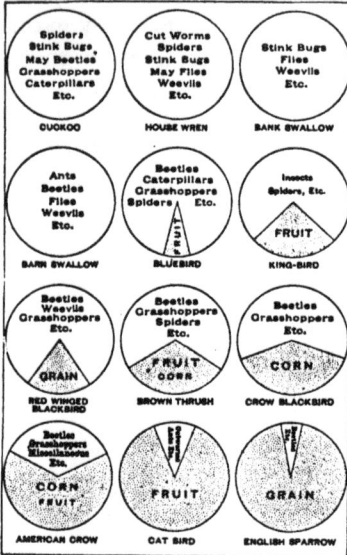

Food of some common birds. Which of the above birds should be protected by man and why?

Birds eat Weed Seeds. — Not only do birds aid man in his battles with destructive insects, but seed-eating birds eat the seeds of weeds. Our native sparrows (not the English sparrow), the mourning dove, bobwhite, and other birds feed largely upon the seeds of many of our common weeds. This fact alone is sufficient to make birds of vast economic importance.

[1] The following quotation from I. P. Trimble, *A Treatise on the Insect Enemies of Fruit and Shade Trees*, bears out this statement: "On the fifth of May, 1864, . . . seven different birds . . . had been feeding freely upon small beetles. . . . There was a great flight of beetles that day; the atmosphere was teeming with them. A few days after, the air was filled with Ephemera flies, and the same species of birds were then feeding upon them."

During the outbreak of Rocky Mountain locusts in Nebraska in 1874–1877, Professor Samuel Aughey saw a long-billed marsh wren carry thirty locusts to her young in an hour. At this rate, for seven hours a day, a brood would consume 210 locusts per day, and the passerine birds of the eastern half of Nebraska, allowing only twenty broods to the square mile, would destroy daily 162,771,000 of the

Not all birds are seed or insect feeders. Some, as the cormorants, ospreys, gulls, and terns, are active fishers. Near large cities gulls especially act as scavengers, destroying much floating garbage that otherwise might be washed ashore to become a menace to health. The vultures of India and semitropical countries are of immense value as scavengers. Birds of prey (owls) eat living mammals, including many rodents; for example, field mice, rats, and other pests.

Extermination of our Native Birds. — Within our own times we have witnessed the almost total extermination of some species of our native birds. The American passenger pigeon, once very abundant in the Middle West, is now extinct. Audubon, the greatest of all American bird lovers, gives a graphic account of the migration of a flock of these birds. So numerous were they that when the flock rose in the air the sun was darkened, and at night the weight of the roosting birds broke down large branches of the trees in which they rested. To-day not a single wild specimen of this pigeon can be found, because they were slaughtered by the hundreds of thousands during the breeding season. The wholesale killing of the snowy egret to furnish ornaments for ladies' headwear is another example of the improvidence of our fellow-countrymen. Charles Dudley Warner said, " Feathers do not improve the appearance of an ugly woman, and a pretty woman needs no such aid." Wholesale killing for plumage, eggs, and food, and, alas, often for mere sport, has reduced the number of our birds more than one half in thirty states and territories within the past fifteen years. Every crusade against indiscriminate killing of our native birds should be welcomed by all thinking

pests. The average locust weighs about fifteen grains, and is capable each day of consuming its own weight of standing forage crops, which at $10 per ton would be worth $1743.26. This case may serve as an illustration of the vast good that is done every year by the destruction of insect pests fed to nestling birds. And it should be remembered that the nesting season is also that when the destruction of injurious insects is most needed; that is, at the period of greatest agricultural activity and before the parasitic insects can be depended on to reduce the pests. The encouragement of birds to nest on the farm and the discouragement of nest robbing are therefore more than mere matters of sentiment; they return an actual cash equivalent, and have a definite bearing on the success or failure of the crops. — *Year Book of the Department of Agriculture.*

Americans. The recent McLane bill which aims at the protection of migrating birds and the bird-protecting clause of the recently passed tariff bill shows that this country is awaking to the value of her bird life. Without the birds the farmer would have a hopeless fight against insect pests. The effect of killing native birds is now well seen in Italy and Japan, where insects are increasing and do greater damage each year to crops and trees.

Of the eight hundred or more species of birds in the United States, only six species of hawks (Cooper's and the sharp-shinned hawk in particular), and the great horned owl, which prey upon useful birds; the sapsucker, which kills or injures many trees because of its fondness for the growing layer of the tree; the bobolink, which destroys yearly $2,000,000 worth of rice in the South; the crow, which feeds on crops as well as insects; and the English sparrow, may be considered as enemies of man.

The English Sparrow. — The English sparrow is an example of a bird introduced for the purpose of insect destruction, that has done great harm because of its relation to our native birds. Introduced at Brooklyn in 1850 for the purpose of exterminating the cankerworm, it soon abandoned an insect diet and has driven out most of our native insect feeders. Investigations by the United States Department of Agriculture have shown that in the country these birds and their young feed to a large extent upon grain, thus showing them to be injurious to agriculture. Dirty and very prolific, it already has worked its way from the East as far as the Pacific coast. In this area the bluebird, song sparrow, and yellowbird have all been forced to give way, as well as many larger birds of great economic value and beauty. The English sparrow has become a pest especially in our cities, and should be exterminated in order to save our native birds. It is feared in some quarters that the English starling which has recently been introduced into this country may in time prove a pest as formidable as the English sparrow.

Food of Snakes. — Probably the most disliked and feared of all animals are the snakes. This feeling, however, is rarely deserved, for, on the whole, our common snakes are beneficial to man. The black snake and the milk snake feed largely on injurious rodents (rats, mice, etc.), the pretty green snake eats injurious insects, and the little DeKay snake feeds partially on slugs. If it were not that the rattlesnake and the copperhead are venomous, they also could be said to be useful, for they live on English sparrows, rats, mice, moles, and rabbits.

Food of Herbivorous Animals. — We must not forget that other animals besides insects and birds help to keep down the rapidly growing weeds. Herbivorous animals the world over destroy, besides the grass which they eat, untold multitudes of weeds, which, if unchecked, would drive out the useful occupants of the pasture, the grasses and grains.

HARM DONE BY ANIMALS

Economic Loss from Insects. — The money value of crops, forest trees, stored foods, and other material destroyed annually by insects is beyond belief. It is estimated that they get one tenth of the country's crops, at the lowest estimate a matter of some $300,000,000 yearly. "The common schools of the country cost in 1902 the sum of $235,000,000, and all higher

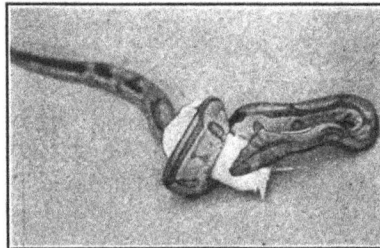

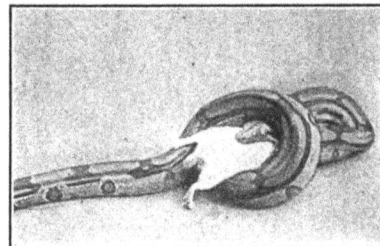

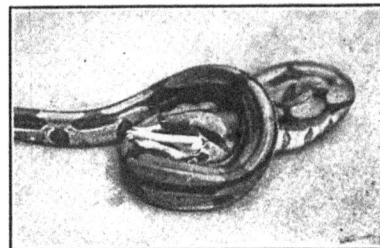

This shows how some snakes (constrictors) kill and eat their prey. (Series photographed by C. W. Beebe and Clarence Halter.)

institutions of learning cost less than $50,000,000, making the total cost of education in the United States considerably less than the farmers lost from insect ravages.

" Furthermore, the yearly losses from insect ravages aggregate nearly twice as much as it costs to maintain our army and navy; more than twice the loss by fire; twice the capital invested in manufacturing agricultural implements; and nearly three times the estimated value of the products of all the fruit orchards, vineyards, and small fruit farms in the country." — SLINGERLAND.

The total yearly value of all farm and forest products in New York is perhaps $150,000,000, and the one tenth that the insects get is worth $15,000,000.

Insects which damage Garden and Other Crops. — The grasshoppers and the larvæ of various moths do considerable harm

Cotton-boll weevil. *a*, larva; *b*, pupa; *c*, adult. Enlarged about four times. (Photographed by Davison.)

here, especially the " cabbage worm," the cutworm, a feeder on all kinds of garden truck, and the corn worm, a pest on corn, cotton, tomatoes, peas, and beans.

Among the beetles which are found in gardens is the potato beetle, which destroys the potato plant. This beetle formerly lived in Mexico upon a wild plant of the same family as the potato, and came north upon the introduction of the potato into Colorado, evidently preferring cultivated forms to wild forms of this family.

The one beetle doing by far the greatest harm in this country is the cotton-boll weevil. Imported from Mexico, since 1892 it has spread over eastern Texas and into Louisiana. The beetle lays its eggs in the young cotton fruit or boll, and the larvæ feed upon the substance within the boll. It is estimated that if unchecked this pest would destroy yearly one half of the cotton crop, causing a loss of $250,000,000. Fortunately, the United States Department of Agriculture is at work on the problem, and, while it has not found any way of exterminating the beetle as yet, it has been found that, by planting more hardy varieties of cotton, the crop matures earlier and ripens before the weevils have increased in sufficient numbers to destroy the crop (see page 126).

The bugs are among our most destructive insects. The most familiar examples of our garden pests are the squash bug; the chinch bug, which yearly does damage estimated at $20,000,000, by sucking the juice from the leaves of grain; and the plant lice, or aphids. One, living on the grape, yearly destroys immense numbers of vines in the vineyards of France, Germany, and California.

Insects which harm Fruit and Forest Trees. — Great damage is annually done trees by the larvæ of moths. Massachusetts has

Female tussock moth which has just emerged from the cocoon at the left, upon which it has deposited over two hundred eggs. (Photograph by Davison.)

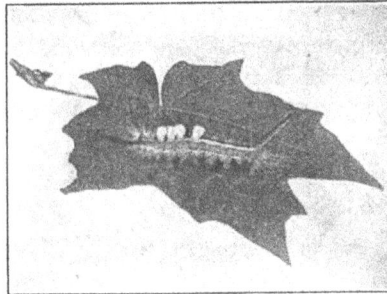

Caterpillar of tussock moth. (Photograph by Davison.)

already spent over $3,000,000 in trying to exterminate the imported gypsy moth. The codling moth, which bores into apples and pears, is estimated to ruin yearly $3,000,000 worth of fruit in New York alone, which is by no means the most important apple region of the United States. Among these pests, the most important to the dweller in a large city is the tussock moth, which destroys our shade trees. The caterpillar may easily be recognized by its hairy, tufted red head. The eggs are laid on the bark of shade trees in what look like masses of foam. (See figure on page 215.) By collecting and burning the egg masses in the fall, we may save many shade trees the following year.

The larvæ of some moths damage the trees by boring into the wood of the tree on which they live. Such are the peach, apple, and other fruit-tree borers common in our orchards. Many beetle larvæ also live in trees and kill annually thousands of forest and shade trees. The hickory borer threatens to kill all the hickory, trees in the Eastern states.

Among the bugs most destructive to trees are the scale insect and the plant lice. The San José scale, a native of China, was introduced into the fruit groves of California about 1870 and has spread all over the country. A ladybird beetle, which has also been imported, is the most effective agent in keeping this pest in check.

Insects of the House or Storehouse. — Weevils are the greatest pests, frequently ruining tons of stored corn, wheat, and other cereals. Roaches will eat almost anything, even clothing; they are especially fond of all kinds of breadstuffs. The carpet beetle is a recognized foe of the housekeeper, the larvæ feeding upon all sorts of woolen material. The larvæ of the clothes moth do an immense amount of damage, especially to stored clothing. Fleas, lice, and particularly bedbugs are among man's personal foes. Besides being unpleasant they are believed to be disease carriers and as such should be exterminated.[1]

Food of Starfish. — Starfish are enormously destructive to young clams and oysters, as the following evidence, collected by Professor A. D. Mead, of Brown University, shows. A single starfish was confined in an aquarium with fifty-six young clams. The largest clam was about the length of one arm of the starfish, the smallest about ten millimeters in length. In six days every clam in the aquarium was devoured. Hundreds of thousands of dollars' damage is done annually to the oysters in Connecticut alone by the ravages of starfish. During the breeding season of the clam and oyster the boats dredge up tons of starfish which are thrown on shore to die or to be used as fertilizer.

[1] Directions for the treatment of these pests may be found in pamphlets issued by the U. S. Department of Agriculture.

THE RELATIONS OF ANIMALS TO DISEASE

The Cause of Malaria. — The study of the life history and habits of the Protozoa has resulted in the finding of many parasitic forms, and the consequent explanation of some kinds of disease. One parasitic protozoan like an amœba is called *Plasmodium malariæ*. It causes the disease known as malaria. When a mosquito (the *anopheles*) sucks the blood from a person having malaria this parasite passes into the stomach of the mosquito. After completing a part of its life history within the mosquito's body the parasite establishes itself within the glands which secrete the saliva of the mosquito. After about eight days, if the infected mosquito bites a person, some of the parasites are introduced into

The life history of the malarial parasite. This cut of the malarial parasite shows parts of the body of the mosquito and of man. To understand the life history begin at the point where the mosquito injects the crescent-shaped bodies into the blood of man. Notice that after the spores are released from the corpuscles of man two kinds of cells *may be formed*. These are probably a sexual stage. Development within the body of the mosquito will only take place when the parasite is taken into its body at this sexual stage.

the blood along with the saliva. These parasites enter the corpuscles of the blood, increase in size, and then form spores. The rapid process of spore formation results in the breaking down of the blood corpuscles and the release of the spores, and the

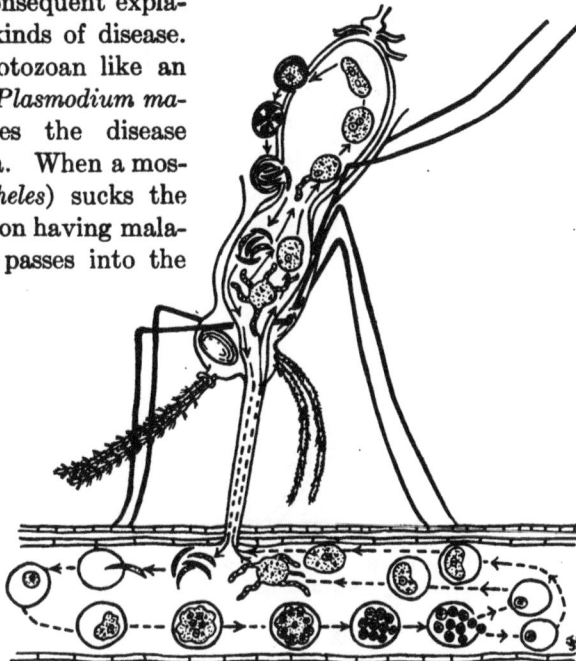

poisons they manufacture, into the blood. This causes the chill followed by the fever so characteristic of malaria. The spores may again enter the blood corpuscles and in forty-eight or seventy-two hours repeat the process thus described, depending on the kind of malaria they cause. The only cure for the disease is *quinine* in rather large doses. This kills the parasites in the blood. But quinine should not be taken except under a physician's directions.

The Malarial Mosquito. — Fortunately for mankind, not all mosquitoes harbor the parasite which causes malaria. The harmless mosquito (*culex*) may be usually distinguished from the mosquito which carries malaria (*anopheles*) by the position taken

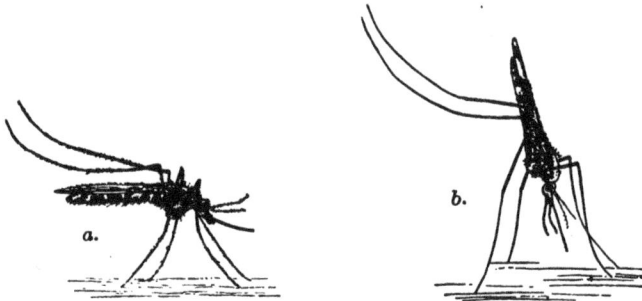

How to distinguish the harmless mosquito (*culex*), a, from the malarial mosquito (*anopheles*), b, when at rest. Notice the position of legs and body.

when at rest. Culex lays eggs in tiny rafts of one hundred or more eggs in any standing water; thus the eggs are distinguished from those of anopheles, which are not in rafts. Rain barrels, gutters, or old cans may breed in a short time enough mosquitoes to stock a neighborhood. The larvæ are known as wigglers. They breathe through a tube in the posterior end of the body, and may be recognized by their peculiar movement when on their way to the surface to breathe. The pupa, distinguished by a large thoracic region, breathes through a pair of tubes on the thorax. The fact that both larvæ and pupæ take air from the surface of the water makes it possible to kill the mosquito during these stages by pouring oil on the surface of the water where they breed. The introduction of minnows, gold fish, or other small fish which feed upon the larvæ in the water where the mosquitoes breed will do much to free a neighborhood from this pest. Draining swamps or low land which holds water after a rain is another method of extermination. Some of the mosquito-infested districts around New York City have been almost freed from mosquitoes by draining the salt marshes where they breed. Long shallow trenches are so built as to tap and drain off any standing water in which the eggs might be laid. In this way the mosquito has been almost exterminated along some parts of our New England coast.

Since the beginning of historical times, malaria has been prevalent in regions infested by mosquitoes. The ancient city of Rome was so greatly troubled by periodic outbreaks of malarial fever that a goddess of fever came to be worshiped in order to lessen the severity of what the inhabitants

Swamps are drained and all standing water covered with a film of oil in order to exterminate mosquitoes. Why is the oil placed on the surface of the water?

believed to be a divine visitation. At the present time the malaria of Italy is being successfully fought and conquered by the draining of the mosquito-breeding marshes. By a little carefully directed oiling of water a few boys may make an almost uninhabitable region absolutely safe to live in. Why not try it if there are mosquitoes in your neighborhood?

Yellow Fever and Mosquitoes. — Another disease carried by mosquitoes is yellow fever. In the year 1878 there were 125,000 cases and 12,000 deaths in the United States, mostly in Alabama, Louisiana, and Mississippi. During the French occupation of the Panama Canal zone the work was at a standstill part of the time because of the ravages of yellow fever. Before the war with Spain thousands of people were ill in Cuba. But to-day this is changed, and yellow fever is under almost complete control, both here and in the Canal zone, where the mosquito (*stegomyia*) which carries yellow fever exists.

YEAR	250 500 750 1000 1250 1500
1871	991
1872	515
1873	1244
1874	1425
1875	1001
1876	1619
1877	1374
1878	1559
1879	1444
1880	645
1881	485
1882	729
1883	849
1884	511
1885	165
1886	167
1887	532
1888	468
1889	303
1890	308
1891	356
1892	357
1893	496
1894	382
1895	533
1896	1282
1897	858
1898	136
1899	103
1900	310
1901	18 Carrier of Yellow Fever discovered
1902	NONE
1903	NONE
1904	NONE
1905	24 First Cuban Rule
1906	12
1907	3
1908	3
1909	NONE
1910	NONE

Notice the difference in the number of yearly deaths from yellow fever before and *after* the American occupation of Havana.

This is due to the experiments during the summer of 1900 of a Commission of United States army officers, headed by Dr. Walter Reed. Of these men one, Dr. Jesse Lazear, gave up his life to prove experimentally that yellow fever was caused by mosquitoes. He allowed himself to be bitten by a mosquito that was known to have bitten a yellow fever patient, contracted the disease, and died a martyr to science. Others, soldiers, volunteered to further test by experiment how the disease was spread, so that in the end Dr. Reed was able to prove to the world that if mosquitoes could be prevented from biting people who had yellow fever the disease could not be spread. The accompanying illustration shows the result of this knowledge for the city of Havana. For years Havana was considered one of the pest spots of the West Indies. Visitors shunned this port and commerce was much affected by the constant menace of yellow fever. At the time of the American occupation after the war with Spain, the experiments referred to above were undertaken. The city was cleaned up, proper sanitation introduced, screens placed in most buildings, and the breeding places of the mosquitoes were so nearly destroyed that the city was practically free from mosquitoes. The result, so far as yellow fever was concerned, was startling, as you can see by reference to the chart. Notice also the rise in the death rate when the young Cuban Republic took control. How do you account for that? We all know what American scientific medicine and sanitation is doing in Panama and in the Philippines.

Other Protozoan Diseases. — Many other diseases of man are probably caused by parasitic protozoans. Dysentery of one kind appears to be caused by the presence of an amœba-like animal in the digestive tract which comes usually through an impure water supply. Smallpox, rabies, and possibly other

Stegomyia, the carrier of yellow fever. (After Howard.)

diseases are caused by protozoans. Smallpox, which was once the most dreaded disease known to man, because of its spread in epidemics, has been conquered by *vaccination*, of which we shall learn more later. The death rate from rabies or hydrophobia has in a like manner been greatly reduced by a treatment founded on the same principles as vaccination and invented by Louis Pasteur.

Another group of protozoan parasites are called *trypanosomes*. These are parasitic in insects, fish, reptiles, birds, and mammals in various parts of the world. They cause various diseases of cattle and other domestic animals, being carried to the animal in most cases by flies. One of this family is believed to live in the blood of native African zebras and antelopes; seemingly it does them no harm. But if one of these parasites is transferred by the dreaded tsetse fly to one of the domesticated horses or cattle of the colonist of that region, death of the animal results.

Another fly carries a species of trypanosome to the natives of Central Africa, which causes "the dreaded and incurable sleeping sickness." This disease carries off more than fifty thousand natives yearly, and many Europeans have succumbed to it. Its ravages are now largely confined to an area near the large Central African lakes and the Upper Nile, for the fly which carries the disease lives near water, seldom going more than 150 feet from the banks of streams or lakes. The British government is now trying to control the disease in Uganda by moving all the villages at least two miles from the lakes and rivers. Among other diseases that may be due to protozoans is kala-agar, a fever in hot Asiatic countries which is probably carried by the bedbug, and African tick fever, probably carried by a small insect called the tick. Bubonic plague, one of the most dreaded of all bacterial diseases, is carried to man by fleas from rats. In this country many fatal diseases of cattle, as "tick," or Texas cattle fever, are probably caused by protozoans.

The Fly a Disease Carrier. — We have already seen that mosquitoes of different species carry malaria and yellow fever. Another rather recent addition to the black list is the house fly or typhoid fly. We shall see later with what reason this name is given. The development of the typhoid fly is extremely rapid. A female may lay from one hundred to two hundred eggs. These are usually deposited in filth or manure. Dung heaps

Life history of house flies, showing from left to right the eggs, larvæ, pupæ, and adult flies. (Photograph, about natural size, by Overton.)

about stables, privy vaults, ash heaps, uncared-for garbage cans, and fermenting vegetable refuse form the best breeding places for flies. In warm weather, the eggs hatch a day or so after they are laid and become larvæ, called maggots. After about one week of active feeding, these wormlike maggots become quiet and go into the pupal stage, whence under favorable conditions they emerge within less than another week as adult flies. The adults breed at once, and in a short summer there may be over ten generations of flies. This accounts for the great number. Fortunately relatively few flies survive the winter. The membranous wings of the adult fly appear to be two in number, a second pair being reduced to tiny knobbed hairs called balancers. The head is freely

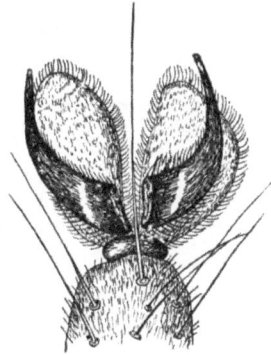

The foot of a fly, showing the hooks, hairs, and pads which collect and carry bacteria. The fly doesn't wipe his feet.

movable, with large compound eyes. The mouth parts form a proboscis, which is tonguelike, the animal obtaining its food by lapping and sucking. The foot shows a wonderful adaptation for clinging to smooth surfaces. Two or three pads, each of which bears tubelike hairs that secrete a sticky fluid, are found on its under surface. It is by this means that the fly is able to walk upside down, and carry bacteria on its feet.

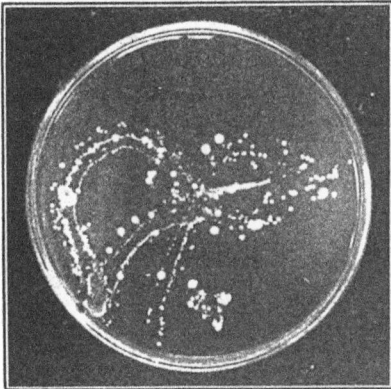

Colonies of bacteria which have developed in a culture medium upon which a fly was allowed to walk.

The Typhoid Fly a Pest. — The common fly is recognized as a pest the world over. Flies have long been known to spoil food through their filthy habits, but it is more recently that the very serious charge of spread of diseases, caused by bacteria, has

Showing how flies may spread disease by means of contaminating food.

been laid at their door. In a recent experiment two young men from the Connecticut Agricultural Station found that a single fly might carry on its feet anywhere from 500 to 6,600,000 bacteria, the average number being over 1,200,000. Not all of these germs are harmful, but they might easily include those of typhoid fever, tuberculosis, summer complaint, and possibly other diseases. A recent pamphlet published by the Merchants' Association in New York City shows that the rapid increase of flies during the summer months has a definite correlation with the increase in the number of cases of summer complaint. Observations in other cities seem to show the increase in number of typhoid cases in the early fall is due, in part at least, to the same cause. A terrible toll of disease and death may be laid at the door of the typhoid fly.

Recently the stable fly has been found to carry the dread disease known as infantile paralysis.

Remedies. — Cleanliness which destroys the breeding place of flies, the frequent removal and destruction of garbage, rubbish, and manure, covering of all food when not in use and especially the *careful* screening of windows and doors during the breeding

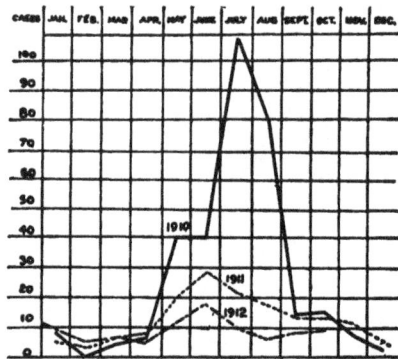

There were 329 typhoid cases in Jacksonville, Florida, in 1910, 158 in 1911, 87 in first 10 months of 1912. 80 to 85 per cent of outdoor toilets were made fly-proof during winter of 1910. Account for the decrease in typhoid after the flies were kept out of the toilets.

season, will all play a part in the reduction of flies. To the motto "swat the fly" should be added, "remove their breeding places!"

Other Insect Disease Carriers. — Fleas and bedbugs have been recently added to those insects proven to carry disease to man. Bubonic plague, which is primarily a disease of rats, is undoubtedly transmitted from the infected rats to man by the fleas. Fleas are also believed to transmit leprosy although this is not proven.

To rid a house of fleas we must first find their breeding places. Old carpets, the sleeping places of cats or dogs or any dirty unswept corner may hold the eggs of the flea. The young breed in cracks and crevices, feeding upon organic matter there. Eventually they come to live as adults on their warm-blooded hosts, cats, dogs, or man. Evidently destruction of the breeding places, careful washing of all infected areas, the use of benzine or gasoline in crevices where the larvæ may be hid are the most effective methods of extermination. Pets which might harbor fleas should be washed frequently with a weak (two to three per cent) solution of creolin.

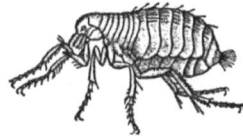
Flea which transmits Bubonic plague from rat to man.

Bedbugs are difficult to prove as an agent in the transmission of disease but their disgusting habits are sufficient reason for their extermination. It has been proven by experiment that they may spread typhoid and relapsing fevers. They prefer human blood to other food and have come to live in bedrooms and beds because this food can be obtained there. They are extremely difficult to exterminate because their flat body allows them to hide in cracks out of sight. Wooden beds are thus better protection for them than iron or brass beds. Boiling water poured over the cracks when they breed or a mixture of strong corrosive sublimate four parts, alcohol four parts and spirits of turpentine one part, are effective remedies.

How the Harm done by Insects is Controlled. — The combating of insects is directed by several bodies of men, all of which have the same end in view. These are the Bureau of Entomology of the United States Department of Agriculture, the various state experiment stations, and medical and civic organizations.

The Bureau of Entomology works in harmony with the other divisions of the Department of Agriculture, giving the time of its experts to the problems of controlling insects which, for good or ill, influence man's welfare in this country. The destruction of the malarial mosquito and control of the typhoid fly; the destruction of harmful insects by the introduction of their natural enemies, plant or animal; the perfecting of the honeybee (see Hodge, *Nature Study and Life*, page 240), and the introduction of new species of insects to pollinate flowers not native to this country (see *Blastophaga*, page 43), are some of the problems to which these men are now devoting their time.

All the states and territories have, since 1888, established state experiment stations, which work in coöperation with the government in the war upon injurious insects. These stations are often connected with colleges, so that young men who are interested in this kind of natural science may have opportunity to learn and to help.

The good done by these means directly and indirectly is very great. Bulletins are published by the various state stations and by the Department of Agriculture, most of which may be obtained free. The most interesting of these from the high school standpoint are the Farmers' Bulletins, issued by the Department of Agriculture, and the Nature Study pamphlets issued by the Cornell University in New York state.

This diagram shows how bubonic plague is carried to man. Explain the diagram.

Animals Other than Insects may be Disease Carriers. — The common brown rat is an example of a mammal, harmful to civilized man, which has followed in his footsteps all over the world. Starting from China, it spread to eastern Europe, thence to western Europe, and in 1775 it had obtained a lodgment in this country. In seventy-five years it reached the Pacific coast, and is now fairly common all over the United States, being one of the most prolific

of all mammals. Rats are believed to carry bubonic plague, the
" Black Death " of the Middle Ages, a disease estimated to have
killed 25,000,000 people during the fourteenth century. The rat,
like man, is susceptible to plague; fleas bite the rat and then biting
man transmit the disease to him. A determined effort is now being
made to exterminate the rat because of its connection with
bubonic plague.

Other Parasitic Animals cause Disease. — Besides parasitic
protozoans other forms of animals have been found that *cause*
disease. Chief among these are certain round and flat worms,
which have come to live as parasites on man and other animals.
A one-sided relationship has thus come into existence where the
worm receives its living from the host, as the animal is called on
which the parasite lives. Consequently the parasite frequently
becomes fastened to its host during adult life and often is reduced
to a mere bag through which the fluid food prepared by its host is
absorbed. Sometimes a complicated life history has arisen from
their parasitic habits. Such is seen in the
life history of the liver fluke, a flatworm
which kills sheep, and in the tapeworm.

Cestodes or Tapeworms. — These parasites infest man and many other vertebrate
animals. The tapeworm (*Tænia solium*)
passes through two stages in its life history,
the first within a pig, the second within the
intestine of man. The developing eggs are
passed off with wastes from the intestine
of man. The pig, an animal with dirty
habits, may take in the worm embryos
with its food. The worm develops within
the intestine of the pig, but soon makes its
way into the muscle or other tissues. It

The life cycle of a tape-
worm. (1) The eggs are
taken in with filthy food
by the pig; (2) man
eats undercooked pork
by means of which
the bladder worm (3) is
transferred to his own
intestine (4).

is here known as a bladderworm. If man eats raw or undercooked
pork containing these worms, *he* may become a host for the tapeworm. Thus during its complete life history it has two hosts.
Another common tapeworm parasitic on man lives part of its life as
an embryo within the muscles of cattle. The adult worm consists

of a round headlike part provided with hooks, by means of which it fastens itself to the wall of the intestine. This head now buds off a series of segmentlike structures, which are practically bags full of sperms and eggs. These structures, called *proglottids*, break off from time to time, thus allowing the developing eggs to escape. The proglottids have no separate digestive systems, but the whole body surface, bathed in digested food, absorbs it and is thus enabled to grow rapidly.

Roundworms. — Still other wormlike creatures called roundworms are of importance to man. Some, as the vinegar eel found in vinegar, or the pinworms parasitic in the lower intestine, particularly of children, do little or no harm. The pork worm or *trichina*, however, is a parasite which may cause serious injury. It passes through the first part of its existence as a parasite in a pig or other vertebrate (cat, rat, or rabbit), where it lies, covered within a tiny sac or *cyst*, in the muscles of its hosts. If raw pork containing these worms is eaten by man, the cyst is dissolved off by the action of the digestive fluids, and the living trichina becomes free in the intestine of man. Here it reproduces and the young bore their way through the intestine walls and enter the muscles, causing inflammation there. This causes a painful and often fatal disease known as *trichinosis*.

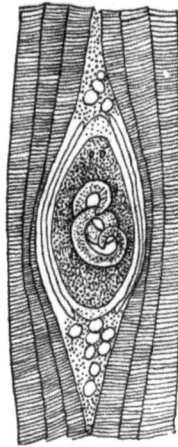

Trichinella spiralis imbedded in human muscle. (After Leuckart.)

The Hookworm. — The discovery by Dr. C. W. Stiles of the Bureau of Animal Industry, that the laziness and shiftlessness of the " poor whites " of the South is partly due to a parasite called the *hookworm*, reads like a fairy tale.

The people, largely farmers, become infected with a larval stage of the hookworm, which develops in moist earth. It enters the body usually through the skin of the feet, for children and adults alike, in certain localities where the disease is common, go barefoot to a considerable extent.

A complicated journey from the skin to the intestine now follows, the larvæ passing through the veins to the heart, from there to the lungs; here they bore into the air passages and eventually work their way by way of the windpipe into the intestine. One result of the injury of the lungs is that many thus infected are subject to tuberculosis. The adult worms, once in the food tube, fasten themselves and feed upon the blood of their host by puncturing the intestine wall. The loss of blood from this cause is not sufficient to account for the bloodlessness of the person infected, but it has been discovered that the hookworm pours out a

A family suffering from hookworm.

poison into the wound which prevents the blood from clotting rapidly (see page 315); hence a considerable loss of blood occurs from the wound after the worm has finished its meal and gone to another part of the intestine.

The cure of the disease is very easy; thymol is given, which weakens the hold of the worm, this being followed by Epsom salts. For years a large area in the South undoubtedly has been retarded in its development by this parasite; hundreds of millions of dollars and thousands of lives have been needlessly sacrificed.

"The hookworm is not a bit spectacular: it doesn't get itself discussed in legislative halls or furiously debated in political campaigns. Modest and unassuming, it does not aspire to such dignity. It is satisfied simply with (1) lowering the working efficiency and the pleasure of living in something like two hundred thousand persons in Georgia and all other Southern states in proportion; with (2) amassing a death rate higher than tuberculosis, pneumonia, or typhoid fever; with (3) stubbornly and quite effectually retarding the agricultural and industrial development of the section; with (4) nullifying the benefit of thousands of dollars spent upon education; with (5) costing the South, in the course of a few decades, several hundred millions of dollars. More serious and closer at hand than the tariff; more costly, threatening, and tangible than the Negro problem; making the menace of the boll weevil laughable in comparison — it is preëminently the problem of the South." — *Atlanta Constitution.*

Animals that prey upon Man. — The toll of death from animals which prey upon or harm man directly is relatively small. Snakes in tropical countries kill many cattle and not a few people.

The bite of the rattlesnake of our own country, although dangerous, seldom kills. The dreaded cobra of India has a record of over two hundred and fifty thousand persons killed in the last thirty-five years. The Indian government yearly pays out large sums for the extermination of venomous snakes, over two hundred thousand of which have been killed during a single year.

A flesh-eating reptile, the alligator.

Alligators and Crocodiles. — These feed on fishes, but often attack large animals, as horses, cows, and even man. They seek their prey chiefly at night, and spend the day basking in the sun. The crocodiles of the Ganges River in India levy a yearly tribute of many hundred lives from the natives.

Carnivorous animals such as lions and tigers still inflict damage in certain parts of the world, but as the tide of civilization advances, their numbers are slowly but surely decreasing so that as important factors in man's welfare they may be considered almost negligible.

REFERENCE BOOKS

ELEMENTARY

Hunter, *Laboratory Problems in Civic Biology*. American Book Company.

Beebe, *The Bird*. Henry Holt and Company.

Bigelow, *Applied Biology*. Macmillan and Company

Davison, *Practical Zoölogy*. American Book Company.

Herrick, *Household Insects* and *Methods of Control*. Cornell Reading Courses.

Hornaday, *Our Vanishing Wild Life*. New York Zoölogical Society.

Hodge, *Nature Study and Life*. Ginn and Company.

Kipling, *Captains Courageous*. Charles Scribner's Sons.

Sharpe, *Laboratory Manual*, pp. 157–158, 182–203, 320–341. American Book Company.

Stone and Cram, *American Animals*. Doubleday, Page and Company.

Toothaker, *Commercial Raw Materials*. Ginn and Company.

ADVANCED

Flower, *The Horse*. D. Appleton and Company.

Hornaday, *The American Natural History*. Macmillan and Company.

Jordan, *Fishes*. Henry Holt and Company.

Jordan and Evermann, *American Food and Game Fishes*. Doubleday, Page and Company.

Schaler, *Domesticated Animals, their Relations to Man and to His Advancement in Civilization*. Charles Scribner's Sons.

XVI. THE FISH AND FROG, AN INTRODUCTORY STUDY OF VERTEBRATES

Problems. — *To determine how a fish and a frog are fitted for the life they lead.*

To determine some methods of development in vertebrate animals.

(a) *Fishes.*

(b) *Frogs.*

(c) *Other animals.*

LABORATORY SUGGESTIONS

Laboratory exercise. — Study of a living fish — adaptations for protection, locomotion, food getting, etc.

Laboratory demonstration. — The development of the fish or frog egg.

Visit to the aquarium. — Study of adaptations, economic uses of fishes, artificial propagation of fishes.

Two Methods of Breathing in Vertebrates. — Vertebrate animals have at least two methods of getting their oxygen. In other respects their life processes are nearly similar. Of all vertebrates fishes are the only ones fitted to breathe all their lives under water. Other vertebrates are provided with lungs and take their oxygen directly from the air.[1] We will next take up the study of a fish to see how it is fitted for its life in the water.

STUDY OF A FISH

The Body. — One of our common fresh-water fish is the bream, or golden shiner. The body of the bream runs insensibly into the head, the neck being absent. The long, narrow body with its smooth surface fits the fish admirably for its life in the water. Certain cells in the skin secrete mucus or slime, another adapta-

---

[1] With the exception of a few lungless salamanders. Most salamanders get much of their supply of oxygen through their moist skins.

tion. The position of the scales, overlapping in a backward direction, is yet another adaptation which aids in passing through the water. Its color, olive above and bright silver and gold below, is protective. Can you see how?

The bream. *A*, dorsal fin; *B*, caudal fin; *C*, anal fin; *D*, pelvic fin; *E*, pectoral fin.

The Appendages and their Uses. — The appendages of the fish consist of paired and unpaired fins. The paired fins are four in number, and are believed to correspond in position and structure with the paired limbs of a man. Note the illustration above and locate the paired *pectoral* and *pelvic* fins. (These are so called because they are attached to the bones forming the pectoral and pelvic girdles. (See page 268.) Find, by comparison with the Figure, the *dorsal*, *anal*, and *caudal* fins. How many unpaired fins are there?

The flattened, muscular body of the fish, tapering toward the caudal fin, is moved from side to side with an undulating motion which results in the forward movement of the fish. This movement is almost identical with that of an oar in sculling a boat. Turning movements are brought about by use of the lateral fins in much the same way as a boat is turned. We notice the dorsal and other single fins are evidently useful in balancing and steering.

The Senses. — The position of the eyes at the side of the head is an evident advantage to the fish. Why? The eye is globular in shape. Such an eye has been found to be very nearsighted. Thus it is unlikely that a fish is able to perceive objects at any great distance from it. The eyes are unprotected by eyelids, but the tough outer covering and their position afford some protection.

Feeding experiments with fishes show that a fish becomes aware of the presence of food by smelling it as well as by seeing it. The nostrils of a fish can be proved to end in little pits, one under each nostril hole. Thus they differ from our own, which are connected with the mouth cavity. In the catfish, for example, the *barbels*, or horns, receive sensations of smell and taste. They do not perceive odors as we do for a fish perceives only substances that are dissolved in the water in which it lives. The senses of taste and touch appear to be less developed than the other senses.

Along each side of most fishes is a line of tiny pits, provided with sense organs and connected with the central nervous system of the fish. This area, called the *lateral line*, is believed to be sensitive to mechanical stimuli of certain sorts. The " ear " of the fish is under the skin and serves partly as a balancing organ.

Food Getting. — A fish must go after its food and seize it, but has no structures for grasping except the teeth. Consequently we find the teeth small, sharp, and numerous, well adapted for holding living prey. The tongue in most fishes is wanting or very slightly developed.

Breathing. — A fish, when swimming quietly or when at rest, seems to be biting when no food is present. A reason for this act is to be seen when we introduce a little finely powdered carmine into the water near the head of the fish. It will be found that a current of water enters the mouth at each of these biting movements and passes out through two slits found on each side of the head of the fish. Investigation shows us that under the broad, flat plate, or *operculum*, forming each side of the head, lie several long, feathery, red structures, the *gills*.

Gills. — If we examine the gills of any large fish, we find that a single gill is held in place by a bony arch, made of several pieces of bone which are hinged in such a way as to give great flexibility to the gill arch, as the support is called. Covering the bony framework, and extending from it, are numerous delicate filaments covered with a very thin membrane or skin. Into each of these filaments pass two blood vessels; in one blood flows downward and in the other upward. Blood reaches the gills and is carried away from these organs by means of two large vessels which pass along the bony arch previously mentioned. In the gill filament the blood comes into contact with the free oxygen of the water bathing the gills. An exchange of gases through the walls of the gill filaments results in the loss of carbon dioxide and a gain of oxygen by the blood. The blood carries oxygen to the cells of the body and (as work is done by the

Diagram of the gills of a fish. (*H*), the heart which forces the blood into the tubes (*V*), which run out into the gill filaments. A gill bar (*G*) supports each gill. The blood after exchanging its carbon dioxide for oxygen is sent out to the cells of the body through the artery (*A*).

cells as a result of the oxidation of food) brings carbon dioxide back to the gills.

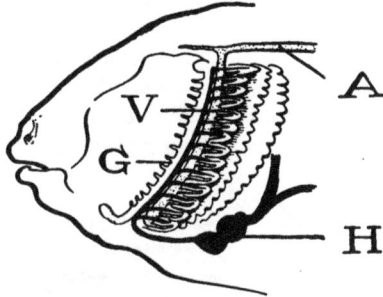

Gill Rakers. — If we open wide the mouth of any large fish and look inward, we find that the mouth cavity leads to a funnel-like opening, the gullet. On each side of the gullet we can see the gill arches, guarded on the inner side by a series of sharp-pointed structures, the *gill rakers*. In some fishes in which the teeth are not well developed, there seems to be a greater development of the gill rakers, which in this case are used to strain out small organisms from the water which passes over the gills. Many fishes make such use of the gill rakers. Such are the shad and menhaden, which feed almost entirely on *plankton*, a name given to the small organisms found by millions near the surface of water.

Digestive System. — The gullet leads directly into a baglike stomach. There are no salivary glands in the fishes. There is, however, a large liver, which appears to be used as a digestive gland. This organ, because of the oil it contains, is in some fishes, as the cod, of considerable economic importance. Many fishes have outgrowths like a series of pockets from the intestine. These structures, called the *pyloric cæca*, are believed to

secrete a digestive fluid.   The intestine ends at the vent, which is usually
located on the under side of the fish, immediately in front of the anal fin.

Swim Bladder. — An organ of unusual significance, called the *swim
bladder*, occupies the region just dorsal to the food tube.   In young fishes
of many species this is connected by a tube with the anterior end of the
digestive tract.   In some forms this tube persists throughout life, but in
other fishes it becomes closed, a thin, fibrous cord taking its place.   The
swim bladder aids in giving the fish nearly the same weight as the water

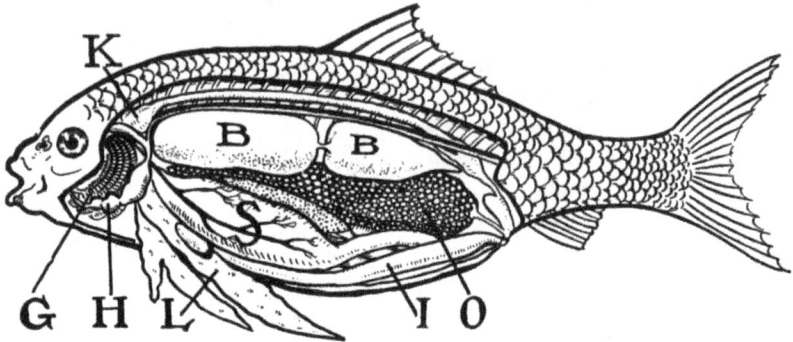

A fish opened to show *H*, the heart; *G*, the gills; *L*, the liver; *S*, the stomach;
I, the intestine; *O*, the ovary; *K*, the kidney, and *B*, the air bladder.

it displaces, thus buoying it up.   The walls of the organ are richly sup-
-plied with blood vessels, and it thus undoubtedly serves as an organ for
supplying oxygen to the blood when all other sources fail.   In some
fishes (the *dipnoi*, page 187) it has come to be used as a lung.

Circulation of the Blood. — In the vertebrate animals the blood is
said to circulate in the body, because it passes through a more or less closed
system of tubes in its course around the body.   In the fishes the heart is
a two-chambered muscular organ, a thin-walled *auricle*, the receiving
chamber, leading into a thick-walled muscular *ventricle* from which the
blood is forced out.   The blood is pumped from the heart to the gills;
there it loses some of its carbon dioxide; it then passes on to other parts
of the body, eventually breaking up into very tiny tubes called *capillaries*.
From the capillaries the blood returns, in tubes of gradually increasing
diameter, toward the heart again.   The body cells lie between the smallest
branches of the capillaries.   Thus they get from the blood food and oxygen and return to the blood the wastes resulting from oxidation within
the cell body.   During its course some of the blood passes through the
kidneys and is there relieved of part of its nitrogenous waste.   Circulation

of blood in the body of the fish is rather slow. The temperature of the blood being nearly that of the surrounding media in which the fish lives, the animal has incorrectly been given the term " cold-blooded."

Nervous System. — As in all other vertebrate animals, the brain and spinal cord of the fish are partially inclosed in bone. The central nervous system consists of a *brain*, with nerves connecting the organs of sight, taste, smell, and hearing, and such parts of the body as possess the sense of touch; a *spinal cord*; and *spinal nerves*. Nerve cells located near the outside of the body send in messages to the central system, which are there received as sensations. Cells of the central nervous system, in turn, send out messages which result in the movement of muscles.

Skeleton. — In the vertebrates, of which the bony fish is an example, the skeleton is under the skin, and is hence called an *endoskeleton*. It consists of a bony framework, the vertebral column which protects the spinal cord and certain attached bones, the ribs, with other spiny bones to which the unpaired fins are attached. The paired fins are attached to the spinal column by two collections of bones, known respectively as the *pectoral* and *pelvic girdles*. The bones in the main skeleton serve in the fish for the attachment of powerful muscles, by means of which locomotion is accomplished. In most fishes, the *exoskeleton*, too, is well developed, consisting usually of scales, but sometimes of bony plates.

Food of Fishes. — We have already seen that in a balanced aquarium the balance of food was preserved by the plants, which furnished food for the tiny animals or were eaten by larger ones, — for example, snails or fish. The smaller animals in turn became food of larger ones. The nitrogen balance was maintained through the excretions of the animals and their death and decay.

The marine world is a great balanced aquarium. The upper layer of water is crowded with all kinds of little organisms, both plant and animal. Some of these are microscopic in size; others, as the tiny crustaceans, are visible to the eye. On these little organisms some fish feed entirely, others in part. Such are the menhaden [1] (bony, bunker, mossbunker of our coast), the shad, and others. Other fishes are bottom feeders, as the blackfish and

[1] It has been discovered by Professor Mead of Brown University that the increase in starfish along certain parts of the New England coast was in part due to overfishing of menhaden, which at certain times in the year feed almost entirely on the young starfish.

the sea bass, living almost entirely upon mollusks and crusta-ceans. Still others are hunters, feeding upon smaller species of fish, or even upon their weaker brothers. Such are the bluefish, squeteague or weakfish, and others.

What is true of salt-water fish is equally true of those inhabiting our fresh-water streams and lakes. It is one of the greatest problems of our Bureau of Fisheries to discover this relation of various fishes to their food supplies so as to aid in the conservation and balance of life in our lakes, rivers, and seas.

Migration of Fishes. — Some fishes change their habitat at different times during the year, moving in vast schools northward in summer and southward in the winter. In a general way such migrations follow the coast lines. Examples of such migratory fish are the cod, menhaden, herring, and bluefish. The migrations are due to temperature changes, to the seeking after food, and to the spawning instinct. Some fish migrate to shallower water in the summer and to deeper water in the winter; here the reason for the migration is doubtless the change in temperature.

The Egg-laying Habits of the Bony Fishes. — The eggs of most bony fishes are laid in great numbers, varying from a few thousand in the trout to many hundreds of thousands in the shad and several millions in the cod. The time of egg-laying is usually spring or early summer. At the time of spawning the male

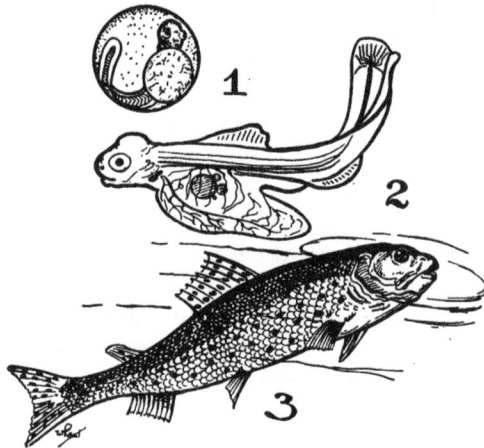

Development of a trout. 1, the embryo within the egg; 2, the young fish just hatched with the yoke sac still attached; 3, the young fish.

usually deposits milt, consisting of millions of sperm cells, in the water just over the eggs, thus accomplishing fertilization. Some

fishes, as sticklebacks, sunfish, toadfish, etc., make nests, but
usually the eggs are left to develop by themselves, sometimes
attached to some submerged object, but more frequently free in
the water. In some eggs a tiny oil drop buoys up the egg to the
surface, where the heat of the sun aids development. They are
exposed to many dangers, and both eggs and developing fish are
eaten, not only by birds, fish of other species, and other water inhabitants, but also by their own relatives, and even parents.
Consequently a very small percentage of eggs ever produce mature fish.

**The Relation of the Spawning Habits to Economic Importance
of Fish.** — The spawning habits of fish are of great importance to
us because of the economic value of fish to mankind, not only
directly as a food, but indirectly as food for other animals in turn
valuable to man. Many of our most desirable food fishes, notably
the salmon, shad, sturgeon, and smelt, pass up rivers from the
ocean to deposit their eggs, swimming against strong currents
much of the way, some species leaping rapids and falls, in order
to deposit their eggs in localities where the conditions of water
and food are suitable, and the water shallow enough to allow
the sun's rays to warm it sufficiently to cause the eggs to develop.
The Chinook salmon of the Pacific coast, the salmon used in the
Western canning industry, travels over a thousand miles up the
Columbia and other rivers, where it spawns. The salmon begin
to pass up the rivers in early spring, and reach the spawning beds,
shallow deposits of gravel in cool mountain streams, before late
summer. Here the fish, both males and females, remain until
the temperature of the water falls to about 54° Fahrenheit. The
eggs and milt are then deposited, and the old fish die, leaving the
eggs to be hatched out later by the heat of the sun's rays.

Need of Conservation. — The instinct of this and other species
of fish to go into shallow rivers to deposit their eggs has been
made use of by man. At the time of the spawning migration the
salmon are taken in vast numbers, for the salmon fisheries net
over $16,000,000 annually.

But the need for conservation of this important national asset
is great. The shad have within recent time abandoned their

breeding places in the Connecticut River, and the salmon have been exterminated along our eastern coast within the past few decades. It is only a matter of a few years when the Western salmon will be extinct if fishing is continued at the present rate. More fish must be allowed to reach their breeding places. To do this a closed season on the rivers of two or three days out of each seven while the shad or the salmon run would do much good.

The sturgeon, the eggs of which are used in the manufacture of the delicacy known as *caviar*, is an example of a fish that is almost extinct in this part of the world. Other food fish taken at the breeding season are also in danger.

Artificial Propagation of Fishes. — Fortunately, the government through the Bureau of Fisheries, and various states by wise protective laws and by artificial propagation of fishes, are beginning to turn the tide. Certain days of the week the salmon are allowed to pass up the Columbia unmolested. Closed breeding seasons protect our trout, bass, and other game fish, also the catching of fish under a certain size is prohibited.

Many fish hatcheries, both government and state, are engaged in artificially fertilizing millions of fish eggs of various species and protecting the young fry until they are of such size that they can take

Artificial fertilization of fish eggs.

care of themselves, when they are placed in ponds or streams. This artificial fertilization is usually accomplished by first squeezing out the ripe eggs from a female into a pan of water; in a similar manner the milt or sperm cells are obtained, and poured over the eggs. The eggs are thus fertilized. They are then placed in receptacles supplied with running water and left to develop under favorable conditions. Shortly after the egg has segmented (divided into many cells) the embryo may be seen developing on one side

of the egg. The rest of the egg is made up of food or yolk,
and when the baby fish hatches it has for some time the yolk
attached to its ventral surface. Eventually the food is absorbed
into the body of the fish. The development of the fish is direct,
the young fish becoming an adult without any great change in
form. The young fry are kept under ideal conditions until later,
when they are shipped, sometimes thousands of miles, to their
new homes.

1   2   3   4   5

Early development of salmon.   Natural size.

NOTE TO TEACHER. — It is suggested that in the spring term the frog be studied,
but if animal biology be taken up during the fall term the fish only might be used.

THE FROG

Adaptations for Life. — The most common frog in the eastern
part of the United States is the leopard frog. It is recognized by
its greenish brown body with dark spots, each spot being outlined
in a lighter-colored background. In spite of the apparent lack of
harmony with their surroundings, their color appears to give
almost perfect protection. In some species of frogs the color of
the skin changes with the surroundings of the frog, another means
of protection.

Adaptations for life in the water are numerous. The ovoid
body, the head merging into the trunk, the slimy covering (for
the frog is provided, like the fish, with mucus cells in the skin),
and the powerful legs with webbed feet, are all evidences of the
life which the frog leads.

Locomotion. — You will notice that the appendages have the
same general position on the body and same number of parts as
do your own (upper arm, forearm, and hand; thigh, shank, and
foot, the latter much longer relatively than your own). Note that
while the hand has four fingers, the foot has five toes, the latter
connected by a web. In swimming the frog uses the stroke we

all aim to make when we are learning to swim. Most of the energy is liberated from the powerful backward push of the hind legs, which in a resting position are held doubled up close to the body. On land, locomotion may be by hopping or crawling.

Sense Organs. — The frog is well provided with sense organs. The eyes are large, globular, and placed at the side of the head. When they are closed, a delicate fold, or third eyelid, called the *nictitating membrane,* is drawn over each eye. Frogs probably see best moving objects at a few feet from them. Their vision is much keener than that of the fish. The external ear (*tympanum*) is located just behind the eye on the side of the body. Frogs hear sounds and distinguish various calls of their own kind, as is proved by the fact that frogs recognize the warning notes of their mates when any one is approaching. The inner ear also has to do with balancing the body as it has in fishes and other vertebrates. Taste and smell are probably not strong sensations in a frog or toad. They bite at moving objects of almost any kind when hungry. The long flexible tongue, which is fastened at the front, is used to catch insects. Experience has taught these animals that moving things, insects, worms, and the like, make good food. These they swallow whole, the tiny teeth being used to hold the food. Touch is a well-developed sense. They also respond to changes in temperature under water, remaining there in a dormant state for the winter when the temperature of the air becomes colder than that of the water.

This diagram shows how the frog uses its tongue to catch insects.

Breathing. — The frog breathes by raising and lowering the floor of the mouth, pulling in air through the two nostril holes. Then the little flaps over the holes are closed, and the frog swallows this air, forcing it down into the baglike lungs. The skin is provided with many tiny blood vessels, and in winter, while the frogs are dormant at the bottom of the ponds, it serves as the only organ of respiration.

The Food Tube and its Glands. — The mouth leads like a funnel into a short tube, the *gullet*. On the lower floor of the mouth can be seen the slitlike *glottis* leading to the lungs. The gullet widens almost at once into a long *stomach*, which in turn leads into a much coiled intestine. This widens abruptly at the lower end to form the *large intestine*. The latter leads into the *cloaca* (Latin, *sewer*), into which open the *kidneys, urinary bladder*, and reproductive organs (*ovaries* or *spermaries*). Several *glands*, the function of which is to produce digestive fluids, open into the food tube. These digestive fluids, by means of the ferments or enzymes contained in them, change insoluble food materials into a soluble form. This allows of the absorption of food material through the walls of the food tube into the blood. The glands (having the same names and uses as those in man) are the *salivary glands*, which pour their juices into the mouth, the *gastric glands* in the walls of the stomach, and the *liver* and *pancreas*, which open into the intestine.

Circulation. — The frog has a well-developed heart, composed of a thick-walled muscular ventricle and two thin-walled auricles. The heart pumps the blood through a system of closed tubes to all parts of the body. Blood enters the right auricle from all parts of the body; it then contains considerable carbon dioxide; the blood entering the left auricle comes

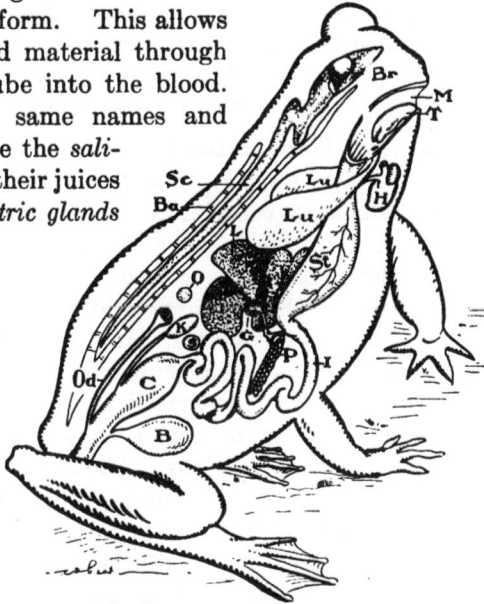

Internal organs of a frog: M, mouth; T, tongue; Lu, lungs; H, heart; St, stomach; I, small intestine; L, liver; G, gall bladder; P, pancreas; C, cloaca; B, urinary bladder; S, spleen; K, kidney, Od, oviduct; O, ovary; Br, brain; Sc, spinal cord; Ba, back bone.

from the lungs, hence it contains a considerable amount of oxygen. Blood leaves the heart through the ventricle, which thus pumps some blood containing much and some containing little oxygen. Before the blood from the tissues and lungs has time to mix, however, it leaves the ventricle and by a delicate adjustment in the vessels leaving the heart most of the blood containing much oxygen is passed to all the various organs of the body, while the blood deficient in oxygen, but containing a large amount of carbon dioxide, is pumped to the lungs, where an exchange of oxygen and carbon dioxide takes place by osmosis.

In the tissues of the body wherever work is done the process of burning or oxidation must take place, for by such means only is the energy necessary to do the work released. Food in the blood is taken to the muscle cells or other cells of the body and there oxidized. The products of the burning — carbon dioxide — and any other organic wastes given off from the tissues must be eliminated from the body. As we know, the carbon dioxide passes off through the lungs and to some extent through the skin of the frog, while the nitrogenous wastes, poisons which must be taken from the blood, are eliminated from it in the kidneys.

Change of Form in Development of the Frog. — Not all vertebrates develop directly into an adult. The frog, for example, changes its form completely before it becomes an adult. This change in form is known as a *metamorphosis*. Let us examine the development of the common leopard frog.

The eggs of this frog are laid in shallow water in the early spring. Masses of several hundred, which may be found attached to twigs or other supports under water, are deposited at a single laying. Immediately before leaving the body of the female they receive a coating of jellylike material, which swells up after the eggs are laid. Thus they are protected from the attack of fish or other animals which might use them as food. The upper side of the egg is dark, the light-colored side being weighted down with a supply of yolk (food). The fertilized egg soon segments (divides into many cells), and in a few days, if the weather is warm, these eggs have each grown into an oblong body which shows the form of a tadpole. Shortly after the tadpole wriggles out of the jellylike case and begins life outside the egg. At first it remains attached to some water weed by means of a pair of suckerlike projections; later a mouth is formed, and the tadpole begins to feed upon algæ or other tiny water plants. At this time, about two weeks after the eggs were laid, gills are

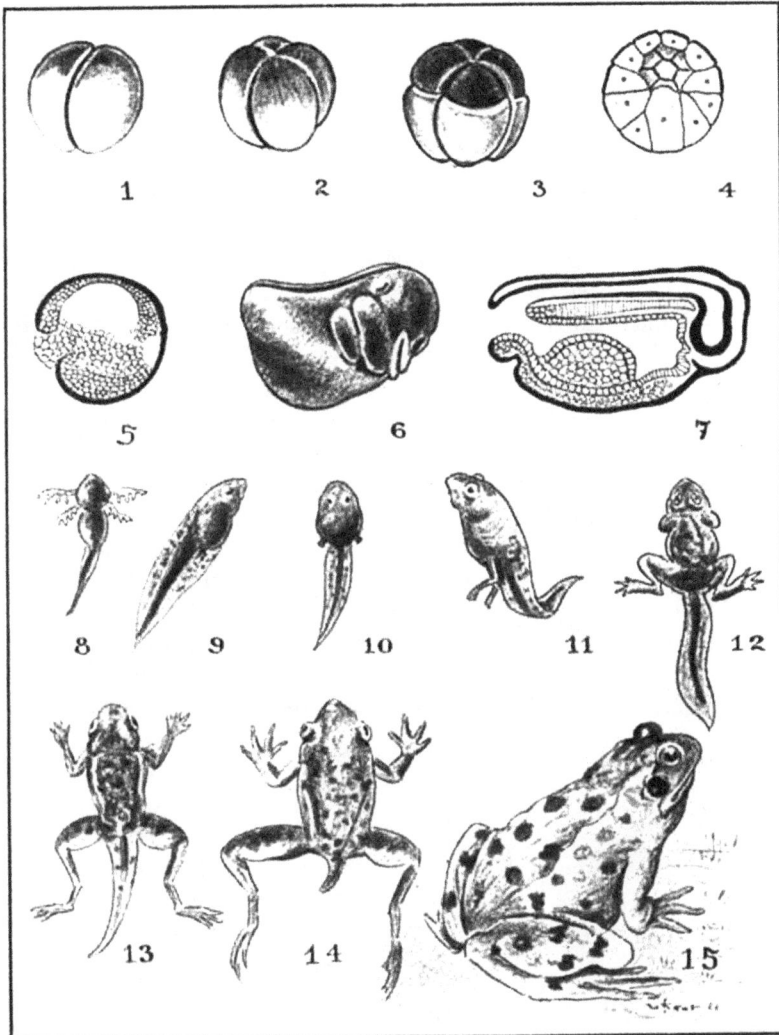

Development of a frog. 1, two cell stage; 2, four cell stage; 3, 8 cells are formed, notice the upper cells are smaller; in (4) the lower cells are seen to be much larger because of the yolk; 5, the egg has continued to divide and has formed a gastrula; 6, 7, the body is lengthening, head is seen at the right hand end; 8, the young tadpole with external gills; 9, 10, the gills are internal, hind legs beginning to form; 11, the hind legs show plainly; 12, 13, 14, later stages in development; 15, the adult frog. Figures 1, 2, 3, 4, 5, 6, and 7 are very much enlarged. (Drawn after Leukart and Kny by Frank M. Wheat.)

present on the outside of the body. Soon after, the external gills are replaced by gills which grow out under a fold of the skin which forms an operculum somewhat as in the fish. Water reaches the gills through the mouth and passes out through a hole on the left side of the body. As the tadpole grows larger, legs appear, the hind legs first, although for a time locomotion is performed by means of the tail. In the leopard frog the change from the egg to adult is completed in one summer. In late July or early August, the tadpole begins to eat less, the tail becomes smaller (being absorbed into other parts of the body), and before long the transformation from the tadpole to the young frog is complete. In the green frog and bullfrog the metamorphosis is not completed until the beginning of the second summer. The large tadpoles of such forms bury themselves in the soft mud of the pond bottom during the winter.

Shortly after the legs appear, the gills begin to be absorbed, and lungs take their place. At this time the young animal may be seen coming to the surface of the water for air. Changes in the diet of the animal also occur; the long, coiled intestine is transformed into a much shorter one. The animal, now insectivorous in its diet, becomes provided with tiny teeth and a mobile tongue, instead of keeping the horny jaws used in scraping off algæ. After the tail has been completely absorbed and the legs have become full grown, there is no further structural change, and the metamorphosis is complete.

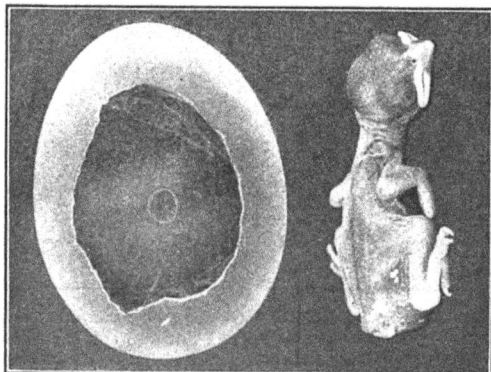

At the left is a hen's egg, opened to show the embryo at the center (the spot surrounded by a lighter area). At the right is an English sparrow one day after hatching.

Development of Birds. — The white of the hen's egg is put on during the passage of the real egg (which is in the **yoke or yellow**

portion) to the outside of the body. Before the egg is laid a shell is secreted over its surface. If the fertilized egg of a hen be broken and carefully examined, on the surface of the yolk will be found a little circular disk. This is the beginning of the growth of an *embryo* chick. If a series of eggs taken from an incubator at periods of twenty-four hours or less apart were examined, this spot would be found at first to increase in size; later the little embryo would be found lying on the surface. Still later small blood vessels could be made out reaching into the yolk for food, the tiny heart beating as early as the second day of incubation. After about three weeks of incubation the little chick hatches; that is, breaks the shell, and emerges in almost the same form as the adult.

Development of a Mammal. — In mammals after fertilization the egg undergoes development within the body of the mother. Instead of blood vessels connecting the embryo with the yolk as in the chick, here the blood vessels are attached to an absorbing organ, known as the *placenta*. This structure sends branchlike processes into the wall of the *uterus* (the organ which holds the embryo) and absorbs nourishment and oxygen by osmosis from the blood of the mother. After a length of time which varies in different species of mammals (from about three weeks in a guinea pig to twenty-two months in an elephant), the young animal leaves the protecting body of the mother, or is born.

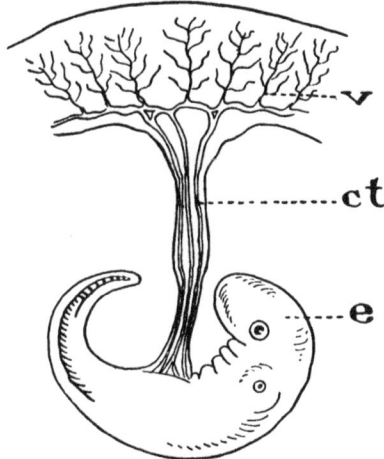

The embryo (*e*) of a rabbit, showing the absorbing organ; the branch-like processes which absorb blood from the mother being shown at (*v*); *ct*, the tube connecting the embryo with the absorbing organ or placenta. (After Haeckel.)

The young, usually, are born in a helpless condition, then nourished by milk furnished by the mother until they are able to take other food.   Thus we see as we go higher in the scale of life fewer eggs formed, but those few eggs are more carefully protected and cared for by the parents.   The chances of their growth into adults are much greater than in the cases when many eggs are produced.

REFERENCE BOOKS

ELEMENTARY

Hunter, *Laboratory Problems in Civic Biology*.   American Book Company.
Bigelow, *Introduction to Biology*.   The Macmillan Company.
Cornell *Nature Study Leaflets*.   Bulletins XVI, XVII.
Davison, *Practical Zoölogy*, pages 185–199.   American Book Company.
Hodge, *Nature Study and Life*, Chaps. XVI, XVII.   Ginn and Company.
Sharpe, *Laboratory Manual*, pp. 195, 204–209.   American Book Company.

ADVANCED

Dickerson, *The Frog Book*.   Doubleday, Page and Company.
Holmes, *The Biology of the Frog*.   The Macmillan Company.
Jordan, *Fishes*.   Henry Holt and Company.
Morgan, *The Development of the Frog's Egg*.   The Macmillan Company.
Needham, *General Biology*.   Comstock Publishing Company.

XVII. HEREDITY, VARIATION, PLANT AND ANIMAL BREEDING

Problems. — *To determine what makes the offspring of animals or plants tend to be like their parents.*

To determine what makes the offspring of animals and plants differ from their parents.

To learn about some methods of plant and animal breeding.

(a) By selection.

(b) By hybridizing.

(c) By other methods.

To learn about some methods of improving the human race.

(a) By eugenics.

(b) By euthenics.

SUGGESTIONS FOR LABORATORY WORK

Laboratory exercise. — On variation and heredity among members of a class in the schoolroom.

Laboratory exercise. — On construction of curve of variation in measurements from given plants or animals.

Laboratory demonstration. — Stained egg cel's (*ascaris*) to show chromosomes.

Laboratory demonstrations. — To illustrate the part played in plant or animal breeding by

(a) selection.

(b) hybridizing.

(c) budding and grafting.

Laboratory demonstration. — From charts to illustrate how human characteristics may be inherited.

HEREDITY AND EUGENICS

Heredity and what it Means. — As I look over the faces of the boys in my class I notice that each boy seems to be more or less like each other boy in the class; he has a head, body, arms, and legs, and even in minor ways he resembles each of the other boys in the room. Moreover, if I should ask him I have no doubt

249

but that he would tell me that he resembled in many respects his mother or father.   Likewise if I should ask his *parents* whom he resembled, they would say, " I can see his grandmother or his grandfather in him."

This wonderful force which causes the likeness of the child to its parents and to *their* parents we call *heredity*.   Heredity causes the plants as well as animals to be like their parents.   If we trace the workings of heredity in our own individual case, we will probably find that we are molded like our ancestors not only in physical characteristics but in mental qualities as well.   The ability to play the piano or to paint is probably as much a case of inheritance as the color of our eyes or the shape of our nose.   We are a complex of physical and mental characters, received in part from all our ancestors.

Variation. — But I notice another thing; no boy in the class before me is *exactly* like any other boy, even twins having minute differences.   In this wonderful mold of nature each one of us

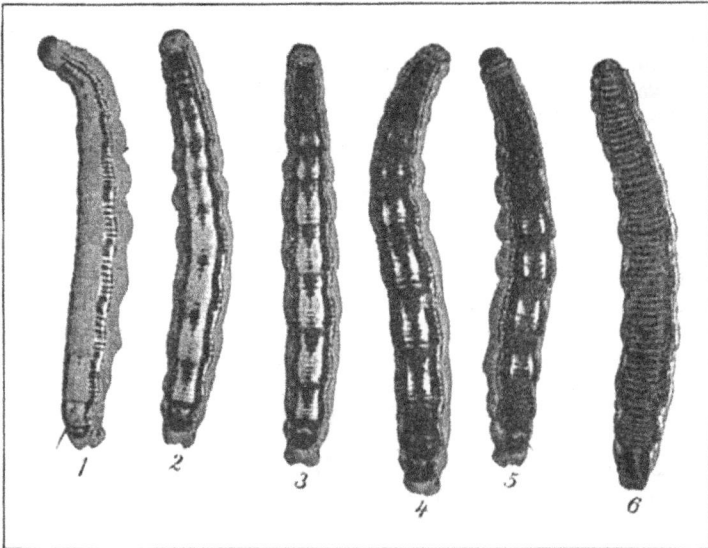

Variations in the Catalpa caterpillar.   (Photographed, natural size, by Davison.)

tends to be slightly different from his or her parents. Each plant, each animal, varies to a greater or lesser degree from its immediate ancestors and may vary to a very great degree. This factor in the lives of plants and animals is called *variation*. Heredity and variation are the cornerstones on which all the work in the improvement of plants and animals, including man himself, is built.

The Bearers of Heredity. — We have seen that somewhere in every living cell is a structure known as a nucleus. In this nucleus, which is a part of the living matter of the cell, are certain very minute structures always present, known as *chromosomes*. These chromosomes (so called because they take up color when stained) are believed to be the structures which contain the *determiners* of the qualities which may be passed from parent plant to offspring or from animal to animal; in other words, the qualities that are inheritable (see page 252).

The Germ Cells. — But it has been found that certain cells of the body, the egg and the sperm cells, before uniting contain only half as many chromosomes as do the body cells. In preparing for the process of fertilization, half of these elements have been eliminated, so that when the egg and sperm cell are united they will have the full number of chromosomes that the other cells have.

If the chromosomes carry the determiners of the characters which are inheritable, then it is easy to see that a fertilized egg must contain an equal number of chromosomes from the bodies of each parent. Consequently characteristics from each parent are handed down to the new individual. This seems to be the way in which nature succeeds in obtaining variation, by providing cell material from two different individuals.

Offspring are Part of their Ancestors. — We can see that if you or I receive characteristics from our parents and they received characteristics from their parents, then we too must have some of the characteristics of the grandparents, and it is a matter of common knowledge that each of us does have some trait or lineament which can be traced back to our grandfather or grandmother. Indeed, as far back as we are able to go, ancestors have added something.

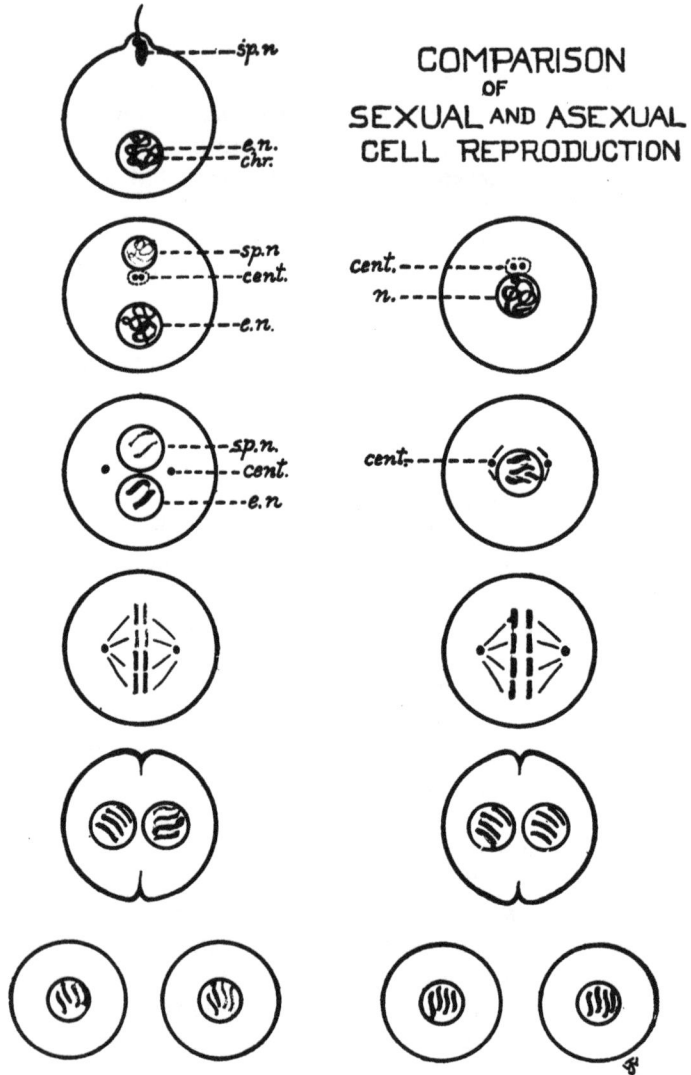

COMPARISON
OF
SEXUAL AND ASEXUAL
CELL REPRODUCTION

Charles Darwin and Natural Selection. — The great English-man Charles Darwin was one of the first scientists to realize how this great force of heredity applied to the development or evolution of plants and animals. He knew that although animals and plants were like their ancestors, they also tended to vary. In nature, the variations which best fitted a plant or animal for life in its own environment were the ones which were handed down because those having variations which were not fitted for life in that particular environment would die. Thus nature seized upon favorable variations and after a time, as the descendants of each of these individuals also tended to vary, a new species of plant or animal, fitted for the place it had to live in, would be gradually evolved.

Mutations. — Recently a new method of variation has been discovered by a Dutch naturalist, named Hugo de Vries. He found that new species of plants and animals arise suddenly by " mutations " or steps. This means that new species instead of arising from very slight variations, continuing during long periods of years (as Darwin believed), might arise very suddenly as a very great variation which would at once breed true. It is easily seen that such a condition would be of immense value to breeders, as new plants or animals quite unlike their parents might thus be formed and perpetuated. It will be one of the future problems of plant and animal breeders to isolate and breed " mutants," as such organisms are called.

Artificial Selection. — Darwin reasoned that if nature seized upon favorable variants, then man by selecting the variations he wanted, could form new varieties of plants or animals much more quickly than nature. And so to-day plant or animal

Improvement in corn by selection. To the left, the corn improved by selection from the original type at the right.

breeders *select* the forms having the characters they wish to perpetuate and breed them together. This method used by plant and animal breeders is known as *selection*.

Selective Planting. — *By selective planting we mean choosing the best plants and planting the seed from these plants with a view of improving the yield.* In doing this we must not necessarily select the most perfect fruits or grains, but must select seeds from the *best plants*. A wheat plant should be selected not from its yield alone, but from its ability to stand disease and other unfavorable conditions. In 1862 a Mr. Fultz, of Pennsylvania, found three heads of beardless or bald wheat while passing through a large field of bearded wheat. These were probably *mutants* which had lost the chaff surrounding the kernel. Mr. Fultz picked them out, sowed them by themselves, and produced a quantity of wheat now known favorably all over the world as the Fultz wheat. In selecting wheat, for example, we might breed for a number of different characters, such as more starch, or more protein in the grain, a larger yield per acre, ability to stand cold or drought or to resist plant disease. Each of these characters would have to be sought for separately and could only be obtained after long and careful breeding. The work of Mendel (see page 257) when applied to plant breeding will greatly shorten the time required to produce better plants of a given kind. By careful seed selection, some Western farmers have increased their wheat production by 25 per cent. This, if kept up all over the United States, would mean over $100,000,000 a year in the pockets of the farmers.

Hybridizing. — We have already seen that pollen from one flower may be carried to another of the same species, thus producing seeds. If pollen from one plant be placed on the pistil of another of an *allied* species or variety, fertilization *may* take place and new plants be eventually produced from the seeds. This process is known as *hybridizing*, and the plants produced by this process known as *hybrids*.

Hybrids are extremely variable, rarely breed from seeds, and often are apparently quite unlike either parent plant. They must be grown for several years, and all plants that do not resemble the desired variety must be killed off, if we expect to produce a hybrid

that will breed more plants like itself. Luther Burbank, the great hybridizer of California, destroys tens of thousands of plants in order to get one or two with the characters which he wishes to preserve. Thus he is yearly adding to the wealth of this country by producing new plants or fruits of commercial value. A number of years ago he succeeded in growing a new variety of potato, which has already en-
riched the farmers of this country about $20,000,000. One of his varieties of black walnut trees, a very valuable hard wood, grows ten to twelve times as rapidly as ordinary black walnuts. With lumber yearly increasing in price, a quick grow-
ing tree becomes a very valuable com-
mercial product. Among his famous hybrids are the plumcot, a cross between an apricot and a plum, his numerous varieties of berries and his splendid "Climax" plum, the result of a cross between a bitter Chinese plum and an edible Japanese plum. But none of Burbank's

In hybridizing, all of the flower is removed at the line (*W*) except the pis-
til (*P*). Then pollen from another flower of a nearly related kind is placed on the pistil and the pollinated flower covered up with a paper bag. Can you explain why?

products grow from seeds; they are all produced *asexually*, from hybrids by some of the processes described in the next paragraph.

The Department of Agriculture and its Methods. — The Depart-
ment of Agriculture is also doing splendid work in producing new varieties of oranges and lemons, of grain and various garden vegetables. The greatest possibilities have been shown by department workers to be open to the farmer or fruit grower through hybridizing, and by budding, grafting, or slipping.

Budding. — If a given tree, for example, produces a kind of fruit which is of excellent quality, it is possible sometimes to attach parts of the tree to another strong tree of the same species that may not bear good fruit. This is done by *budding*. A T-shaped incision is cut in the bark; a bud from the tree bearing the desired fruit is placed in the cut and bound in place. When a shoot from the

embedded bud grows out the following spring, it is found to have all the characters of the tree from which it was taken.

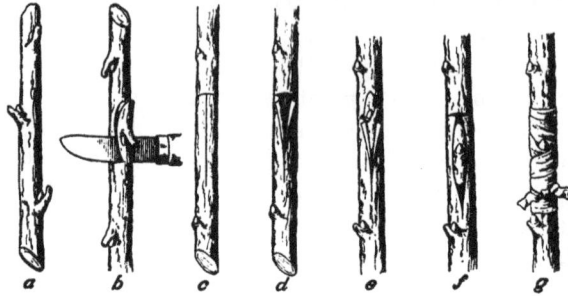

Steps in budding. *a*, twig having suitable buds to use; *b*, method of cutting out bud; *c*, how the bark is cut; *d*, how the bark is opened; *e*, inserting the bud; *f*, the bud in place; *g*, the bud properly bound in place.

Grafting. — Of much the same nature is grafting. Here, however, a small portion of the stem of the closely allied tree is fas-

Steps in tongue grafting. *a*, the two branches to be formed; *b*, a tongue cut in each; *c*, fitted together; *d*, method of wrapping.

tened into the trunk of the growing tree in such a manner that the two cut layers just under the bark will coincide. This will allow of the passage of food into the grafted part and insure the ultimate growth of the twig. Grafting and budding are of considerable economic value to the fruit grower, as it enables him to produce at will, trees bearing choice varieties of fruit.[1]

Other Methods. — Other methods of plant propagation are by means of runners, as when strawberry plants strike root from long stems that run along the ground; layering, where roots may develop on covered up branches of blackberry or raspberry plants; slips, roots developing from stems which are cut off and placed in moist sand; from tubers, as in planting potatoes; and by means of

[1] For full directions for budding and grafting, see Goff and Mayne, *First Principles of Agriculture*, Chap. XIX, Mayne and Hatch, *High School Agriculture*, pp. 159–165, or Hodge, *Nature Study and Life*, pages 169–179.

bulbs, as the tulip or hyacinth.   All of the above means of propagation are asexual and are of importance in our problem of
plant breeding.

Plant breeding plots.   (Minnesota Experiment Station.)

The Work of Gregor Mendel. — Fifty years ago, an Austrian·
monk, Gregor Mendel, found in breeding garden peas that these
plants passed on certain *fixed characters*, as the shape of the
seed, the color of the pod when ripe, and others, and that when
two pea plants of different characters were crossed, one of these

characters would be likely to appear in the offspring of the second generation in the ratio of three to one. Such characters as would appear to the exclusion of others in the first crossing of the plants were called *dominant,* the ones not appearing, *recessive* characteristics. When these seeds were again sown the ones bearing a recessive characteristic would produce only peas with this recessive characteristic, but the ones with a dominant characteristic might give rise to a pure dominant or to offspring having partly a dominant and partly a recessive character; pure dominants being to the mixed offspring in the ratio of 1 to 2. The pure dominants if bred with others like themselves would produce only pure dominants, but the cross breeds would again produce mixed offspring of three kinds in the ratio of one dominant to two cross breeds and one recessive. The feature of this work that interests us is that *unit* characters are passed along by heredity in the germ cells *pure,* that is, unchanged, from one generation to another, and independently of each other.

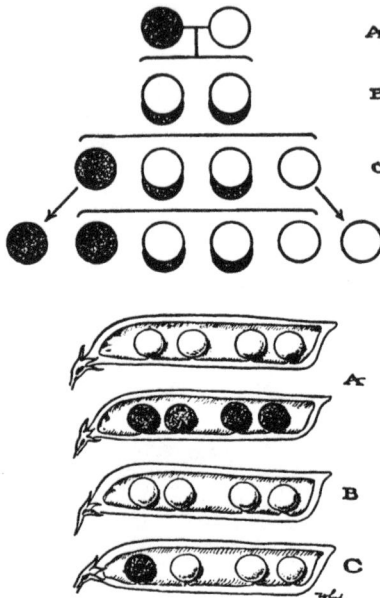

Illustration of Mendel's Law.

Determiners of Character. — A child then resembles his parents in some definite particulars because certain *determiners* of characters have been present in the germ cells of one of the parents. If the determiner of a certain character is *absent* from the germ cells of both parents, it will be *absent* in *all* of their offspring.

These discoveries of Mendel are of the greatest importance in plant and animal breeding because they enable the breeder to isolaté certain characters and by proper selection to breed varieties which have these desired characters, instead of waiting for a *chance* union of the desired characters by nature.

Animal Breeding. — It has been pointed out that the domestication of wild animals, the horse, cattle, sheep, goats, and the dog, marked a great advance in civilization in the history of the earth's peoples. As the young of these animals came to be bred in captivity the peoples owning them would undoubtedly pick out the strongest and best of the offspring, killing off the others for food. Thus they came unconsciously to select and aid nature in producing a stronger and better stock. Later man began to recognize certain characters that he wished to have in horses, dogs, or cattle, and so by slow processes of breeding and " crossing " or hybridizing one nearly allied form with another the numerous groups of domesticated animals began to appear.

What has resulted from artificial selection among dogs. (After Romanes.)

In Darwin's time animal breeding was so far advanced that he got his ideas of selection by nature in evolution from the artificial selection practiced by animal breeders. A glance at the pictures will give some idea of the changes that have taken place in the form of some animals since man began to breed them a few thousand years ago.

Some Domesticated Animals. — Our domesticated dogs are descended from a number of wolflike forms in various parts of the world. All the present races of cats, on the other hand, seem to be traced back to Egypt. Modern horses are first noted in Europe and Asia, but far older forms flourished on the earth in former geologic periods. It is interesting to note that America was the original home of the horse, although at the time of the earliest explorers the horse was unknown here, the wild horse of the Western plains having arisen from horses introduced by the Spaniards. Long ages ago, the first ancestors of the horse were probably little animals about the size of a fox. The earliest horse we have knowledge of had four toes on the fore and three toes on the hind foot. Thousands of years later we find a larger horse, the size of a sheep, with a three-toed foot. By gradual changes, caused by the tendency of the animals to vary and by the action of the surroundings upon the animal in preserving these variations, there was eventually produced our present horse, an animal with legs adapted for rapid locomotion, with feet particularly fitted for the life in open fields, and with teeth which serve well to seize and grind herbage. Knowledge of this sort was also used by Darwin to show that constant changes in the form of animals have been taking place since life began on the earth.

The four-toed ancestor of the present horse, restored from a study of its fossil skeleton. (After Knight in American Museum of Natural History.)

The horse, which for some reason disappeared in this country, continued to exist in Europe, and man, emerging from his early savage condition, began to make use of the animal. We know the horse was domesticated in early Biblical times, and that he soon became one of man's most valued servants. In more recent times, man has begun to change the horse by breeding for certain desired characteristics. In this manner have been established and improved the various types of horses familiar to us as draft horses, coach horses, hackneys, and the trotters.

It is needless to say that all the various domesticated animals have been tremendously changed in a similar manner since civilized man has come to live on the earth. When we realize the very great amount of money invested in domesticated animals; that there are over 60,000,000 each of sheep, cattle, and swine and over 20,000,000 horses owned in this country, then we may see how very important a part the domestic animals play in our lives.

Improvement of Man. — If the stock of domesticated animals can be improved, it is not unfair to ask if the health and vigor of the future generations of men and women on the earth might not be improved by applying to them the laws of selection. This improvement of the future race has a number of factors in which we as individuals may play a part. These are personal hygiene, selection of healthy mates, and the betterment of the environment.

Personal Hygiene. — In the first place, good health is the one greatest asset in life. We may be born with a poor bodily machine, but if we learn to recognize its defects and care for it properly, we may make it do its required work effectively. If certain muscles are poorly developed, then by proper exercise we may make them stronger. If our eyes have some defect, we can have it remedied by wearing glasses. If certain drugs or alcohol lower the efficiency of the machine, we can avoid their use. With proper *care* a poorly developed body may be improved and do effective work.

Eugenics. — When people marry there are certain things that the individual as well as the race should demand. The most important of these is freedom from germ diseases which might be handed down to the offspring. Tuberculosis, that dread white plague which is still responsible for almost one seventh of all deaths, epilepsy, and feeble-mindedness are handicaps which it is not only unfair but criminal to hand down to posterity. The science of being well born is called *eugenics*.

The Jukes. — Studies have been made on a number of different families in this country, in which mental and moral defects were present in one or both of the original parents. The "Jukes" family is a notorious example. The first mother is known as

" Margaret, the mother of criminals." In seventy-five years the
progeny of the original generation has cost the state of New York
over a million and a quarter of dollars, besides giving over

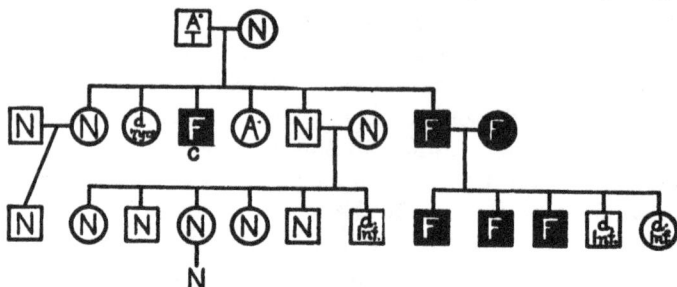

In this and the following diagrams the circle represents a female, the square a
male. Ⓝ means normal; 🅵 means feeble-minded; A, alcoholic; T, tubercular. This chart shows the record of a certain family for three generations.
A normal woman married an alcoholic and tubercular man. He must have
been feeble-minded also, as two of his children were born feeble-minded. One
of these children married another feeble-minded woman, and of their five
children two died in infancy and three were feeble-minded. (After Davenport.)

to the care of prisons and asylums considerably over a hundred feeble-minded, alcoholic, immoral, or criminal persons.
Another case recently studied is the " Kallikak " family.[1] This
family has been traced to the union of Martin Kallikak, a young
soldier of the War of the Revolution, with a feeble-minded girl.

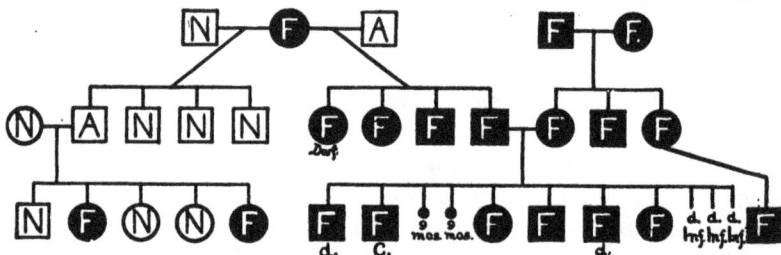

This chart shows that feeble-mindedness is a characteristic sure to be handed
down in a family where it exists. The feeble-minded woman at the top left
of the chart married twice. The first children from a normal father are all
normal, but the other children from an alcoholic father are all feeble-minded.
The right-hand side of the chart shows a terrible record of feeble-mindedness.
Should feeble-minded people be allowed to marry ? (After Davenport.)

[1] The name Kallikak is fictitious.

She had a feeble-minded son from whom there have been to the present time 480 descendants. Of these 33 were sexually immoral, 24 confirmed drunkards, 3 epileptics, and 143 *feeble-minded*. The man who started this terrible line of immorality and feeble-mindedness later married a normal Quaker girl. From this couple a line of 496 descendants have come, with *no* cases of feeble-mindedness. The evidence and the moral speak for themselves !

Parasitism and its Cost to Society. — Hundreds of families such as those described above exist to-day, spreading disease, immorality, and crime to all parts of this country. The cost to society of such families is very severe. Just as certain animals or plants become parasitic on other plants or animals, these families have become parasitic on society. They not only do harm to others by corrupting, stealing, or spreading disease, but they are actually protected and cared for by the state out of public money. Largely for them the poorhouse and the asylum exist. They take from society, but they give nothing in return. They are true parasites.

The Remedy. — If such people were lower animals, we would probably kill them off to prevent them from spreading. Humanity will not allow this, but we do have the remedy of separating the sexes in asylums or other places and in various ways preventing intermarriage and the possibilities of perpetuating such a low and degenerate race. Remedies of this sort have been tried successfully in Europe and are now meeting with success in this country.

Blood Tells. — Eugenics show us, on the other hand, in a study of the families in which are brilliant men and women, the fact that the descendants have received the *good* inheritance from their ancestors. The following, taken from Davenport's *Heredity in Relation to Eugenics*, illustrates how one family has been famous in American History.

In 1667 Elizabeth Tuttle, " of strong will, and of extreme intellectual vigor, married Richard Edwards of Hartford, Conn., a man of high repute and great erudition. From their one son descended another son, Jonathan Edwards, a noted divine, and president of Princeton College. Of the descendants of Jonathan Edwards much has been written ; a brief catalogue must suffice : Jonathan Edwards, Jr., president of Union College; Timothy

Dwight, president of Yale; Sereno Edwards Dwight, president of Hamilton College; Theodore Dwight Woolsey, for twenty-five years president of Yale College; Sarah, wife of Tapping Reeve,

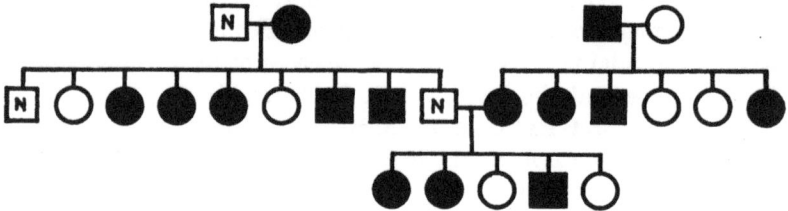

This record shows the inheritance of artistic ability (black circles and squares). (After Davenport.)

founder of Litchfield Law School, herself no mean lawyer; Daniel Tyler, a general in the Civil War and founder of the iron industries of North Alabama; Timothy Dwight, second, president of Yale University from 1886 to 1898; Theodore William Dwight, founder and for thirty-three years warden of Columbia Law School; Henrietta Frances, wife of Eli Whitney, inventor of the cotton gin, who, burning the midnight oil by the side of her ingenious husband, helped him to his enduring fame; Merrill Edwards Gates, president of Amherst College; Catherine Maria Sedgwick of graceful pen; Charles Sedgwick Minot, authority on biology and embryology in the Harvard Medical School; Edith Kermit Carow, wife of Theodore Roosevelt; and Winston Churchill, the author of *Coniston* and other well-known novels."

Of the daughters of Elizabeth Tuttle distinguished descendants also came. Robert Treat Paine, signer of the Declaration of Independence; Chief Justice of the United States Morrison R. Waite; Ulysses S. Grant and Grover Cleveland, presidents of the United States. These and many other prominent men and women can trace the characters which enabled them to occupy the positions of culture and learning they held back to Elizabeth Tuttle.

Euthenics. — Euthenics, the betterment of the environment, is another important factor in the production of a stronger race. The strongest physical characteristics may be ruined if the surroundings are unwholesome and unsanitary. The slums of a city are " at once symptom, effect, and cause of evil." A city which allows foul tenements, narrow streets, and crowded slums to exist will spend too much for police protection, for charity, and for hospitals.

Every improvement in surroundings means improvement of the chances of survival of the race. In the spring of 1913 the health department and street-cleaning department of the city of New York coöperated to bring about a " clean up " of all filth, dirt, and rubbish from the houses, streets, and vacant lots in that city. During the summer of 1913 the health department reported a smaller percentage of deaths of babies than ever before. We must draw our own conclusions. Clean streets and houses, clean milk and pure water, sanitary housing, and careful medical inspection all do their part in maintaining a low rate of illness and death, thus reacting upon the health of the citizens of the future. It will be the purpose of the following pages to show how we may best care for our own bodies and how we may better the environment in which we are placed.

REFERENCE BOOKS

ELEMENTARY

Hunter, *Laboratory Problems in Civic Biology.* American Book Company.
Bailey, *Plant Breeding.* Macmillan and Company.
Harwood, *New Creations in Plant Life.* The Macmillan Company.
Jordan, *The Heredity of Richard Roe.* American Unitarian Association.
Sharpe, *Laboratory Manual,* pp. 64–72, 345–347. American Book Company.

ADVANCED

Allen, *Civics and Health.* Ginn and Company.
Coulter, Castle, East, Tower, and Davenport, *Heredity and Eugenics.* University of Chicago Press.
Davenport, *Heredity in Relation to Eugenics.* Henry Holt and Company.
De Vries, *Plant Breeding.* Open Court Publishing Company.
Goddard, *The Kallikak Family.* The Macmillan Company.
Kellicott, *The Social Direction of Human Evolution.* Appleton Company.
Punnet, *Mendelism.* The Macmillan Company.
Richards, Helen M. *Euthenics, the Science of Controllable Environment.*
Walter, *Genetics.* The Macmillan Company.

XVIII. THE HUMAN MACHINE AND ITS NEEDS

Problem. — *To obtain a general understanding of the parts and uses of the bodily machine.*

Man and his Environment. — In the last chapter we saw that one factor in the improvement of man lies in giving him better surroundings. It will be the purpose of the following chapters to show how man is fitted to live in the environment in which he is placed. He comes in contact with air, light, water, soil, food, and shelter which make his somewhat artificial environment; he must adapt himself to get the best he can out of this environment.

The Needs of Living Things. — We have already found that the primary needs of plants and animals are the same. They both need food, they both need to digest their food and to have it circulate in a fluid form to the cells where it will be used. They both need oxygen so as to release the energy locked up in their food. And they both need to reproduce so that their kind may be continued on the earth. What is true of plants and other animals is true of man.

The Needs of Simple and Complex Animals the Same. — The simplest animal, a single cell, has the same needs as the most complex. The *cell* paramœcium feeds, digests, oxidizes its food, and releases energy. The *cells* of the human body built up into tissues have the same needs and perform the same functions as the paramœcium. It is the *cells* of the body working together

266

in groups as tissues and organs that make the complicated actions of man possible. Division of labor has arisen because of the complex needs and work of the organism.

The Human Body a Machine. — In all animals, and the human animal is no exception, the body has been likened to a machine in that it turns over the *latent* or potential energy stored up in food into *kinetic* energy (mechanical work and heat), which is manifested when we perform work. One great difference exists between an engine and the human body. The engine uses fuel unlike the substance out of which it is made. The human body, on the other hand, uses for fuel the same substances out of which it is formed; it may, indeed, use part of its own substance for food. It must as well do more than purely mechanical work. The human organism must be so delicately adjusted to its surroundings that it will react in a ready manner to stimuli from without; it must be able to utilize its fuel (food) in the most economical manner; it must be fitted with machinery for transforming the energy received from food into various kinds of work; it must properly provide the machine with oxygen so that the fuel will be oxidized, and the products of oxidation must be carried away, as well as other waste materials

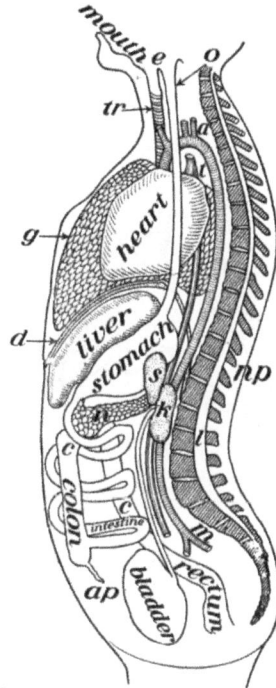

The human body seen from the side in longitudinal section.

which might harm the effectiveness of the machine. Most important of all, the human machine must be able to repair itself.

In order to understand better this complicated machine, the human body, let us briefly examine the structure of its parts and thus get a better idea of the interrelation of these parts and of their functions.

The Skin. — Covering the body is a protective structure called the skin. Covered on the outside with dead cells, yet it is provided with delicate sense organs, which give us perception of touch, taste, smell, pressure, and temperature. It also aids in getting wastes out of the body by means of its sweat glands and plays an important part in equalizing the temperature of the body.

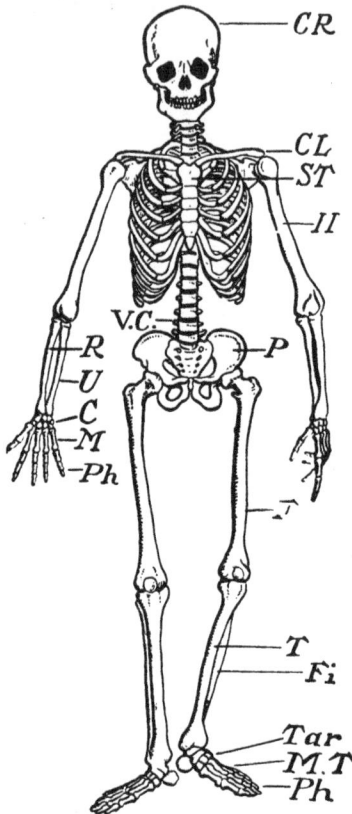

Bones and Muscles. — The body is built around a framework of bones. These bones, which are bound together by tough *ligaments*, fall naturally into two great groups, the bones of the body proper, vertebral column, ribs, breast bone, and skull, which form the *axial* skeleton, and the appendages, two sets of bones which form the framework of the arms and legs, which with the bones which attach them to the axial skeleton form the *appendicular* skeleton.

To the bones are attached the muscles of the body. Movement is accomplished by contraction of muscles, which are attached so as to cause the bones to act as levers. Bones also protect the nervous system and other delicate organs. They also help to give form and rigidity to the body.

Hygiene of Muscles and Bones. —Young people especially need to know how to prevent certain defects which are largely the result of bad habits of posture. Standing erect

Skeleton of a man. *CR.*, cranium; *CL.*, clavicle; *ST.*, sternum; *H.*, humerus; *V.C.*, vertebral column; *R.*, radius; *U.*, ulna; *P.*, pelvic girdle; *C.*, carpals; *M.*, metacarpals; *Ph.*, phalanges; *F.*, femur; *Fi.*, fibula; *T.*, tibia; *Tar.*, tarsals; *MT.*, metatarsals.

is an example of a good habit, round shoulders a bad habit of this sort. The habit of a wrong position of bones and muscles once formed is very hard to correct. This can best be done by certain corrective exercises at home or in the gymnasium.

Round shoulders is most common among people whose occupation causes them to stoop. Drawing, writing, and a wrong position when at one's desk are among the causes. Exercises which strengthen the back muscles and cause the head to be kept erect are helpful in forming the habit of erect carriage.

Slight curvature of the spine either backward or forward is helped most by exercises which tend to straighten the body, such as stretching up with the

Diagram showing action of biceps muscle. *a*, contracted; *b*, extended; *h*, humerus; *s*, scapula.

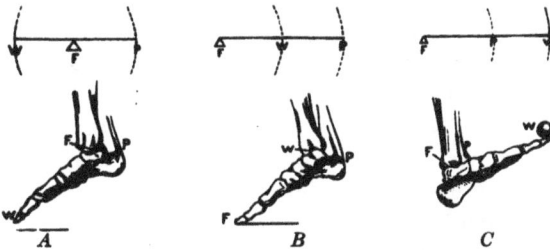

hands above the head. Lateral curvature of the spine, too often caused by a "hunched-up" position at the school desk, may also

Three classes of levers in the human body; bones and muscles act together. *A*, a lever of the first class; *B*, a lever of the second class; *C*, a lever of the third class.

be corrected by exercises which tend to lengthen the spinal column.

It is the duty of every girl and boy to have good posture and

Bad posture in the schoolroom may cause permanent injury to the spine.

erect carriage, not only because of the better state of health which comes with it, but also because one's self-respect demands that each one of us makes the best of the gifts that nature has given us. An erect head, straight shoulders, and elastic carriage go far toward making their owner both liked and respected.

Other Body Structures. — In spaces between the muscles are found various other structures, — blood vessels, which carry blood to and from the great pumping station, the heart, and thence to all parts of the body; connective tissue, which holds groups of muscle or other cells together; fat cells, scattered in various parts of the body; various gland cells, which manufacture enzymes; and the cells of the nervous system, which aid in directing the body parts.

Body Cavity. — Within the body is a cavity, which in life is almost completely. filled with various organs. A thin wall of muscle called the *diaphragm* divides the body cavity into two unequal spaces. In the upper space are found the *heart* and *lungs*, in the lower, the digestive tract with its glands, the *liver, kidneys*, and other structures (see page 267).

Digestion, Absorption, and Excretion. — Running through the body is a food tube in which undigested food is placed and from which digested or liquid food is absorbed into the blood so that the cells of the various organs which do the work may receive food. Emptying into this food tube are various groups of gland cells, which pour digestive fluids over the solid foods, thus aiding in changing them to liquids. Solid wastes are passed out through the posterior end of the food tube, while liquid wastes are excreted by means of glands called *kidneys*.

Work done by Cells. — Food, prepared in the digestive tract, and oxygen from the lungs are taken by the blood to the cells. Bathed in liquid food, the cells do their work; they promote the oxidization of food and the exchange of carbon dioxide for oxygen in the blood, while other wastes of the cells are given off, to pass eventually through the kidneys and out of the body.

The Nervous System. — The smooth working of the bodily machine is due to another set of structures which direct the working of the parts so that they will act in unison. This director is the nervous system. We have seen that, in the simplest of animals, one cell performs the functions necessary to its existence. In the more complex animals, where groups of cells form tissues, each having a different function, a nervous system is developed. *The functions of the human nervous system are:* (1) *the providing of man with sensation, by means of which he gets in touch with the world about him;* (2) *the connecting of organs in different parts of the body so that they act as a united and harmonious whole;* (3) *the giving to the human being a will, a provision for thought.* Coöperation in word and deed is the end attained. We are all familiar with examples of the coöperation of organs. You see food; the thought comes that it is good to eat; you reach out, take it, raise it to the mouth; the jaws move in response to your will; the food is chewed and swallowed. While digestion and absorption of the food are taking place, the nervous system is still in control. The nervous system also regulates pumping of blood over the body, respiration, secretion of glands, and, indeed, every bodily function. Man is the highest of all animals because of the extreme development of the nervous system. Man is the thinking animal, and as such is master of the earth.

REFERENCE READING FOR THIS AND SUCCEEDING CHAPTERS ON HUMAN BIOLOGY

ELEMENTARY

Hunter, *Laboratory Problems in Civic Biology.* American Book Company.
Davison, *The Human Body and Health.* American Book Company.
Gulick, *The Gulick Hygiene Series.* Ginn and Company.
Overton, *General Hygiene.* American Book Company.
Ritchie, *Human Physiology.* World Book Company.
Sharpe, *Laboratory Manual in Botany,* pages 218–225. American Book Company.

ADVANCED

Halliburton, *Kirk's Handbook of Physiology.* P. Blakiston's Son and Company.
Hough and Sedgwick, *The Human Mechanism.* Ginn and Company.
Howell, *Physiology,* 3d edition. W. B. Saunders Company.
Schafer, *Textbook of Physiology.* The Macmillan Company.
Stiles, *Nutritional Physiology.* W. B. Saunders Company.
Verworn, *General Physiology.* The Macmillan Company.

XIX. FOODS AND DIETARIES

Problems. — *A study of foods to determine:* —

(a) *Their nutritive value.*

(b) *The relation of work, environment, age, sex, and digestibility of foods to diet.*

(c) *Their relative cheapness.*

(d) *The daily Calorie requirement.*

(e) *Food adulteration.*

(f) *The relation of alcohol to the human system.*

LABORATORY SUGGESTIONS

Laboratory exercise. — Composition of common foods. The series of food charts supplied by the United States Department of Agriculture makes an excellent basis for a laboratory exercise to determine common foods rich in (a) water, (b) starch, (c) sugar, (d) fats or oils, (e) protein, (f) salts, (g) refuse.

Demonstration. — Method of using bomb calorimeter.

Laboratory and home exercise. — To determine the best individual balanced dietary (using standard of Atwater, Chittenden, or Voit) as determined by the use of the 100-Calorie portion.

Demonstration. — Tests for some common adulterants.

Demonstration. — Effect of alcohol on protein, *e.g.* white of egg.

Demonstration. — Alcohol in some patent medicines.

Demonstration. — Patent medicines containing acetanilid. Determination of acetanilid.

Why we Need Food. — A locomotive engine takes coal, water, oxygen, from its environment. A living plant or animal takes organic food, water, and oxygen from its environment. Both the living and nonliving machine does the same thing with this fuel or food. They oxidize it and release the energy in it. But the living organism in addition may use the food to repair parts that have broken down or even build new parts. *Thus food may be defined as something that releases energy or that forms material for* *the growth or repair of the body of a plant or animal.* The millions of cells of which the body is composed must be given material which will form more living matter or material which can be oxidized to release energy when muscle cells move, or gland cells secrete, or brain cells think.

Nutrients. — Certain nutrient materials form the basis of food of both plants and animals. These have been stated to be *proteins* (such as lean meat, eggs, the gluten of bread), *carbohydrates* (starches, sugars, gums, etc.), *fats* and *oils* (both animal and vegetable), *mineral matter* and *water.*

The composition of milk. Why is it considered a good food?

Proteins. — Protein substances contain the element nitrogen. Hence such foods are called nitrogenous foods. Man must form the protoplasm of his body (that is, the muscles, tendons, nervous system, blood corpuscles, the living parts of the bone and the skin, etc.) in part at least from nitrogenous food. Some of this he obtains by eating the flesh of animals, and some he obtains directly from plants (for example, peas and beans). Proteins are the only foods directly available for tissue building. They may be oxidized to release energy if occasion requires it.

Fats and Oils. — Fats and oils, both animal and vegetable, are the materials from which the body derives part of its energy. The chemical formula of a fat shows that, compared with other food substances, there is very little oxygen present; hence the greater capacity of this substance for uniting with oxygen. The rapid burning of fat compared with the slower combustion of a piece of meat or a piece of bread illustrates this. A pound of butter releases over twice as much energy to the body as does a pound of sugar or a pound of steak. Human fatty tissue is formed in part from fat eaten, but carbohydrate or even protein food may be changed and stored in the body as fat.

Carbohydrates. — We see that the carbohydrates, like the fats, contain carbon, hydrogen, and oxygen. *Carbohydrates are essentially energy-producing foods.* They are, however, of use in building up or repairing tissue. It is certainly true that in both plants and animals such foods pass directly, together with foods containing nitrogen, to repair waste in tissues, thus giving the needed proportion of carbon, oxygen, and hydrogen to unite with the nitrogen in forming the protoplasm of the body.

Inorganic Foods. — Water forms a large part of almost every food substance. It forms about five sixths of a normal daily diet.

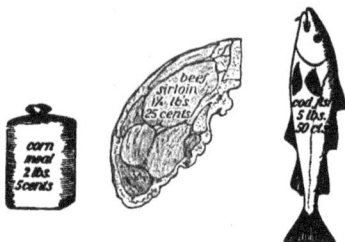

Three portions of foods, each of which furnishes about the same amount of nourishment.

The human body, by weight, is about two thirds water. About 90 per cent of the blood is water. Water is absolutely essential in passing off waste of the body. When we drink water, we take with it some of the inorganic salts used by the body in the making of bone and in the formation of protoplasm. Sodium chloride (table salt), an important part of the blood, is taken in as a flavoring upon our meats and vegetables. Phosphate of lime and potash are important factors in the formation of bone.

Phosphorus is a necessary substance for the making of living matter, milk, eggs, meat, whole wheat, and dried peas and beans containing small amounts of it. Iron also is an extremely important mineral, for it is used in the building of red blood cells. Meats, eggs, peas and beans, spinach, and prunes, are foods containing some iron.

Some other salts, compounds of calcium, magnesium, potassium, and phosphorus, have been recently found to aid the body in many of its most important functions. The beating of the heart, the contraction of muscles, and the ability of the nerves to do their work appear to be due to the presence of minute quantities of these salts in the body.

Uses of Nutrients. — The following table sums up the uses of nutrients to man : [1] —

[1] Adapted from Atwater, *Principles of Nutrition and Nutritive Value of Food* U.S. Department of Agriculture, 1902.

Protein . . . . Forms tissue (mus-
    White of eggs (albumen),   cles, tendon,
    curd of milk (casein), lean   and probably
    meat, gluten of wheat, etc.   fat).
Fats . . . . Form fatty tissue.
    Fat of meat, butter, olive oil,
    oils of corn and wheat, etc.
Carbohydrates . . . . . Transformed into
    Sugar, starch, etc.   fat.
Mineral matters (ash). . . Aid in forming bone,
    Phosphates of lime, potash,   assist in diges-
    soda, etc.   tion, aid in ab-
        sorption and in
        other ways help
        the body parts
        do their work.

All serve as *fuel* and yield *energy* in form of heat and muscular strength.

Water used as a vehicle to carry nutrients, and enters into the composition of living matter.

Common Foods contain the Nutrients. — We have already found in our plant study that various plant foods are rich in different nutrients, carbohydrates forming the chief nutrient in the foods we call cereals, breads, cake, fleshy fruits, sugars, jellies, and the like. Fats and oils are most largely found in nuts and some grains. Animal foods are our chief supply of protein. White of egg and lean meat are almost pure protein and water. Proteins are most abundant, as we should expect, in those plants which are richly supplied with nitrogen; peas and beans, and in grains and nuts. Fats, which are melted into oils at the temperature of the body, are represented by the fat in meats, bacon, pork, lard, butter, and vegetable oils.

Water. — Water is, as we have seen, a valuable part of food. It makes up a very high percentage of fresh fruits and vegetables; it is also present in milk and eggs, less abundant in meats and fish, and is lowest in dried foods and nuts. The amount of water in a given food is often a decided factor in the cost of the given food, as can easily be seen by reference to the chart on page 283.

Refuse. — Some foods bought in the market may contain a certain unusable portion. This we call refuse. Examples of

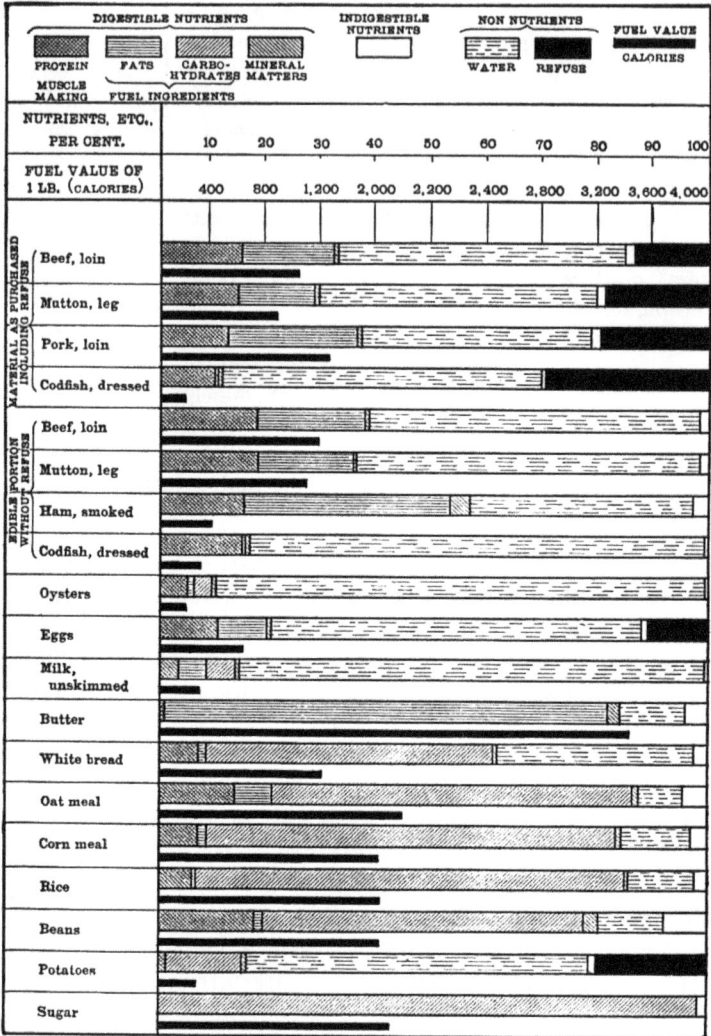

Table of food values. Determine the percentage of water in codfish, loin of beef, milk, potatoes. Percentage of refuse in leg of mutton, codfish, eggs, and potatoes. What *is* the refuse in each case? Find three foods containing a high percentage of protein; of fat; of carbohydrate. Find some food in which the proportions of protein, fat, and carbohydrate are combined in a good proportion.

refuse are bones in meat, shells of eggs or of shellfish, the covering
of plant cells which form the skins of potatoes or other vegetables.
The amount of refuse present also plays an important part in the
values of foods for the table. The table[1] on page 276 gives the per-
centages of organic nutrients, water, and refuse present in some
common foods.

Fuel Values of Nutrients. — In experiments performed by
Professor Atwater and others, and in the appended tables, the
value of food as a source of energy is stated in heat units called
Calories. *A Calorie is the amount of heat required to raise the temperature of one kilogram of water from zero to one degree Centigrade.*
This is about equivalent to raising one pound four degrees Fahrenheit. The fuel value of different foods may be computed in a
definite manner. This is done by burning a given portion of a
food (say one gram) in the apparatus known as a *calorimeter*.
By this means may be determined the number of degrees the
temperature of a given amount of water is raised during the process
of burning. It has thus been found that a gram of fat will liber-
ate 9.3 Calories of heat, while a gram of starch or sugar only about
4 Calories. The burning value of fat is, therefore, over twice that
of carbohydrates. In a similar manner protein has been shown to
have about the same fuel value as carbohydrates, *i.e.* 4 Calories
to a gram.[1]

The Relation of Work to Diet. — It has been shown experimentally that a man doing hard, muscular work needs more food
than a person doing light work. The mere exercise gives the
individual a hearty appetite; he eats more and needs more of
all kinds of food than a man or boy doing light work. Especially
is it true that the person of sedentary habits, who does brain work,
should be careful to eat less food and food that will digest easily.
His protein food should also be reduced. Rich or hearty foods
may be left for the man who is doing hard manual labor out
of doors, for any extra work put on the digestive organs takes
away just so much from the ability of the brain to do its
work.

[1] W. O. Atwater, *Principles of Nutrition and Nutritive Value of Food*, U.S. Department of Agriculture, 1902.

COMPOSITION OF FOOD MATERIALS.

U. S. Department of Agriculture
Office of Experiment Stations
A. C. True: Director

Prepared by
C. F. Langworthy
Expert in Charge of Nutrition Investigations

Protein Fat Carbohydrates Ash Water Fuel Value 1/16 Sq. In. Equals 1000 Calories

SHELLED BEAN, FRESH.
Water: 58.9
Fat: 0.6
Carbohydrates: 29.1
Protein: 9.4
Ash: 2.0
Fuel value: 720 CALORIES PER POUND

NAVY BEAN, DRY.
Water: 12.6
Protein: 22.5
Fat: 1.8
Carbohydrates: 59.6
Ash: 3.5
Fuel value: 1560 CALORIES PER POUND

STRING BEAN, GREEN.
Carbohydrates: 7.4
Ash: 0.8
Water: 89.2
Fat: 0.3
Protein: 2.3
Fuel value: 190 CALORIES PER POUND

CORN, GREEN.
SINGLE PORTION
Water: 75.4
Protein: 3.1
Carbohydrates: 19.7
Ash: 0.7
Fuel value: 460 CALORIES PER POUND
Fat: 1.1

COMPOSITION OF FOOD MATERIALS.

U. S. Department of Agriculture
Office of Experiment Stations
A. C. True: Director

Prepared by
C. F. Langworthy
Expert in Charge of Nutrition Investigations

WHITE BREAD
Water: 35.3
Protein: 9.2
Fat: 1.3
Carbohydrates: 53.1
Ash: 1.1
Fuel value: 1180 CALORIES PER POUND

WHOLE WHEAT BREAD
Water: 38.4
Protein: 9.7
Fat: 0.9
Carbohydrates: 49.7
Ash: 1.3
Fuel value: 1110 CALORIES PER POUND

OAT BREAKFAST FOOD
COOKED
Water: 84.5
Protein: 2.8
Fat: 0.5
Carbohydrates: 11.5
Ash: 0.7
Fuel value: 280 CALORIES PER POUND

TOASTED BREAD
Water: 24.0
Protein: 11.5
Fat: 1.6
Ash: 1.7
Carbohydrates: 61.2
Fuel value: 1380 CALORIES PER POUND

CORN BREAD
Water: 58.9
Protein: 7.9
Fat: 4.7
Carbohydrates: 46.3
Ash: 2.2
Fuel value: 1175 CALORIES PER POUND

MACARONI
COOKED
Fat: 1.5
Protein: 3.0
Water: 78.4
Ash: 1.3
Carbohydrates: 15.8
Fuel value: 400 CALORIES PER POUND

COMPOSITION OF FOOD MATERIALS.

U. S. Department of Agriculture
Office of Experiment Stations
A. C. True: Director

Prepared by
C. F. Langworthy
Expert in Charge of Nutrition Investigations

Protein Fat Carbohydrates Ash Water Fuel Value 1/16 Sq. In. Equals 1000 Calories

ONION
Water: 87.6
Protein: 1.6
Fat: 0.3
Carbohydrates: 9.9
Ash: 0.6
Fuel value: 220 CALORIES PER POUND

PARSNIP
Water: 83.0
Protein: 1.6
Fat: 0.5
Carbohydrates: 13.5
Ash: 1.4
Fuel value: 295 CALORIES PER POUND

CELERY

POTATO
Protein: 2.2
Fat: 0.1
Ash: 1.0
Carbohydrates: 18.4
Water: 78.3
Fuel value: 375 CALORIES PER POUND

Water: 94.5
Protein: 1.1
Carbohydrates: 3.4
Ash: 1.0
Fuel value: 80 CALORIES PER POUND

COMPOSITION OF FOOD MATERIALS.

U. S. Department of Agriculture
Office of Experiment Stations
A. C. True: Director

Prepared by
C. F. Langworthy
Expert in Charge of Nutrition Investigations

Protein Fat Carbohydrates Ash Water Fuel Value 1/16 Sq. In. Equals 1000 Calories

SUGAR
GRANULATED
Carbohydrates 100.0
Fuel value: 1810 CALORIES PER POUND

MOLASSES
Water: 25.1
Protein: 2.4
Carbohydrates: 69.3
Ash: 3.2
Fuel value: 1300 CALORIES PER POUND

STICK CANDY
Carbohydrates: 96.5
Water: 3.0
Ash: 0.5
Fuel value: 1745 CALORIES PER POUND

MAPLE SUGAR
Water: 16.3
Carbohydrates: 82.8
Ash: 0.9
Fuel value: 1500 CALORIES PER POUND

HONEY
Water: 18.2
Protein: 0.4
Carbohydrates: 81.2
Ash: 0.2
Fuel value: 1475 CALORIES PER POUND

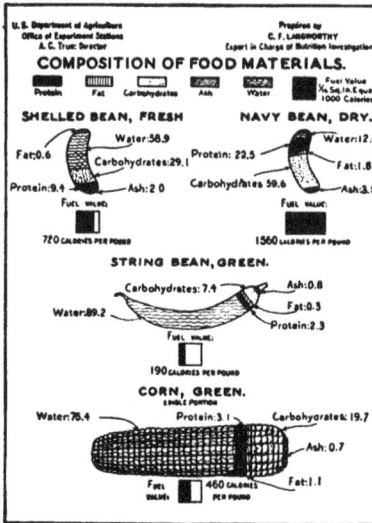

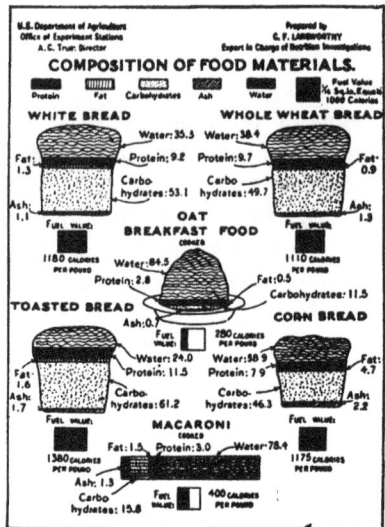

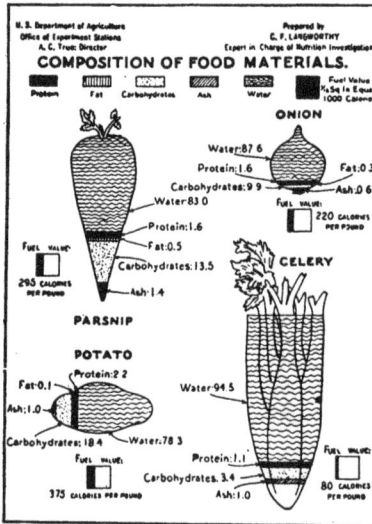

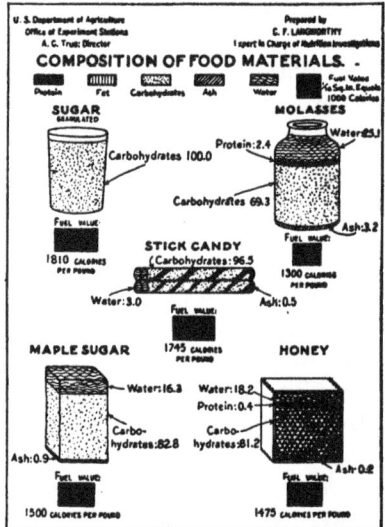

Foods of plant origin. Select 5 foods containing a high percentage of protein, 5 with a high percentage of carbohydrates, 5 with a high percentage of water. Do vegetable foods contain much fat? Which of the above-mentioned foods have the highest burning value?

COMPOSITION OF FOOD MATERIALS.

U.S. Department of Agriculture
Office of Experiment Stations
A.C. True-Director

Prepared by
C.F. Langworthy
Expert in Charge of Nutrition Investigations

Protein Fat Carbohydrates Ash Water
Fuel Value
½ Sq. In. Equals
1000 Calories

COD
Lean Fish

SALT COD

FUEL VALUE

Water:82.6
300 CALORIES PER POUND
Protein:15.8

Water:53.5

FUEL VALUE
400 CALORIES PER POUND
Protein:27.5
Fat:0.3
Ash:24.7

OYSTER

Water:86.9

Fat:0.4
Ash:1.2
Carbohydrates:3.7

MACKEREL
Fat Fish

Protein:6.2
Fat:1.2
Ash:2.0

FUEL VALUE
230 CALORIES PER POUND

SMOKED HERRING

Water:34.6
Protein:36.4

Water:73.4
Protein:18.3

FUEL VALUE
620 CALORIES PER POUND

FUEL VALUE
1305 CALORIES PER POUND

Fat:15.8
Ash:13.2

Fat:7.1
Ash:1.2

COMPOSITION OF FOOD MATERIALS.

U.S. Department of Agriculture
Office of Experiment Stations
A.C. True-Director

Prepared by
C.F. Langworthy
Expert in Charge of Nutrition Investigations

Protein Fat Carbohydrates Ash Water
Fuel Value
½ Sq. In. Equals
1000 Calories

LAMB CHOP
EDIBLE PORTION

PORK CHOP
EDIBLE PORTION

Water:53.1
Protein:17.6
Fat:28.3
Ash:1.0

Water:52.0
Protein:16.9
Fat:30.1
Ash:1.0

SMOKED HAM
EDIBLE PORTION

Water:40.3
Protein:16.1
Fat:38.8
Ash:4.8

FUEL VALUE
1475 CALORIES PER POUND

FUEL VALUE
1535 CALORIES PER POUND

BEEF STEAK
EDIBLE PORTION

DRIED BEEF
EDIBLE PORTION

Water:61.9
Protein:18.6
Fat:18.5
Ash:1.0

Water:54.3
Protein:30.0
Fat:6.6
Ash:9.1

FUEL VALUE
1090 CALORIES PER POUND

FUEL VALUE
810 CALORIES PER POUND

1875 CALORIES PER POUND

COMPOSITION OF FOOD MATERIALS

U.S. Department of Agriculture
Office of Experiment Stations
A.C. True-Director

Prepared by
C.F. Langworthy
Expert in Charge of Nutrition Investigations

Protein Fat Carbohydrates Ash Water
Fuel Value
½ Sq. In. Equals
1000 Calories

WHOLE EGG

(YOLK)

EGG
WHITE AND YOLK

Water:73.7
Protein:14.8
Fat:10.5
Ash:1.0

Water:49.5
Protein:16.1
Fat:33.3
Ash:1.1

(WHITE)
Water:86.2
Protein:13.0
Fat:0.2
Ash:0.6

FUEL VALUE OF WHOLE EGG
695 CALORIES PER POUND

FUEL VALUE OF YOLK
1650 CALORIES PER POUND

FUEL VALUE OF WHITE
245 CALORIES PER POUND

CREAM CHEESE

COTTAGE CHEESE

Water:34.2
Protein:25.9
Fat:33.7
Ash:3.8

Carbo-hydrates:2.4

Water:72.0
Protein:20.9
Fat:1.0
Ash:1.8

Carbo-hydrates:4.3

FUEL VALUE
1885 CALORIES PER POUND

FUEL VALUE
495 CALORIES PER POUND

COMPOSITION OF FOOD MATERIALS.

U.S. Department of Agriculture
Office of Experiment Stations
A.C. True Director

Prepared by
C.F. Langworthy
Expert in Charge of Nutrition Investigations

Protein Fat Carbohydrates Ash Water
Fuel Value
½ Sq. In. Equals
1000 Calories

VEGETABLE OILS, AS
OLIVE,
PEANUT,
COTTONSEED

BACON

Protein:9.4
Fat:67.4
Water:18.8
Ash:6.4

Fat:100.0

FUEL VALUE
3090 CALORIES PER POUND

FUEL VALUE
4080 CALORIES PER POUND

Fat:81.8

BEEF SUET

Water:13.2
Protein:4.7
Ash:0.3

FUEL VALUE
3425 CALORIES PER POUND

BUTTER

LARD

Fat:83.0
Water:13.0
Ash:3.0
Protein:1.0

Fat:100.0

FUEL VALUE
3405 CALORIES PER POUND

FUEL VALUE
4080 CALORIES PER POUND

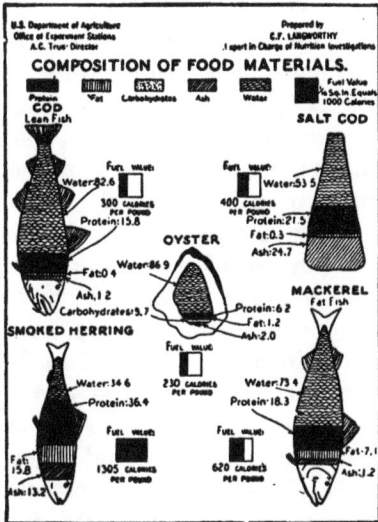

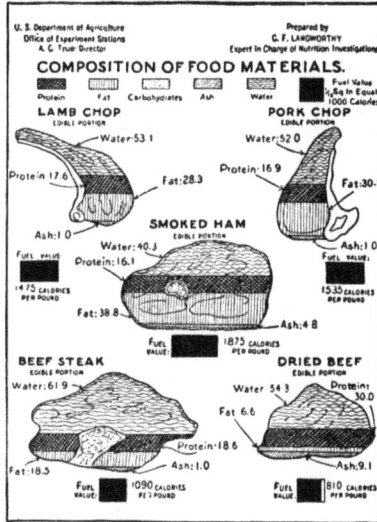

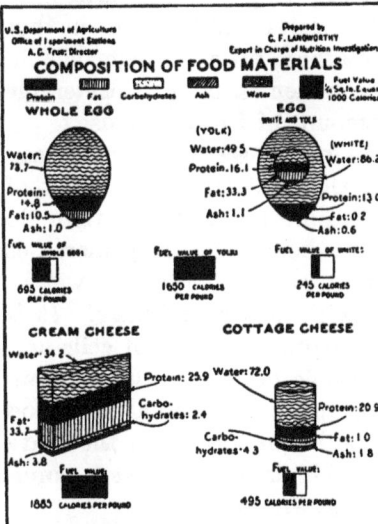

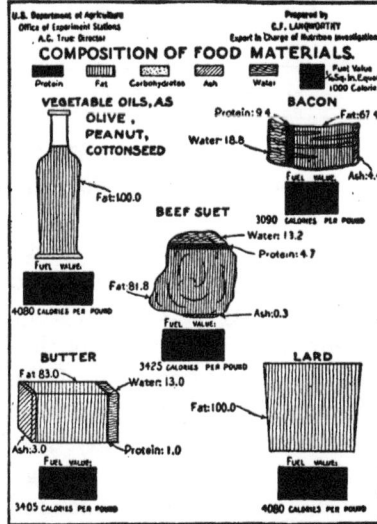

Foods largely of animal origin. Compare with the previous chart with reference to amount of protein, carbohydrate, and fat in foods. Compare the burning value of plant and animal foods. Compare the relative percentage of water in both kinds of foods.

The Relation of Environment to Diet. — We are all aware of the fact that the body seems to crave more food in winter than in summer. The temperature of the body is maintained at 98.6° in winter as in summer, but much more heat is lost from the body in cold weather. Hence feeding in winter should be for the purpose of maintaining our fuel supply. We need heat-producing food, and we need *more* food in winter than in summer. We may use carbohydrates for this purpose, as they are economical and digestible. The inhabitants of cold countries get their heat-releasing foods largely from fats. In tropical countries and in hot weather little protein should be eaten and a considerable amount of fresh fruit used.

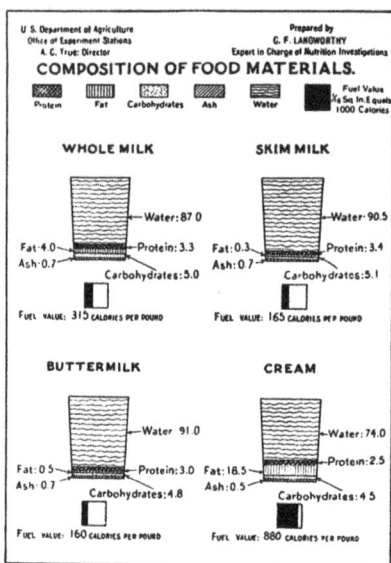

U S. Department of Agriculture
Office of Experiment Stations
A. C. True: Director

Prepared by
G. F. LANGWORTHY
Expert in Charge of Nutrition Investigations

COMPOSITION OF FOOD MATERIALS.

Protein Fat Carbohydrates Ash Water

Fuel Value
⅛ Sq. In. Equals
1000 Calories

WHOLE MILK

Water: 87.0
Fat: 4.0 Protein: 3.3
Ash: 0.7
Carbohydrates: 5.0

FUEL VALUE: 315 CALORIES PER POUND

SKIM MILK

Water: 90.5
Fat: 0.3 Protein: 3.4
Ash: 0.7
Carbohydrates: 5.1

FUEL VALUE: 165 CALORIES PER POUND

BUTTERMILK

Water 91.0
Fat: 0.5 Protein: 3.0
Ash: 0.7
Carbohydrates: 4.8

FUEL VALUE: 160 CALORIES PER POUND

CREAM

Water: 74.0
Protein: 2.5
Fat: 18.5
Ash: 0.5
Carbohydrates: 4.5

FUEL VALUE: 880 CALORIES PER POUND

The composition of milk.

The Relation of Age to Diet. — As we will see a little later, age is a factor not only in determining the kind but the amount of food to be used. Young children require far less food than do those of older growth or adults. The body constantly increases in weight until young manhood or womanhood, then its weight remains nearly stationary, varying with health or illness. It is evident that food in adults simply repairs the waste of cells and is used to supply energy. Elderly people need much less protein than do younger persons. But inasmuch as the amount of food to be taken into the body should be in proportion to the body weight, it is also evident that growing children do not, as is popularly supposed, need as much food as grown-ups.

The Relation of Sex to Diet. — As a rule boys need more food than girls, and men than women. This seems to be due to, first, the more active muscular life of the man and, secondly, to the greater amount of fat in the tissues of the woman, making loss of heat less. Larger bodies, because of greater surface, give off more heat than smaller ones. Men are usually larger in bulk than are women, — another reason for more food in their case.

The Relation of Digestibility to Diet. — Animal foods in general may be said to be more completely digested within the body than plant foods. This is largely due to the fact that plant cells have woody walls that the digestive juices cannot act upon. Cereals and legumes are less digestible foods than are dairy products, meat, or fish. This does not mean necessarily that these foods would not agree with you or me but that in general the body would get less nourishment out of the total amount available.

The agreement or disagreement of food with an individual is largely a personal matter. I, for example, cannot eat raw tomatoes without suffering from indigestion, while some one else can digest tomatoes but not strawberries. Each individual should learn early in life the foods that disagree with him personally and leave such foods out of his dietary. For " what is one man's meat may be another man's poison."

The Relation of Cost of Food to Diet. — It is a mistaken notion that the best foods are always the most expensive. A glance at the table (page 283) will show us that both fuel value and tissue-building value is present in some foods from vegetable sources, as well as in those from animal sources, and that the vegetable foods are much cheaper. The American people are far less economical in their purchase of food than most other nations. Nearly one half of the total income of the average workingman· is spent on food. Not only does he spend a large amount on food, but he wastes money in purchasing the wrong kinds of food. A comparison of the daily diets of persons in various occupations in this and other countries shows that as a rule we eat more than is necessary to supply the necessary fuel and repair, and that our workingmen eat more than those of other countries. Another waste of money by the American is in the false notion that a large proportion of the daily dietary should be meat. Many people think that the most expensive cuts of meat are the most nutritious.

The falsity of this idea may be seen by a careful study of the tables on pages 283 and 286.

The Best Dietary. — Inasmuch as all living substance contains nitrogen, it is evident that protein food must form a part of the dietary; but protein alone is not usable. If more protein is eaten than the body requires, then immediately the liver and kidneys have to work overtime to get rid of the excess of protein which forms a poisonous waste harmful to the body. We must take foods that will give us, as nearly as possible, the proportion of the different chemical elements as they are contained in protoplasm. It has been found, as a result of studies of Atwater and others, that a man who does muscular work requires a little less than one quarter of a pound of protein, the same amount of fat, and about one pound of carbohydrate to provide for the growth, waste, and repair of the body and the energy used up in one day.

The Daily Calorie Requirement. — Put in another way, Atwater's standard for a man at light exercise is food enough to yield 2816 Calories; of these, 410 Calories are from protein, 930 Calories from fat, and 1476 Calories from carbohydrate. That is, for every 100 Calories furnished by the food, 14 are from protein, 32 from fat, and 54 from carbohydrate. In exact numbers, the day's ration as advocated by Atwater would contain about 100 grams or 3.7 ounces protein, 100 grams or 3.7 ounces fat, and 360 grams or 13 ounces carbohydrate. Professor Chittenden of Yale University, another food expert, thinks we need proteins, fats, and carbohydrates in about the proportion of 1 to 3 to 6, thus differing from Atwater in giving less protein in proportion. Chittenden's standard for the same man is food to yield a total of 2360 Calories, of which protein furnishes 236 Calories, fat 708 Calories, and carbohydrates 1416 Calories. For every 100 Calories furnished by the food, 10 are from protein, 30 from fat, 60 from carbohydrate. In actual amount the Chittenden diet would contain 2.16 ounces protein, 2.83 ounces fat, and 13 ounces carbohydrate. A German named Voit gives as ideal 25 Calories from proteins, 20 from fat, and 55 from carbohydrate, out of every 100 Calories; this is nearer our actual daily ration. In addition, an ounce of salt and nearly one hundred ounces of water are used in a day.

	PROTEIN	FATS	CARBOHYDRATES	FUED VALUE

FOOD MATERIALS	PRICE PER POUND	TEN CENTS WILL BUY	POUNDS OF NUTRIENTS AND CALORIES OF FUEL VALUE IN 10 CENTS WORTH		
			1 LB.	2 LBS.	3 LBS.
	CENTS	POUNDS	2,000 CAL.	4,000 CAL.	6,000 CAL.
Beef, round	14	.71			
Beef, sirloin	20	.50			
Beef, shoulder	12	.83			
Mutton, leg	16	.63			
Pork, loin	12	.83			
Pork, salt, fat	12	.83			
Ham, smoked	18	.56			
Codfish, fresh, dressed	10	1.00			
Oysters 35 cents per quart	18	.56			
Milk, 6 cents per quart	3	3.33			
Butter	25	.40			
Cheese	16	.63			
Eggs, 24 cents per dozen	16	.63			
Wheat bread	5	2.00			
Corn meal	2½	4.00			
Oat meal	4	2.50			
Beans, white, dried	5	2.00			
Rice	8	1.25			
Potatoes, 60 cents per bushel	1	10.00			
Sugar	6	1.67			

Table showing the cost of various foods. Using this table, make up an economical dietary for one day, three meals, for a man doing moderate work. Give reasons for the amount of food used and for your choice of foods. Make up another dietary in the same manner, using expensive foods. What is the difference in your bill for the day?

A Mixed Diet Best. — Knowing the proportion of the different food substances required by man, it will be an easy matter to determine from the tables and charts shown you the best foods for use in a mixed diet. Meats contain too much nitrogen in proportion to the other substances. In milk, the proportion of proteins, carbohydrates, and fats is nearly right to make protoplasm; a considerable amount of mineral matter being also present. For these reasons, milk is extensively used as a food for children, as it combines food material for the forming of protoplasm with mineral matter for the building of bone. Some vegetables (for example, peas and beans) contain a large amount of nitrogenous material but in a less digestible form than is found in some other foods. Vegetarians, then, are correct in theory when they state that a diet of vegetables may contain everything necessary to sustain life. But a mixed diet containing meat is healthier. A purely vegetable diet contains much waste material, such as the cellulose forming the walls of plant cells, which is indigestible. It has been recently discovered that the outer coats of some grains, as rice, contain certain substances (enzymes) which aid in digestion. In the case of polished rice, when this outer coat is removed the grain has much less food value.

Daily Fuel Needs of the Body. — It has been pointed out that the daily diet should differ widely according to age, occupation, time of year, etc. The following table shows the daily fuel needs for several ages and occupations : —

Daily Calorie Needs (Approximately)

1. For child under 2 years . . . . 900 Calories
2. For child from 2–5 years . . . . 1200 Calories
3. For child from 6–9 years . . . . 1500 Calories
4. For child from 10–12 years . . . . 1800 Calories
5. For child from 12–14 (woman, light work, also) . . 2100 Calories
6. For boy (12–14), girl (15–16), man, sedentary . . . 2400 Calories
7. For boy (15–16) (man, light muscular work) . . . 2700 Calories
8. For man, moderately active muscular work . . . . 3000 Calories
9. For farmer (busy season) . . . . 3200 to 4000 Calories
10. For ditchers, excavators, etc.. . . . . . . 4000 to 5000 Calories
11. For lumbermen, etc. . . . . 5000 and more Calories

Normal Heat Output. — The following table gives the result of some experiments made to determine the hourly and daily expenditure of energy of the average normal grown person when asleep and awake, at work or at rest: —

AVERAGE NORMAL OUTPUT OF HEAT FROM THE BODY

CONDITIONS OF MUSCULAR ACTIVITY	AVERAGE CALORIES PER HOUR
Man at rest, sleeping	65 Calories
Man at rest, awake, sitting up	100 Calories
Man at light muscular exercise	170 Calories
Man at moderately active muscular exercise	290 Calories
Man at severe muscular exercise	450 Calories
Man at very severe muscular exercise	600 Calories

It is very simple to use such a table in calculating the number of Calories which are spent in twenty-four hours under different bodily conditions. For example, suppose the case of a clerk or school teacher leading a relatively inactive life, who

sleeps for 9 hours . . . . × 65 Calories = 585
works at desk 9 hours . . . . × 100 Calories = 900
reads, writes, or studies 4 hours . . × 100 Calories = 400
walks or does light exercise 2 hours . × 170 Calories = 340
                                                        ————
                                                        2225

This comes out, as we see, very close to example 6 of the table [1] on page 284.

How we may Find whether we are Eating a properly Balanced Diet. — We already know approximately our daily Calorie needs and about the proportion of protein, fat, and carbohydrate needed. Dr. Irving Fisher of Yale University has worked out a very easy method of determining whether one is living on a proper diet. He has made up a number of tables, in which he has designated portions of food, each of which furnishes 100 Calories of energy.

[1] The above tables have been taken from the excellent pamphlet of the Cornell Reading Course, No. 6, *Human Nutrition*.

TABLE OF 100 CALORIE PORTIONS — MODIFIED FROM FISHER

Food	Port. containing 100 Calories	Wt. in Oz. 100 Cal. Port.	Cal. Furnished Protein	Fat	Carbo-hydrate	Price 1 Lb.	100 Cal. Por.
Oysters	1 doz.	6.8	49	22	29	.175	.07
Bean soup	½ small serving	2.6	24	12	64		.007
Cream of corn	⅔ ordin. serv.	3.1	11	58	31		.02
Vegetable soup	½ ordin. serv.	2.4	8	89	3		.01
Cod fish (fresh)	ordin. serv.	5	95	5	0	.12	.04
Salmon (canned)	small serv.	1.75	45	55	0	.22	.03
Chicken	½ large serv.	1.75	39	56	5	.22	.05
Veal cutlet	⅔ large serv.	2.4	54	46	0	.28	.045
Beef, corned	½ large serv.	1.0	15	85	0	.16	.01
Beef, sirloin	small serv.	1.6	33	67	0	.34	.04
Beef, round	small serv.	1.8	39	61	0	.24	.025
Ham, lean	ordin. serv.	1.1	28	72	0	.22	.015
Lamb chops	½ ordin. serv.	1.0	24	76	0	.20	.013
Mutton, leg	ordin. serv.	1.2	35	65	0	.20	.015
Eggs, boiled	1 large egg	2.1	32	68	0	.30 doz.	.025
Eggs, scrambled	1⅓ ordin. serv.	2.5	37	58	5	.30 doz.	.03
Beans, baked	side dish	2.66	21	18	61	.08	.013
Potatoes, mashed	ordin. serv.	3.2	10	25	65	.02	.005
Macaroni	¼ large serv.	.95	15	3	82	.10	.01
Potato salad	ordin. serv.	2.25	10	57	33	.20	.025
Tomatoes, sliced	4 large serv.	15.	15	16	69	.10	.10
Rolls, plain	1 large roll	1.2	12	7	81	.10 doz.	.01
Butter	ordin. pat	.44	5	99.5		.35	.01
Wheat bread	1 small slice	.96	15	5	80	.07	.005
Chocolate cake	½ ord. sq. piece	.98	7	22	71	.32	.02
Gingerbread	½ ord. sq. piece	.96	6	23	71	.16	.01
Custard pudding	ordin. serv.	3.25	18	42	40	.15	.03
Rice pudding	very small serv.	2.65	8	13	79	.13	.02
Apple pie	⅓ piece	1.3	5	32	63		.013
Cheese, American	1½ cu. in.	.77	25	73	2	.19	.01
Crackers (soda)	2 crackers	.9	10	20	70	.10	.007
Currant jelly	2 heap. spoons	1.1	2	0	98	.40	.025
Sugar	3 teaspoons	.86	0	0	100	.06	.003
Milk as bought	small glass	4.9	19	52	29	.05	.015
Milk, cond., sweet	4 teaspoons	1.06	10	23	67		.01
Oranges	1 large one	9.4	6	3	91		.025
Peanuts	13 double ones	.62	20	63	17		.004
Almonds, shelled	8–15	.53	13	77	10		.025

The tables show the proportion of protein, fat, and carbohydrate in each food, so that it is a simple matter by using such a table to estimate the proportions of the various nutrients in our dietary. We may depend upon taking somewhere near the proper amount of food if we take a diet based upon either Atwater's, Chittenden's, or Voit's standard. One of the most interesting and useful pieces of home work that you can do is to estimate your own personal dietary, using the tables giving the 100-Calorie portion to see if you have a properly balanced diet. From the table on page 286 make out a simple dietary for yourself for one day, estimating your own needs in Calories and then picking out 100-Calorie portions of food which will give you the proper proportions of protein, fat, and carbohydrate.

From the preceding table plan a well-balanced and cheap dietary for one day for a family of five, two adults and three children. Make a second dietary for the same time and same number of people which shall give approximately the same amount of tissue and energy producing food from more expensive materials.

Food Waste in the Kitchen. — Much loss occurs in the improper cooking of foods. Meats especially, when overdone, lose much of their flavor and are far less easily digested than when they are cooked rare. The chief reasons for cooking meats are that the muscle fibers may be loosened and softened, and that the bacteria or other parasites in the meat may be killed by the heat. The common method of frying makes foods less digestible. Stewing is an economical as well as healthful method. A good way to prepare meat, either for stew or soup, is to place the meat, cut in small pieces, in cold water, and allow it to simmer for several hours. Rapid boiling toughens the muscle fibers by the too rapid coagulation of the albuminous matter in them, just as the white of egg becomes tough when boiled too long. Boiling and roasting are excellent methods of cooking meat. In order to prevent the loss of the nutrients in roasting, it is well to baste the meat frequently; thus a crust is formed on the outer surface of the meat, which prevents the escape of the juices from the inside.

Vegetables are cooked in order that the cells containing starch grains may be burst open, thus allowing the starch to be more easily attacked by the digestive fluids. Inasmuch as water may dissolve out nutrients from vegetable tissues, it is best to boil them rapidly in a small amount of water. This gives less time for the solvent action to take place. Vegetables should be cooked with the outer skin left on when it is possible.

Adulterations in Foods. — The addition of some cheaper substance to a food, or the subtraction of some valuable substance from a food, with the view to cheating the purchaser, is known as *adulteration*. Many foods which are artificially manufactured have been adulterated to such an extent as to be almost unfit for food, or even harmful. One of the commonest adulterations is the substitution of grape sugar (glucose) for cane sugar. Glucose, however, is not a harmful adulterant. It is used largely in candy making. Flour and other cereal foods are sometimes adulterated with some cheap substitutes, as bran or sawdust. Alum is sometimes added to make flour whiter. Probably the food which suffers most from adulteration is milk, as water can be added without the average person being the wiser. By means of an inexpensive instrument known as a *lactometer*, this cheat may easily be detected. In most cities, the milk supply is carefully safeguarded, because of the danger of spreading typhoid fever from impure milk (see Chapter XX). Before the pure food law was passed in 1906, milk was frequently adulterated with substances like formalin to make it keep sweet longer. Such preservatives are harmful, and it is now against the law to add anything whatever to milk.

Coffee, cocoa, and spices are subject to great adulteration; cottonseed oil is often substituted for olive oil; butter is too frequently artificial; while honey, sirups of various kinds, cider and vinegar, have all been found to be either artificially made from cheaper substitutes or to contain such substitutes.

Pure Food Laws. — Thanks to the National Pure Food and Drug Law passed by Congress in 1906, and to the activity of various city and state boards of health, the opportunity to pass adulterated foods on the public is greatly lessened. This law compels manufacturers of foods or medicines to state the composition of their products on the labels placed on the jars or bottles.

So if a person reads the label he can determine exactly what he is getting for his money.

Impure Water. — Great danger comes from drinking impure water. This subject has already been discussed under Bacteria, where it was seen that the spread of typhoid fever in particular is due to a contaminated water supply. As citizens, we must aid all legislation that will safeguard the water used by our towns and cities. Boiling water for ten minutes or longer will render it safe from all organic impurities.

Stimulants. — We have learned that food is anything that supplies building material or releases energy in the body; but some materials used by man, presumably as food, do not come under this head. Such are tea and coffee. When taken in moderate quantities, *they produce a temporary increase in the vital activities* of the person taking them. This is said to be a stimulation; and material taken into the digestive tract, producing this, is called a *stimulant*. In moderation, tea and coffee appear to be harmless. Some people, however, cannot use either without ill effects, even in small quantity. It is the *habit* formed of relying upon the stimulus given by tea or coffee that makes them a danger to man. Cocoa and chocolate, although both contain a stimulant, are in addition good foods, having from 12 per cent to 21 per cent of protein, from 29 per cent to 48 per cent fat, and over 30 per cent carbohydrate in their composition.

Is Alcohol a Food? — The question of the use of alcohol has been of late years a matter of absorbing interest and importance among physiologists. A few years ago Dr. Atwater performed a series of very careful experiments by means of the respiration calorimeter, to ascertain whether alcohol is of use to the body as food.[1] In these experiments the subjects were given, instead of their daily allotment of carbohydrates and fats, enough alcohol to supply the same amount of energy that these foods would have given. The amount was calculated to be about two and one half ounces per day, about as much as would be contained in

---

[1] Alcohol is made up of carbon, oxygen, and hydrogen. It is very easily oxidized, but it cannot, as is shown by the chemical formula, be of use to the body in tissue building, because of its lack of nitrogen.

a bottle of light wine.[1] This alcohol was administered in small
doses six times during the day. Professor Atwater's results may
be summed up briefly as follows : —

1. The alcohol administered was almost all oxidized in the body.

2. The potential energy in the alcohol was transformed into heat
or muscular work.

3. The body did about as well with the rations including alcohol as it did without it.

The committee of fifty eminent men appointed to report on the
physiological aspects of the drink problem reported that a large
number of scientific men state that they are in the habit of taking
alcoholic liquor in small quantities, and many report that they do
not *feel* harm thereby. A number of scientists seem to agree
that within limits alcohol may be a kind of food, although a very
poor food.

On the other hand, we know that although alcohol may technically be considered as a food, it is a very unsatisfactory food and,
as the following statements show, it has an effect on the body
tissues which foods do not have.

Professor Chittenden of Yale College, · in discussing the food
problem of alcohol, writes as follows : " It is true that alcohol
in moderate quantities may serve as a food, *i.e.* it can be oxidized
with the liberation of heat. It may to some extent take the
place of fat and carbohydrates, but it is not a perfect substitute
for them, and for this reason alcohol has an action that cannot be ignored. It reduces liver oxidation. It therefore presents a dangerous side wholly wanting in carbohydrates and fat.
The latter are simply burned up to carbonic acid and water or are
transformed to glycogen and fat, but alcohol, although more easily
oxidized, is at all times liable to obstruct, in a measure at least, the
oxidative processes of the liver and probably of other tissues also,
thereby throwing into the circulation bodies, such as uric acid,
which are harmful to health, a fact which at once tends to draw a
distinct line of demarcation between alcohol and the two non-

[1] Alcoholic beverages contain the following proportions of alcohol: beer, from
2 to 5 per cent; wine, from 10 to 20 per cent; liquors, from 30 to 70 per cent. Patent medicines frequently contain as high as 60 per cent alcohol. (See page 294.)

nitrogenous foods, fat and carbohydrates. Another matter must be emphasized, and it is that the form in which alcohol is taken is of importance. Port wine, for instance, has more influence on the amount of uric acid secreted than an equivalent amount of alcohol has in some other form. To conclude: as an adjunct to the ordinary daily diet of the healthy man alcohol cannot be considered as playing the part of a true nonnitrogenous food." — Quoted in *American Journal of Inebriety*, Winter, 1906.

Effect of Alcohol on Living Matter. — If we examine raw white of egg, we find a protein which closely resembles protoplasm in its chemical composition; it is called albumen. Add to a little albumen in a test tube some 95 per cent alcohol and notice what happens. As soon as the alcohol touches the albumen the latter coagulates and becomes hard like boiled white of egg. Shake the alcohol with the albumen and the entire mass soon becomes a solid. This is because the alcohol draws the water out of the albumen. It has been shown that albumen is somewhat like protoplasm in structure and chemical composition. Strong alcohol acts in a similar manner on living matter when it is absorbed by the living body cells. It draws water from them and hardens them. It has a chemical and physical action upon living matter.

Alcohol a Poison. — But alcohol is also in certain quantities a poison. *A commonly accepted definition of a poison is that it is any substance which, when taken into the body, tends to cause serious detriment to health, or the death of the organism.* That alcohol may do this is well known by scientists.

It is a matter of common knowledge that alcohol taken in small quantities does not do any *apparent* harm. But if we examine the vital records of life insurance companies, we find a large number of deaths directly due to alcohol and a still greater number due in part to its use. In the United States every year there are a third more deaths from alcoholism and cirrhosis of the liver (a disease *directly* caused by alcohol) than there are from typhoid fever. The poisonous effect is not found in small doses, but it ultimately shows its harmful effect. Hardening of the arteries, an old-age disease, is rapidly becoming in this country a disease of the middle aged.

From it there is no escape. It is chiefly caused by the cumulative effect of alcohol. The diagram following, compiled by two English life insurance companies that insure moderate drinkers and

COMPANIES	← 100 EXPECTED DEATHS →	
Sceptre Life Insurance Company 1884-1909.	MODERATE DRINKERS	79.7
	ABSTAINERS	52.9
United Kingdom Temperance AND General Provident Institution 1866-1909.	MODERATE DRINKERS	93
	ABSTAINERS	70

Abstainers live longer than moderate drinkers.

abstainers, shows the death rate to be considerably higher among those who use alcohol.

Dr. Kellogg, the founder of the famous Battle Creek Sanitarium, points out that strychnine, quinine, and many other drugs are oxidized in the body but surely cannot be called foods. The following reasons for not considering alcohol a food are taken from his writings: —

"1. A habitual user of alcohol has an intense craving for his accustomed dram. Without it he is entirely unfitted for business. One never experiences such an insane craving for bread, potatoes, or any other particular article of food.

"2. By continuous use the body acquires a tolerance for alcohol. That is, the amount which may be imbibed and the amount required to produce the characteristic effects first experienced gradually increase until very great quantities are sometimes required to satisfy the craving which its habitual use often produces. This is never the case with true foods. . . . Alcohol behaves in this regard just as does opium or any other drug. It has no resemblance to a food.

"3. When alcohol is withdrawn from a person who has been accustomed to its daily use, most distressing effects are experienced. . . . Who ever saw a man's hand trembling or his nervous system unstrung because he could not get a potato or a piece of cornbread for breakfast? In this respect, also, alcohol behaves like opium, cocaine, or any other enslaving drug.

"4. Alcohol lessens the appreciation and the value of brain and nerve activity, while food reënforces nervous and mental energy.

"5. Alcohol as a protoplasmic poison lessens muscular power, whereas food increases energy and endurance.

"6. Alcohol lessens the power to endure cold. This is true to such a marked degree that its use by persons accompanying Arctic expeditions is absolutely prohibited. Food, on the other hand, increases ability to endure cold. The temperature after taking food is raised. After taking alcohol, the temperature, as shown by the thermometer, is lowered.

"7. Alcohol cannot be stored in the body for future use, whereas all food substances can be so stored.

"8. Food burns slowly in the body, as it is required to satisfy the body's needs. Alcohol is readily oxidized and eliminated, the same as any other oxidizable drug."

The Use of Tobacco. — A well-known authority defines a narcotic as a substance "*which directly induces sleep, blunts the senses, and, in large amounts, produces complete insensibility.*" Tobacco, opium, chloral, and cocaine are examples of narcotics. Tobacco owes its narcotic influence to a strong poison known as nicotine. Its use in killing insect parasites on plants is well known. In experiments with jellyfish and other lowly organized animals, the author has found as small a per cent as one part of nicotine to one hundred thousand parts of sea water to be sufficient to profoundly affect an animal placed within it. The illustration here given shows the

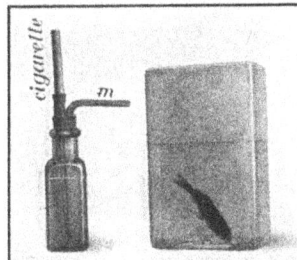

Experiment (by Davison) to show how the nicotine in six cigarettes was sufficient to kill this fish. The smoke from the cigarettes was passed through the water in which the fish is swimming.

effect of nicotine upon a fish, one of the vertebrate animals. Nicotine in a pure form is so powerful a poison that two or three drops would be sufficient to cause the death of a man by its action upon the nervous system, especially the nerves controlling the beating of the heart. This action is well known among boys training for athletic contests. The heart is affected; boys become "short-winded" as a result of the action on the heart. It has been demonstrated that tobacco has, too, an important effect on muscular development. The stunted appearance of the young smoker is well known.

Use and Abuse of Drugs. — The American people are addicted to the use of drugs, and especially patent medicines. A glance at

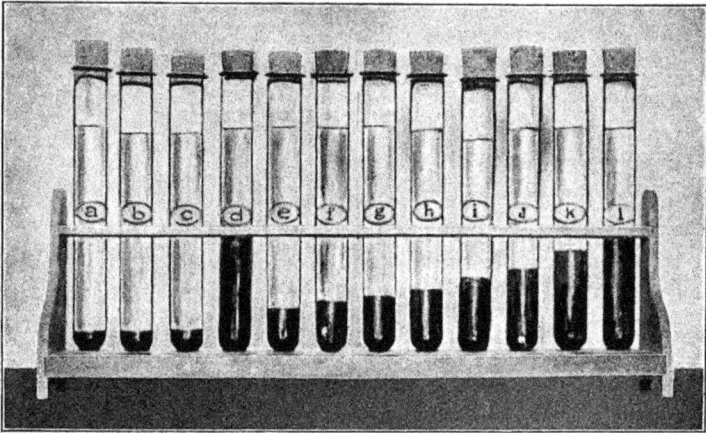

The amounts of alcohol in some liquors and in some patent medicines. *a*, beer, 5 %; *b*, claret, 8 %; *c*, champagne, 9 %; *d*, whisky, 50 %; *e*, well-known sarsaparilla, 18 %; *f, g, h*, much-advertised nerve tonics, 20 %, 21 %, 25 %; *i*, another much-advertised sarsaparilla, 27 %; *j*, a well-known tonic, 28 %; *k, l*, bitters, 37 %, 44 % alcohol.

the street-car advertisements shows this. Most of the medicines advertised contain alcohol in greater quantity than beer or wine, and many of them have opium, morphine, or cocaine in their composition. Paregoric and laudanum, medicines sometimes given to young children, are examples of dangerous drugs that contain opium. Dr. George D. Haggard of Minneapolis has shown

by many analyses that a large number of the so-called " malts,"
" malt extracts," and " tonics," including several of the best known
and most advertised on the market, are simply disguised beers
and, frequently, very poor beers at that. These drugs, in addition
to being harmful, affect the person using them in such a manner
that he soon feels the need for the drug. Thus the drug habit is
formed, — a condition which has wrecked thousands of lives. A
number of articles on patent medicines recently appeared in a
leading magazine and have been collected and published under the
title of *The Great American Fraud.* In this booklet the author
points out a number of different kinds of " cures " and patent
medicines. The most dangerous are those headache or neuralgia
cures containing *acetanilid.* This drug is a heart depresser and
should not be used without medical advice. Another drug which
is responsible for habit formation is *cocaine.* This is often found in
catarrh or other cures. Alcohol is the basis of all tonics or
" bracers." Every boy and girl should read this booklet so as to
be forearmed against evils of the sort just described.

REFERENCE READING ON FOODS

Hunter, *Laboratory Problems in Civic Biology.* American Book Company.
Allen, *Civics and Health.* Ginn and Company.
Bulletin 13, American School of Home Economics, Chicago.
Cornell University Reading Course, Buls. 6 and 7, *Human Nutrition.*
Davison, *The Human Body and Health.* American Book Company.
Jordan, *The Principles of Human Nutrition.* The Macmillan Company.
Kebler, L. F., *Habit-forming Agents.* Farmers' Bulletin 393, U.S. Dept. of Agri.
Lusk, *Science and Nutrition.* W. B. Saunders Company.
Norton, *Foods and Dietetics.* American School of Home Economics.
Olsen, *Pure Foods.* Ginn and Company.
Sharpe, *A Laboratory Manual for the Solution of Problems in Biology,* pp. 226–240.
    American Book Company.
Stiles, *Nutritional Physiology.* W. B. Saunders Company.
The Great American Fraud. American Medical Association, Chicago.
The Propaganda for Reform in Proprietary Medicines. Am. Medical Association.
Farmers' Bulletin: numbers 23, 34, 42, 85, 93, 121, 128, 132, 142, 182, 249, 295,
    298.
Reprint from Yearbook, 1901, Atwater, *Dietaries in Public Institutions.*
Reprint from Yearbook, 1902, Milner, *Cost of Food related to its Nutritive Value.*
Experiment Station, Circular 46, Langworthy, *Functions and Uses of Food.*

XX. DIGESTION AND ABSORPTION

Problems. — *To determine where digestion takes place by examining :* —

(a) *The functions of glands.*

(b) *The work done in the mouth.*

(c) *The work done in the stomach.*

(d) *The work done in the small intestine.*

(e) *The function of the liver.*

To discover the absorbing apparatus and how it is used.

LABORATORY SUGGESTIONS

Demonstration of food tube of man (manikin). — Comparison with food tube of frog. Drawing (comparative) of food tube and digestive glands of frog and man.

Demonstration of simple gland. — (Microscopic preparation.)

Home experiment and laboratory demonstration. — The digestion of starch by saliva. Conditions favorable and unfavorable.

Demonstration experiment. — The digestion of proteins with artificial gastric juice. Conditions favorable and unfavorable.

Demonstration. — An emulsion as seen under the compound microscope.

Demonstration. — Emulsification of fats with artificial pancreatic fluid. Digestion of starch and protein with artificial pancreatic fluid.

Demonstration of "tripe" to show increase of surface of digestive tube.

Laboratory or home exercise. — Make a table showing the changes produced upon food substances by each digestive fluid, the reaction (acid or alkaline) of the fluid, when the fluid acts, and what results from its action.

Purpose of Digestion. — We have learned that starch and protein food of plants are formed in the leaves. A plant, however, is unable to make use of the food in this condition. Before it can be transported from one part of the plant body to another, it is changed into a soluble form. In this state it can be passed from cell to cell by the process of osmosis. Much the same condition exists in animals. In order that food may be of use to man, it must be changed into a state that will allow of its passage in a soluble form through the walls of the alimentary canal, or food tube.

This is done by the enzymes which cause digestion. It will be the purpose of this chapter to discover where and how digestion takes place in our own body.

Alimentary Canal. — In all vertebrate animals, including man, food is taken in the mouth and passed through a *food tube* in which it is digested. This tube is composed of different portions, named, respectively, as we pass from the *mouth* downward, the *gullet, stomach, small* and *large intestine,* and *rectum.*

Comparison of Food Tube of a Frog and Man. — If we compare the food tube of a dissected frog with the food tube of man (as shown by a manikin or chart), we find part for part they are much the same. But we notice that the intestines of man, both small and large, are relatively longer than in the frog.

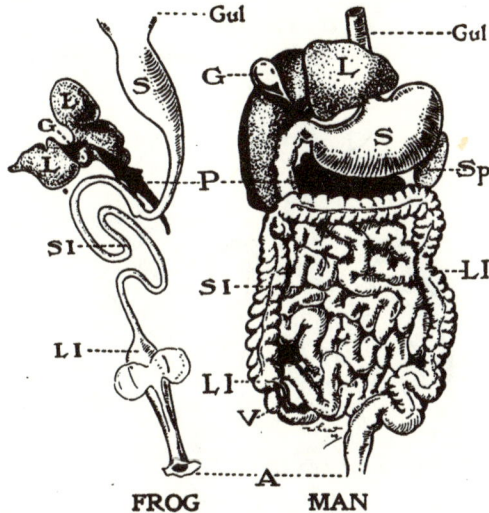

FROG   MAN

The digestive tract of the frog and man. *Gul,* gullet; *S,* stomach; *L,* liver; *G,* gall bladder; *P,* pancreas; *Sp,* spleen; *SI,* small intestine; *LI,* large intestine; *V,* appendix; *A,* anus.

We also notice in man the body cavity or space in which the internal organs rest is divided in two parts by a wall of muscle, the *diaphragm,* which separates the heart and lungs from the other internal organs. In the frog no muscular diaphragm exists. In the frog we can see plainly the silvery transparent *mesentery* or double fold of the lining of the body cavity in which the organs of digestion are suspended. Numerous blood vessels can be found especially in the walls of the food tube.

Glands. — In addition to the alimentary canal proper, we find a number of *digestive glands,* varying in size and position, connected with the canal.

What a Gland Does. Enzymes. — In man there are the saliva gland of the mouth, the gastric glands of the stomach, the pancreas and liver, the two latter connected with the small intestine, and the intestinal glands in the walls of the intestine. Besides glands which aid in digestion there are several others of which we will speak later. As we have already learned, a gland is a collection of cells which takes up material from within the body and manufactures from it something which is later poured out as a secretion. An example of a gland in plants is found in the nectar-secreting cells of a flower.

food tube

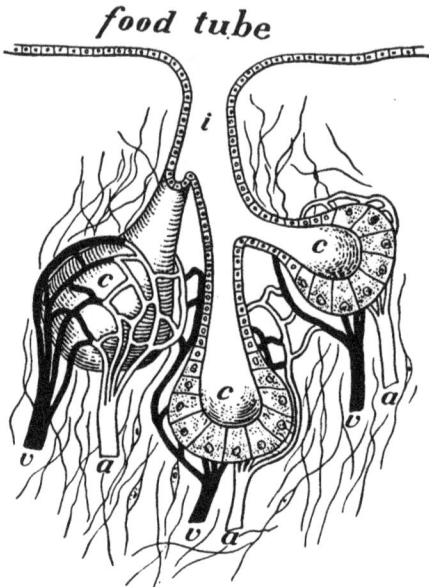

Diagram of a gland. *i*, the common tube which carries off the secretions formed in the cells lining the cavity *c*; *a*, arteries carrying blood to the glands; *v*, veins taking blood away from the glands.

Certain substances, called *enzymes*, formed by glands cause the digestion of food. The enzymes secreted by the cells of the glands and poured out into the food tube act upon insoluble foods so as to change them to a soluble form. They are the product of the activity of the cell, although they are not themselves alive. We do not know much about enzymes themselves, but we can observe what they do. Some enzymes render soluble different foods, others work in the blood, still others probably act within any cell of the body as an aid to oxidation, when work is done. Enzymes are very sensitive to changes in temperature and to the degree of *acidity* or *alkalinity* [1] of the material in which they act. We will find that the enzymes found in glands in the mouth will not act long in the

[1] The teacher should explain the meaning of these terms.

stomach because of the change from an alkaline surrounding in the mouth to that of an acid in the stomach. Enzymes seem to be able to work indefinitely, providing the surroundings are favorable. A small amount of digestive fluid, if it had long enough to work, could therefore digest an indefinite amount of food.

Gland Structure. — The entire inner surface of the food tube is covered with a soft lining of *mucous membrane*. This is always moist because certain cells, called *mucus cells*, empty out their contents into the food tube, thus lubricating its inner surface. When a large number of cells which have the power to secrete fluids are collected together, the surface of the food tube may become indented at this point to form a pitlike *gland*. Often such depressions are branched, thus giving a greater secreting surface, as is seen in the figure on page 298. The cells of the gland are always supplied with blood vessels and nerves, for the secretions of the glands are under the control of the nervous system.

How a Gland Secretes. — We must therefore imagine that as the blood goes to the cells of a gland it there loses some substances which the gland cells take out and make over into the particular enzyme that they are called upon to manufacture. Under certain conditions, such as the sight or smell of food, or even the desire for it, the activity of the gland is stimulated. It then pours out its secretion containing the digestive enzyme. Thus a gland does its work.

Salivary Glands. — We are all familiar with the substance called *saliva* which acts as a lubricant in the mouth. Saliva is manufactured in the cells of three pairs of glands which empty into the mouth, and which are called, according to their position, the *parotid* (beside the ear), the *submaxillary* (under the jawbone), and the *sublingual* (under the tongue).

Digestion of Starch. — If we collect some saliva in a test tube, add to it a little starch paste, place the tube containing the mixture for a few minutes in tepid water, and then test with Fehling's solution, we shall find grape sugar present. Careful tests of the starch paste and of the saliva made separately will usually show no grape sugar in either.

If another test be made for grape sugar, in a test tube containing

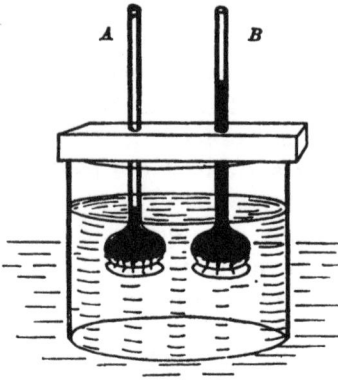

Experiment showing non-osmosis of starch in tube *A*, and osmosis of sugar in tube *B*.

starch paste, saliva, and a few drops of any weak acid, the starch will be found not to·have changed. The digestion or change of starch to grape sugar is caused by the presence in the saliva of an *enzyme*, or *digestive ferment*. You will remember that starch in the growing corn grain was changed to grape sugar by an enzyme called *diastase*. Here a similar action is caused by an enzyme called *ptyalin*. This ferment acts *only* in an alkaline medium at about the temperature of the body.

Mouth Cavity in Man. — In our study of a frog we find that the mouth cavity has two unpaired and four paired tubes leading from it. These are (*a*) the *gullet* or food tube, (*b*) the *windpipe* (in the frog opening through the *glottis*), (*c*) the paired nostril holes (*posterior nares*), (*d*) the paired *Eustachian tubes*, leading to the ear. All of these openings are found in man.

In man the mouth cavity, and all internal surfaces of the food tube, are lined with a *mucous membrane*. The *mucus* secreted from gland cells in this lining makes a slippery surface so

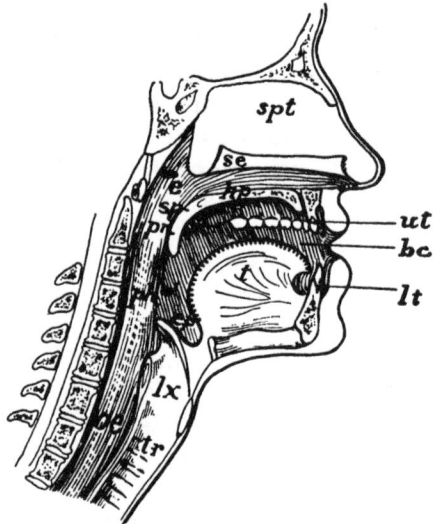

The mouth cavity of man. *e*, Eustachian tube; *hp*, hard palate; *sp*, soft palate; *ut*, upper teeth; *bc*, buccal cavity; *lt*, lower teeth; *t*, tongue; *ph*, pharynx; *ep*, epiglottis; *lx*, voice box; *oe*, gullet; *tr*, trachea.

that the food may slip down easily. The roof of the mouth is formed in front by a plate of bone called the *hard palate*, and a softer continuation to the back of the mouth, the *soft palate*. These separate the nose cavity from that of the mouth proper. The part of the space back of the soft palate is called the *pharynx*, or throat cavity. From the pharynx lead off the *gullet* and *windpipe*, the former back of the latter. The lower part of the mouth cavity is occupied by a muscular tongue. Examination of its surface with a looking-glass shows it to be almost covered in places by tiny projections called *papillæ*. These papillæ contain organs known as *taste buds*, the sensory endings of which determine the taste of substances. The tongue is used in moving food about in the mouth, and in starting it on its way to the gullet; it also plays an important part in speaking.

The Teeth. — In man the teeth, unlike those of the frog, are used in the mechanical preparation of the food for digestion. Instead of holding prey, they crush, grind, or tear food so that more surface may be given for the action of the digestive fluids. The teeth of man are divided, according to their functions, into four groups.

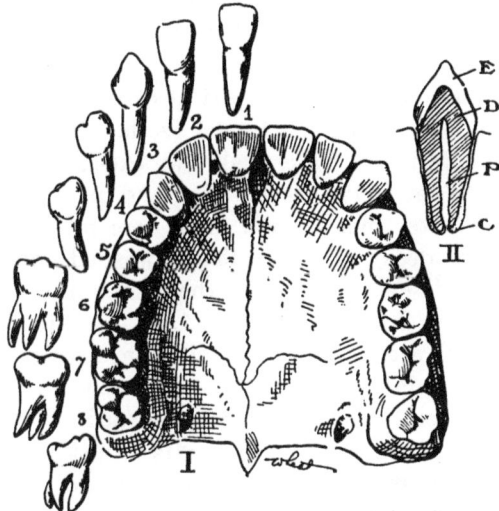

I. Teeth of the upper jaw, from below. *1, 2,* incisors; *3,* canine; *4, 5,* premolars; *6, 7, 8,* molars. II. longitudinal section of a tooth. *E,* enamel; *D,* dentine; *C,* cement; *P,* pulp cavity.

In the center of both the upper and lower jaw in front are found eight teeth with chisel-like edges, four in each jaw; these are the *incisors*, or cutting teeth. Next is found a single tooth on each side (four in all); these have rather sharp points and are called the *canines*. Then come two teeth on each side, eight in all, called *premolars*. Lastly, the *flat-top molars*, or grinding teeth, of which there are six in each jaw. Food is caught between irregular projections on the surface of the molars and crushed to a pulpy mass.

Hygiene of the Mouth. — Food should simply be chewed and relished, with no thought of swallowing. There should be no more effort to prevent than to force swallowing. It will be found that if you attend only to the agreeable task of extracting the flavors of your food, Nature will take care of the swallowing, and this will become, like breathing, involuntary. The instinct by which most people eat is perverted through the " hurry habit " and the use of abnormal foods. Thorough mastication takes time, and therefore one must not feel hurried at meals if the best results are to be secured. The stopping point for eating should be at the *earliest* moment after one is really satisfied.

Care of the Teeth. — It has been recently found that fruit acids are very beneficial to the teeth. Vinegar diluted to about half strength with water makes an excellent dental wash. Clean your teeth carefully each morning and before going to bed. Use dental silk after meals. We must remember that the bacteria which cause decay of the teeth are washed down into the stomach and may do even more harm there than in the mouth.

How Food is Swallowed. — After food has been chewed and mixed with saliva, it is rolled into little balls and pushed by the tongue into such position that the muscles of the throat cavity may seize it and force it downward. Food, in order to reach the gullet from the mouth cavity, must pass over the opening into the windpipe. When food is in the course of being swallowed, the upper part of this tube forms a trapdoor over the opening. When this trapdoor is not closed, and food " goes down the wrong way," we choke, and the food is expelled by coughing.

The Gullet, or Esophagus. — Like the rest of the food tube the gullet is lined by soft and moist mucous membrane. The wall is made up of two sets of muscles, — the inside ones running around the tube ; the outer layer of muscle taking a longitudinal course. After food leaves the mouth cavity, it gets beyond our direct control, and the muscles of the gullet, stimulated to activity by the presence of food in the tube, push the food down to the stomach by a series of contractions until it reaches the stomach. These wavelike movements (called *peristaltic* movements) are characteristic of other parts of the food tube, food being pushed along in the stomach and the small intestine by a series of slow-moving muscular waves. Peristaltic movement is caused by muscles which are not under voluntary nervous

bolus of food.

Peristaltic waves on the gullet of man.
(A bolus means little ball.)

control, although anger, fear, or other unpleasant emotions have the effect of slowing them up or even stopping them entirely.

Stomach of Man. — The stomach is a pear-shaped organ capable of holding about three pints. The end opposite to the gullet, which empties into the small intestine, is provided with a ring of muscle forming a valve called the *pylorus*. There is also another ring of muscle guarding the entrance to the stomach.

Gastric Glands. — If we open the stomach of the frog, and remove its contents by carefully washing, its wall is seen to be thrown into folds internally. Between the folds in the stomach of man, as well as in the frog, are located a number of tiny pits. These form the mouths of the *gastric glands*, which pour into the stomach a secretion known as the *gastric juice*. The gastric glands are little tubes, the lining of which secretes the fluid. When we think of or see appetizing food, this secretion is given out in considerable quantity. The stomach, like the mouth, " waters " at the sight of food. Gastric juice is slightly acid in its chemical reaction, containing about .2 per cent *free hydrochloric acid*. It also contains two very important enzymes, one called *pepsin*, and another less important one called *rennin*.

Action of Gastric Juice. — If protein is treated with artificial gastric juice at the temperature of the body, it will be found to become swollen and then gradually to change to a substance which is soluble in water. This is like the action of the gastric juice upon proteins in the stomach.

The other enzyme of gastric juice, called *rennin*, curdles or coagulates a protein found in milk; after the milk is curdled, the pepsin is able to act upon it. "Junket" tablets, which contain rennin, are used in the kitchen to cause this change.

The hydrochloric acid found in the gastric juice acts upon lime and some other salts taken into the stomach with food, changing them so that they may pass into the blood and eventually form the mineral part of bone or other tissue. The acid also has a decided antiseptic influence in preventing growth of bacteria which cause decay, and some of which might cause disease.

Movement of Walls of Stomach. — The stomach walls, provided with three layers of muscle which run in an oblique, circular, and longitudinal direction (taken from the inside outward), are well fitted for the constant churning of the food in that organ. Here, as elsewhere in the digestive tract, the muscles are involuntary, muscular action being under the control of the so-called *sympathetic nervous system*. Food material in the stomach makes several complete circuits during the process of digestion in that organ. Contrary to common belief, the greatest amount of food is digested

A peptic gland, from the stomach, very much magnified. *A*, central or chief cell, which makes pepsin; *B*, border cells, which make acid. (From Miller's *Histology*.)

after it leaves the stomach. But this organ keeps the food in it in almost constant motion for a considerable time, a meal of meat and vegetables remaining in the stomach for three or four hours. While movement is taking place, the gastric juice acts upon proteins, softening them, while the constant churning movement tends to separate the bits of food into finer particles. Ultimately the semifluid food, much of it still undigested, is allowed to pass in small amounts through the pyloric valve, into the small intestine. This is allowed by the relaxation of the ringlike muscles of the pylorus.

Experiments on Digestion in the Stomach. — Some very interesting experiments have recently been made by Professor Cannon of Harvard with reference to movements of the stomach contents. Cats were fed with material having in it bismuth, a harmless chemical that would be visible under the X-ray. It was found that shortly after food reached the stomach a series of waves began which sent the food toward the pyloric end of the stomach. If the cat was feeling happy and well, these contractions continued regularly, but if the cat was cross or bad tempered, the movements would stop. This shows the importance of *cheerfulness* at meals. Other experiments showed that food which was churned into a soft mass was only permitted to leave the stomach when it became thoroughly permeated by the gastric juice. It is the *acid* in the partly digested food that causes the stomach valve to open and allow its contents to escape little by little into the small intestine.

The partly digested food in the small intestine almost immediately comes in contact with fluids from two glands, the liver and pancreas. We shall first consider the function of the pancreas.

Position and Structure of the Pancreas. — The most important digestive gland in the human body is the pancreas. The gland is a rather diffuse structure ; its duct empties by a common opening with the bile duct into the small intestine, a short distance below the pylorus. In internal structure, the pancreas resembles the salivary glands.

Work done by the Pancreas. — Starch paste added to artificial pancreatic fluid and kept at blood heat is soon changed to sugar. Protein, under the same conditions, is changed to a peptone.

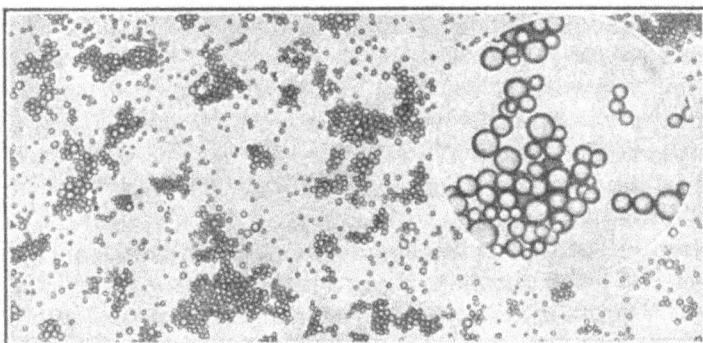

Appearance of milk under the microscope, showing the natural grouping of the fat globules. In the circle a single group is highly magnified. Milk is one form of an emulsion. (S. M. Babcock, Wis. Bul. No. 61.)

Fats, which so far have been unchanged except to be melted by the heat of the body, are changed by the action of the pancreas into a form which can pass through the walls of the food tube. If we test pancreatic fluid, we find it strongly *alkaline* in its reaction. If two test tubes, one containing olive oil and water, the other olive oil and a weak solution of caustic soda, an *alkali*, be shaken violently and then allowed to stand, the oil and water will quickly separate, while the oil, caustic soda, and water will remain for some time in a milky *emulsion*. If this emulsion be examined under the microscope, it will be found to be made of millions of little droplets of fat, floating in the liquid. The presence of the caustic soda helped the forming of the emulsion. Pancreatic fluid similarly emulsifies fats and changes them into soft soaps and fatty acids. Fat in this form may be absorbed. The process of this transformation is not well understood.

Conditions under which the Pancreas does its Work. — The secretion from this gland seems to be influenced by the overflow of acid material from the stomach. This acid, on striking the lining of the small intestine, causes the formation in its walls of a substance known as *secretin*. This secretin reaches the blood and seems to stimulate all the glands pouring fluid into the intestine to do more work. A pint or more of pancreatic fluid is secreted every day.

The Intestinal Fluid. — Three different pancreatic enzymes do the work of digestion, one acting on starch, another on protein, and a third on fats. It has been found that some of these enzymes will not do their work unless aided by the *intestinal* fluid, a secretion formed in glands in the walls of the small intestine. This fluid, though not much is known about it, is believed to play an important part in the digestion of all kinds of foods left undigested in the small intestine.

Liver. — The liver is the largest gland in the body. In man, it hangs just below the diaphragm, a little to the right side of the body. During life, its color is deep red. It is divided into three lobes, between two of which is found the *gall bladder*, a thin-walled sac which holds the *bile*, a secretion of the liver. Bile is a strongly alkaline fluid of greenish color. It reaches the intestine through the same opening as the pancreatic fluid. Almost one quart of bile is passed daily into the digestive canal. The color of bile is due to certain waste substances which come from the destruction of worn-out red corpuscles of the blood. This destruction takes place in the liver.

Functions of Bile. — The action of bile is not very well known. It has the very important faculty of aiding the pancreatic fluid in digestion, though alone it has slight if any digestive power. Certain substances in the bile aid especially in the absorption of fats. Bile seems to be mostly a waste product from the blood and as such incidentally serves to keep the contents of the intestine in a more or less soft condition, thus preventing extreme constipation.

The Liver a Storehouse. — Perhaps the most important function of the liver is the formation within it of a material called *glycogen*, or animal starch. The liver is supplied by blood from two sources. The greater amount of blood received by the liver comes directly

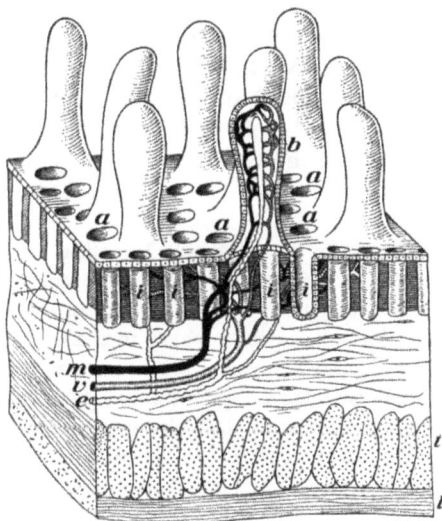

Diagram of a bit of the wall of the small intestine, greatly magnified. *a*, mouths of intestinal glands; *b*, villus cut lengthwise to show blood vessels and lacteal (in center); *e*, lacteal sending branches to other villi; *i*, intestinal glands; *m*, artery; *v*, vein; *l, t,* muscular coats of intestine wall.

from the walls of the stomach and intestine to this organ. It normally contains about one fifth of all the blood in the body. This blood is very rich in food materials, and from it the cells of the liver take out sugars to form glycogen.[1] Glycogen is stored in the liver until such a time as a food is needed that can be quickly

---

[1] It is known that glycogen *may* be formed in the body from protein, and possibly from fatty foods.

oxidized; then it is changed to sugar and carried off by the blood to the tissue which requires it, and there used for this purpose. Glycogen is also stored in the muscles, where it is oxidized to release energy when the muscles are exercised.

The Absorption of Digested Food into the Blood. — The object of digestion is to change foods from an insoluble to a soluble form. This has been seen in the study of the action of the various digestive fluids in the body, each of which is seen to aid in dissolving solid foods, changing them to a fluid, and, in case of the bile, actually assisting them to pass through the wall of the intestine. A small amount of digested food may be absorbed by the blood in the blood vessels of the walls of the stomach. Most of the absorption, however, takes place through the walls of the small intestine.

Structure of the Small Intestine. — The small intestine in man is a slender tube nearly twenty feet in length and about one inch in diameter. If the chief function of the small intestine is that of absorption, we must look for adaptations which increase the absorbing surface of the tube. This is gained in part by the inner surface of the tube being thrown into transverse folds which not only retard the rapidity with which food passes down the intestine, but also give more absorbing surface. But far more important for absorption are millions of little projections which cover the inner surface of the small intestine.

The Villi. — So numerous are these projections that the whole surface presents a velvety appearance. Collectively, these structures are called the *villi* (singular *villus*). They form the chief organs of absorption in the intestine, several thousand being distributed over every square inch of surface. By means of the folds and villi the small intestine is estimated to have an absorbing surface equal to twice that of the surface of the body. Between the villi are found the openings of the *intestinal glands*.

Internal Structure of a Villus. — The internal structure of a villus is best seen in a longitudinal section. We find the outer wall made up of a thin layer of cells, the *epithelial* layer. It is the duty of these cells to absorb the semifluid food from within the intestine. Underneath these cells lies a network of very tiny blood vessels, while inside of these, occupying the core of the villus, are found spaces which, because of their white appearance after absorption of fats, have been called *lacteals*. (See figure, page 207.)

Absorption of Foods. — Let us now attempt to find out exactly how foods are passed from the intestines into the blood. Food substances in solution may be soaked up as a sponge would take up water, or they may pass by osmosis into the cells lining the villus.

These cells break down the peptones into a substance that will pass into and become part of the blood. Once within the villus, the sugars and digested proteins pass through tiny blood vessels into the larger vessels comprising the *portal circulation*. These pass through the liver, where, as we have seen, sugar is taken from the blood and stored as glycogen. From the liver, the food within the blood is sent to the heart, from there is pumped to the lungs, from there returns to the heart, and is pumped to the tissues of the body. A large amount of water and some salts are also absorbed through the walls of the stomach and intestine as the food passes on its course. The fats in the form of soaps and fatty acids pass into the space in the center of the villus. Later they are changed into fats again, probably in certain groups of gland cells

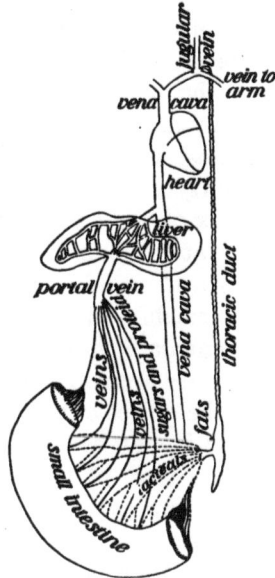

Diagram to show how the nutrients reach the blood.

known as *mesenteric* glands, and eventually reach the blood by way of the thoracic duct *without* passing through the liver.

Large Intestine. — The large intestine has somewhat the same structure as the small intestine, except that it lacks the villi and has a greater diameter. Considerable absorption, however, takes place through its walls as the mass of food and refuse material is slowly pushed along by the muscles within its walls.

Vermiform Appendix. — At the point where the small intestine widens to form the large intestine, a baglike pouch is formed. From one side of this pouch is given off a small tube about four inches long, closed at the lower end. This tube, the rudiment of what is an important part of the food tube in the lower vertebrates, is called the *vermiform appendix*. It has come to have unpleasant notoriety in late years, as the site of serious inflammation.

· **Constipation.** — In the large intestine live millions of bacteria, some of which make and give off poisonous substances known as toxins. These substances are easily absorbed through the walls of the large intestine, and, when they pass into the blood, cause headaches or sometimes serious trouble. Hence it follows that the lower bowel should be emptied of this matter as frequently as possible, at least once a day. Constipation is one of the most serious evils the American people have to deal with, and it is largely brought about by the artificial life which we lead, with its lack of exercise, fresh air, and sleep. Fruit with meals, especially at breakfast, plenty of water between meals and before breakfast, exercise, particularly of the abdominal muscles, and regular habits will all help to correct this evil.

Hygienic Habits of Eating; the Causes and Prevention of Dyspepsia. — From the contents of the foregoing chapter it is evident that the object of the process of digestion is to break up solid food so that it may be absorbed to form part of the blood. Any habits we may form of thoroughly chewing our food will evidently aid in this process. Undoubtedly much of the distress known as dyspepsia is due to too hasty meals with consequent lack of proper mastication of food. The message of Mr. Horace Fletcher in bringing before us the need of proper mastication of food and the attendant evils of overeating is one which we cannot afford to ignore. It is a good rule to go away from the table feeling a little hungry. Eating too much overtaxes the digestive organs and prevents their working to the best advantage. Still another cause of dyspepsia is eating when in a *fatigued* condition. It is always a good plan to rest a short time before eating, especially after any hard manual work. We have seen how great a part unpleasant emotions play in preventing peristaltic movements of the food tube. Conversely, pleasant conversation, laughter, and fun will help you to digest your meal. Eating between meals is condemned by physicians because it calls the blood to the digestive organs at a time when it should be more active in other parts of the body.

Effect of Alcohol on Digestion. — It is a well-known fact that alcohol extracts water from tissues with which it is in contact. This fact works much harm to the interior surface of the food tube, especially the walls of the stomach, which in the case of a hard drinker are likely to become irritated and much toughened. In very small amounts alcohol stimulates the secretion of the salivary and gastric glands, and thus appears to aid in digestion.

The following results of experiments on dogs, published in the *American Journal of Physiology*, Vol. I, Professor Chittenden of Yale University gives as " strictly comparable," because " they were carried out in succession on the same day." They show that alcohol retards rather than aids in digestion : —

NUMBER OF EXPERIMENT	$\frac{1}{10}$ LB. MEAT WITH WATER	$\frac{1}{10}$ LB. MEAT WITH DILUTE ALCOHOL
XVII α 9 : 15 A.M.	Digested in 3 hours	
XVII β 3 : 00 P.M.		Digested in 3 : 15 hours
XVIII α 8 : 30 A.M.	Digested in 2 : 30 hours	
XVIII β 2 : 10 P.M.		Digested in 3 : 00 hours
XIX α 9 : 00 A.M.	Digested in 2 : 30 hours	
XIX β 2 : 30 P.M.		Digested in 3 : 00 hours
XX α 9 : 15 A.M.		Digested in 2 : 45 hours
XX β 2 : 30 P.M.	Digested in 2 : 15 hours	
VI α 9 : 15 A.M.		Digested in 3 : 45 hours
VI β 1 : 00 P.M.	Digested in 3 : 15 hours	
Average	2 : 42 hours	3 : 09 hours

As a result of his experiments, Professor Chittenden remarks : " We believe that the results obtained justify the conclusion that gastric digestion as a whole is not materially modified by the introduction of alcoholic fluids with the food. In other words, the unquestionable acceleration of gastric secretion which follows the ingestion of alcoholic beverages is, as a rule, counterbalanced by the inhibitory effect of the alcoholic fluids upon the chemical process of gastric digestion, with perhaps at times a tendency towards preponderance of inhibitory action." Others have come to the same or stronger conclusions as to the undesirable action of alcohol on digestion, as a result of their own experiments.

Effect of Alcohol on the Liver. — The effect of heavy drinking upon the liver is graphically shown in the following table prepared by the Scientific Temperance Federation of Boston, Mass.: —

DEATHS BY:	10	20	30	40	50	60	70	80	90
Accidents.	516			4587					
Pul. Tuberculosis	4068				29,832				
Heart Disease.	3662				19,225				
Apoplexy	2585			9163					
Paralysis	512			1817					
Pneumonia	3090			10,954					
Arterial Disease	645			2158					
Suicide	1388			4647					
Bright's Disease.	5254			12,259					
Cirrhosis of Liver.		2762					1361		
Alcoholism.			2025						
TOTAL	26,507			96,063					

Proportion of deaths from disease in a certain area due to alcohol. The black area shows deaths due to alcohol.[1]

" Alcoholic indulgence stands almost if not altogether in the front rank of the enemies to be combated in the battle for health." — PROFESSOR WILLIAM T. SEDGWICK.

[1] Does not include deaths from general alcoholic paralysis or other organic diseases due to alcohol. Liver cirrhosis due to alcohol conservatively estimated at 75 per cent of total cases.

XXI. THE BLOOD AND ITS CIRCULATION

Problems. — *To discover the composition and uses of the different parts of the blood.*

To find out the means by which the blood is circulated about the body.

LABORATORY SUGGESTIONS

Demonstration. — Structure of blood, fresh frog's blood and human blood. Drawings.

Demonstration. — Clotting of blood.

Demonstration. — Use of models to demonstrate that the heart is a force pump.

Demonstration. — Capillary circulation in web of frog's foot or tadpole's tail. Drawing.

Home or laboratory exercise. — On relation of exercise on rate of heart beat.

Function of the Blood. — The chief function of the digestive tract is to change foods to such form that they can be absorbed through the walls of the food tube and become part of the blood.[1]

If we examine under the microscope a drop of blood taken from the frog or man, we find it made up of a fluid called *plasma* and two kinds of bodies, the so-called *red corpuscles* and *colorless corpuscles*, floating in this plasma.

Composition of Plasma. — The plasma of blood is found to be largely (about 90 per cent) water. It also contains a considerable amount of protein, some sugar, fat, and mineral material. It is, then, the medium which holds the fluid food that has been absorbed from within the intestine. This food is pumped to the body cells where, as work is performed, oxidation takes place and heat is given off as a form of energy. The almost constant temperature

---

[1] This change is due to the action of certain enzymes upon the nutrients in various foods. But we also find that peptones are changed back again to proteins when once in the blood. This appears to be due to the *reversible* action of the enzymes acting upon them. (See page 307.)

313

of the body is also due to the blood, which brings to the surface of
the body much of the heat given off by oxidation of food in the

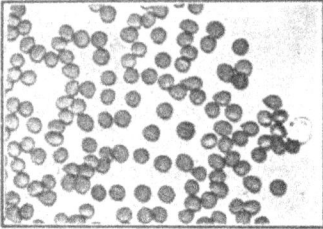

Human blood as seen under the high power of the compound microscope; at the extreme right is a colorless corpuscle.

muscles and other tissues. When
the blood returns from the tissues
where the food is oxidized, the
plasma brings back with it to the
lungs part of the carbon dioxide
liberated where oxidation has taken
place. Some waste products, to be
spoken of later, are also found in
the plasma.

**The Red Blood Corpuscle; its
Structure and Functions.** — The
red corpuscle in the blood of the
frog is a true cell of disklike form, containing a nucleus. The red
corpuscle of man is made in the red marrow of bones and in
its young stages has a nucleus. In its adult form, however,
it lacks a nucleus. Its form is that of a biconcave disk. So
small and so numerous are these corpuscles that about five
million are found in a cubic millimeter of normal blood. They
make up almost one half the total volume of the blood. The
color, which is found to be a dirty yellow when separate corpuscles are viewed under the microscope, is due to a protein
material called *hæmoglobin.* Hæmoglobin contains a large amount
of iron. It has the power of uniting very readily with oxygen
whenever that gas is abundant, and, after having absorbed it,
of giving it up to the surrounding media, when oxygen is there
present in smaller amounts than in the corpuscle. This function
of carrying oxygen is the most important function of the red
corpuscle, although the red corpuscle also removes part of the
carbon dioxide from the tissues on their return to the lungs. The
taking up of oxygen is accompanied by a change in color of the
mass of corpuscles from a dull red to a bright scarlet.

Clotting of Blood. — If fresh beef blood is allowed to stand overnight,
it will be found to have separated into two parts, a dark red, almost solid
clot and a thin, straw-colored liquid called *serum.* Serum is found to
be made up of about 90 per cent water, 8 per cent protein, 1 per cent

other organic foods, and 1 per cent mineral substances. In these respects it very closely resembles the fluid food that is absorbed from the intestines.

If another jar of fresh beef blood is poured into a pan and briskly whipped with a bundle of little rods (or with an egg beater), a stringy substance will be found to stick to the rods. This, if washed carefully, is seen to be almost colorless. Tested with nitric acid and ammonia, it is found to contain a protein substance which is called *fibrin*.

Blood plasma, then, is made up of a fluid portion of serum, and fibrin, which, although in a fluid state in the blood vessels within the body, coagulates when blood is removed from the blood vessels. This coagulation aids in making a blood clot. A clot is simply a mass of fibrin threads with a large number of corpuscles tangled within. The clotting of blood is of great physiological importance, for otherwise we might bleed to death even from a small wound.

Blood Plates. — In blood within the circulatory system of the body, the fibrin is held in a fluid state called *fibrinogen*. An enzyme, acting upon this fibrinogen, the soluble protein in the blood, causes it to change to an insoluble form, the fibrin of the clot. This change seems to be due to the action of minute bodies in the blood known as *blood plates*. Under abnormal conditions these blood plates break down, releasing some substances which eventually cause this enzyme to do its work.

The Colorless Corpuscle; Structure and Functions. — A colorless corpuscle is a cell irregular in outline, the shape of which is constantly changing. These corpuscles

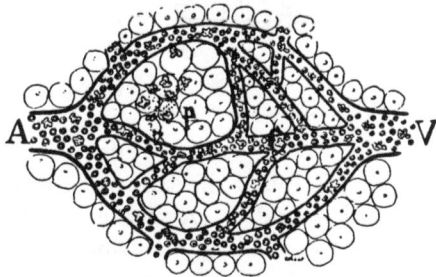

A small artery (A) breaking up into capillaries (c) which unite to form a vein (V). Note at (P) several colorless corpuscles, which are fighting bacteria at that point.

are somewhat larger than the red corpuscles, but less numerous, there being about one colorless corpuscle to every three hundred red ones. They have the power of movement, for they are found not only inside but outside the blood vessels, showing that they have worked their way between the cells that form the walls of the blood tubes.

A Russian zoölogist, Metchnikoff, after studying a number of simple animals, such as medusæ and sponges, found that in such animals some of the cells lining the inside of the food cavity take up or engulf minute bits of food. Later, this food is changed into the protoplasm of the cell. Metchnikoff believed that the colorless corpuscles of the blood have somewhat the same function. This he later proved to be true. Like the amœba, they feed by engulfing their prey. This fact has a very important bearing on the relation of colorless corpuscles to certain diseases caused by bacteria within the body. If, for example, a cut becomes infected by bacteria, inflammation may set in. Colorless corpuscles at once surround the spot and attack the bacteria which cause the inflammation. If the bacteria are few in number, they are quickly eaten by certain of the colorless corpuscles, which are known as *phagocytes*. If bacteria are present in great quantities, they may prevail and kill the phagocytes by poisoning them. The dead bodies of the phagocytes thus

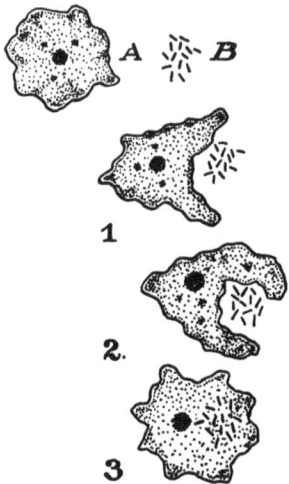

A colorless corpuscle catching and eating germs.

killed are found in the pus, or matter, which accumulates in infected wounds. In such an event, we must come to the aid of nature by washing the wound with some antiseptic, as weak carbolic acid or hydrogen peroxide.

Antibodies and their Uses. — In case of disease where, for example, fever is caused by poison given off from bacteria we find the cells of the body manufacture and pour into the blood a substance known as an *antibody*. This substance does not of necessity kill the harmful germs or even stop their growth. It does, however, unite with the toxin or poison given off by the germs and renders it entirely harmless.

Function of Lymph. — The tissues and organs of the body are traversed by a network of tubes which carry the blood. Inside these tubes is the blood proper, consisting of a fluid plasma, the colorless corpuscles, and the red corpuscles. Outside the blood tubes, in spaces between the cells which form tissues, is found another fluid, which is in chemical composition very much like plasma of the blood. This is the *lymph*. It is, in fact, fluid food in which some colorless amœboid corpuscles are found Blood gives up its food material to the lymph. This it does by passing it through the walls of the capillaries. The food is in turn given up to the tissue cells, which are bathed by the lymph.

Some of the amœboid corpuscles from the blood make their way between the cells forming the walls of the capillaries. *Lymph, then, is practically blood plasma plus some colorless corpuscles. It acts as the medium of exchange between the blood proper and the cells in the tissues of the body.* By means of the food supply thus brought, the cells of the body are able to grow, the fluid food being changed

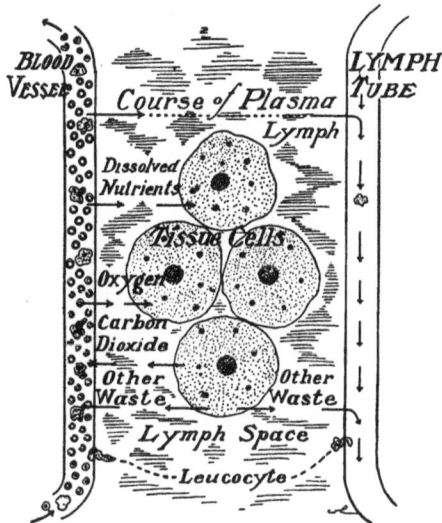

The exchange between blood and the cells of the body.

to the protoplasm of the cells. By means of the oxygen passed over by the lymph, oxidation may take place within the cells. Lymph not only gives food to the cells of the body, but also takes away carbon dioxide and other waste materials, which are ultimately passed out of the body by means of the lungs, skin, and kidneys.

Internal Secretions. — In addition to all the functions given above, the blood has recently been shown to carry the secretions of a number of glands through which it passes, although these glands have no ducts to carry off their secretions.   These internal secretions seem *absolutely necessary* for the health of the body. Several glands, the thyroid, adrenal bodies, the testes, and ovaries, as well as the pancreas, give off these remarkable substances.

The Amount of Blood and its Distribution. — Blood forms, by weight, about one sixteenth of the body.   This would be about four quarts to a body weight of 130 pounds.   Normally, about one half of the blood of the body is found in or near the organs lying in the body cavity below the diaphragm, about one fourth in the muscles, and the rest in the head, heart, lungs, large arteries, and veins.

Blood Temperature. — The temperature of blood in the human body is normally about 98.6° Fahrenheit when tested under the tongue by a thermometer, although the temperature drops almost two degrees after we have gone to sleep at night.   It is highest about 5 P.M. and lowest about 4 A.M.   In fevers, the temperature of the body sometimes rises to 107°; but unless this temperature is soon reduced, death follows.   Any considerable drop in temperature below the normal also means death. Body heat results from the oxidation of food, and the circulation of blood keeps the temperature nearly uniform in all parts of the body.

Cold-blooded Animals. — In animals which are called cold-blooded, the blood has no fixed temperature, but varies with the temperature of the medium in which the animal lives.   Frogs, in the summer, may sit for hours in water with a temperature of almost 100°.   In winter, they often endure freezing so that the blood and lymph within the spaces under the loose skin are frozen into ice crystals.   This change in body temperature is evidently an adaptation to the mode of life.

Circulation of the Blood in Man. — The blood is the carrying agent of the body.   Like a railroad or express company, it takes materials from one part of the human organism to another.   This it does by means of the organs of circulation, — the heart and blood vessels.   These blood vessels are called *arteries* where they carry blood away from the heart, *veins* where they bring blood back to the heart, and *capillaries* where they connect the larger blood vessels.   The organs of circulation thus form a system of connected tubes through which the blood flows.

The Heart; Position, Size, Protection. — The heart is a cone-shaped muscular organ about the size of a man's fist.   It is located immediately above the diaphragm, and lies so that the muscular apex, which points downward, moves while beating against the fifth and sixth ribs, just a little to the left of the midline of the body. This fact gives rise to the notion that the heart is on the left side of the body. The heart is surrounded by a loose membranous bag called the *pericardium*, the inner lining of which secretes a fluid in which the heart lies. When, for any reason, the pericardial fluid is not secreted, inflammation arises in that region.

Internal Structure of Heart. — If we should cut open the heart of a mammal down the midline, we could divide it into a right and a left side, *each of which would have no internal connection with the other*. Each side is made up of an upper thin-walled portion with a rather large internal cavity, the *auricle*, which opens into a lower smaller portion with heavy muscular walls, the *ventricle*. Communication between auricles and ventricles is guarded by little flaps or *valves*. The auricles receive blood from the veins. The ventricles pump the blood into the arteries.

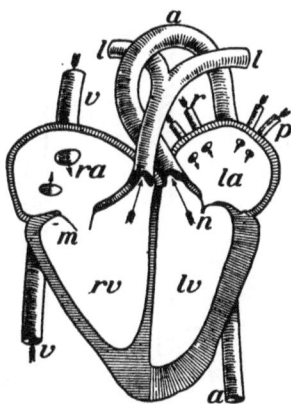

Diagram showing the front half of the heart cut away: *a*, aorta; *l*, arteries to the lungs; *la*, left auricle; *lv*, left ventricle; *m*, tricuspid valve open; *n*, bicuspid or mitral valve closed; *p* and *r*, veins from the lungs; *ra*, right auricle; *rv*, right ventricle; *v*, vena cava. Arrows show direction of circulation.

The Heart in Action. — The heart is constructed on the same plan as a force pump, the valves preventing the reflux of blood into the auricle when it is forced out of the ventricle. Blood enters the auricles from the veins because the muscles of that part of the heart relax; this allows the space within the auricles to fill. Almost immediately the muscles of the ventricles relax, thus allowing blood to pass into the chambers within the ventricles. Then, after a short pause, during which time the muscles of the heart are resting, a wave of muscular contraction begins in the auricles and ends in the ventricles, with a sudden strong contraction which forces the blood out into the arteries. Blood is kept on its course by the valves, which act in the same manner as do the valves in a pump. The blood is thus made to pass into the arteries upon the contraction of the ventricle walls.

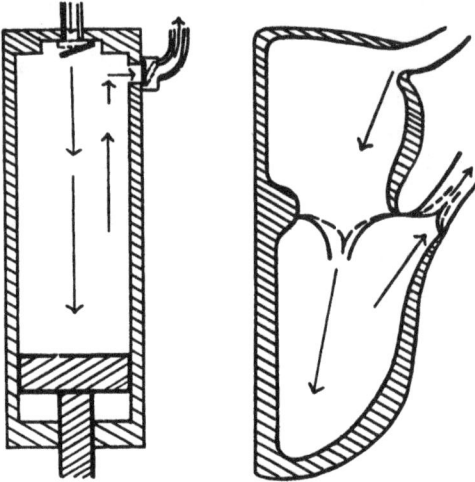

The Course of the Blood in the Body. — Although the two sides of the heart are separate and distinct from each other, yet every drop of blood that passes through the right heart likewise passes later through the left heart. There are two distinct systems of circulation in the body. The *pulmonary circulation* takes the blood through the right auricle and ventricle, to the lungs, and passes it back to the left auricle. This is a relatively short circulation, the blood receiving in the lungs its supply of oxygen, and there giving up some of its carbon dioxide. The greater circulation is known as the *systemic circulation;* in this system, the blood leaves the left ventricle through the great dorsal *aorta.* A large part of the blood passes directly to the muscles; some of it goes to the nervous system, kidneys, skin, and other organs of the body. It gives up its supply of food and oxygen in these tissues, receives the waste products of oxidation while passing through the capillaries, and returns to

The heart is a force pump; prove it from these diagrams.

I. Circulation in a fish. *G*, gills; *C*, capillaries of the body. Notice the two-chambered heart.

II. The circulation in a frog. *L*, the lungs; *C*, the capillaries. Notice the heart has three chambers. What is the condition of blood leaving the ventricle to go to the cells of the body?

III. The circulation in man. *H*, head; *A*, arms; *L*, lungs; *S*, stomach; *Li*, liver; *K*, kidney: *S.I.*, small intestine; *L.I.*, large intestine; *Le*, legs; *1*, right auricle; *2*, right ventricle; *3*, left ventricle; *4*, left auricle; *5*, dorsal aorta; *6*, vein to lungs.

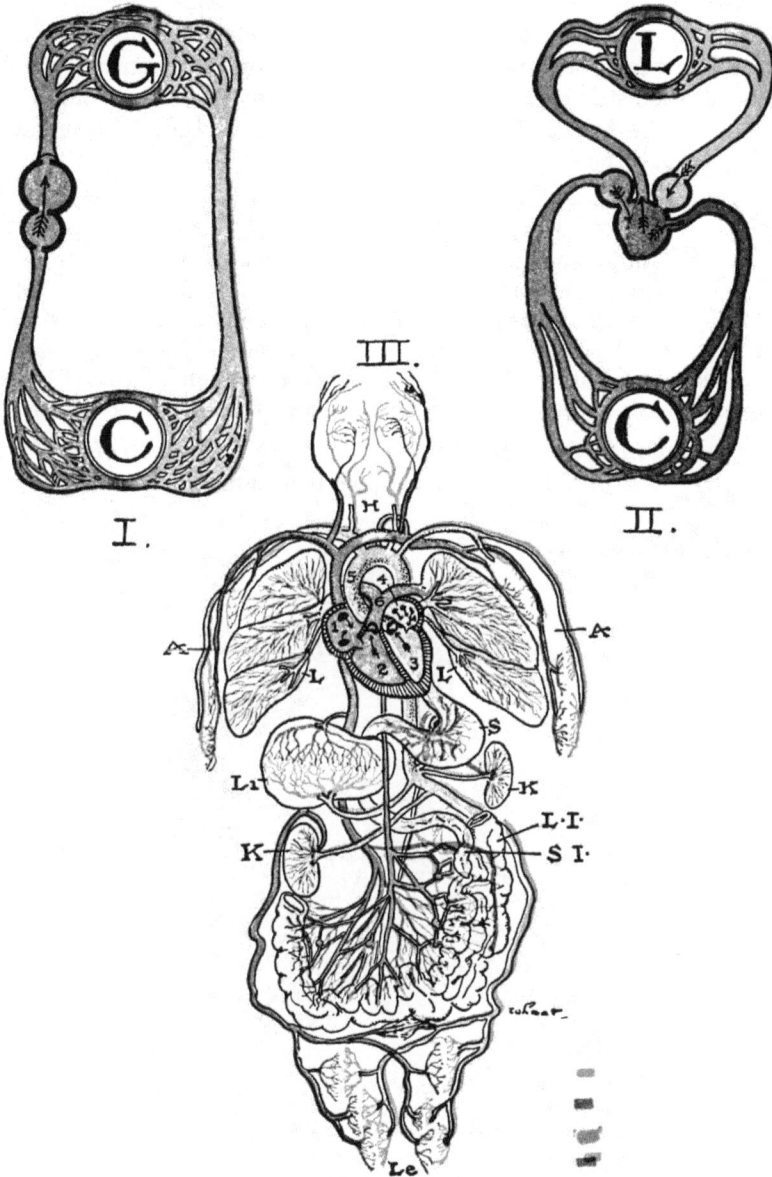

the right auricle through two large vessels known as the *venæ cavæ*. It requires only from twenty to thirty seconds for the blood to make the complete circulation from the ventricle back again to the starting point. This means that the entire volume of blood in the human body passes three or four thousand times a day through the various organs of the body.[1]

Portal Circulation. — Some of the blood, on its way back to the heart, passes to the walls of the food tube and to its glands. From there it is sent with its load of absorbed food to the liver. Here the vein which carries the blood (called the portal vein) breaks up into capillaries around the cells of the liver, when it gives up sugar to be stored as glycogen. From the liver, blood passes directly to the right auricle. The *portal circulation*, as it is called, is the only part of the circulation where the blood passes through two sets of capillaries on its way from auricle to auricle.

Circulation in the Web of a Frog's Foot. — If the web of the foot of a live frog or the tail of a tadpole is examined under the com-

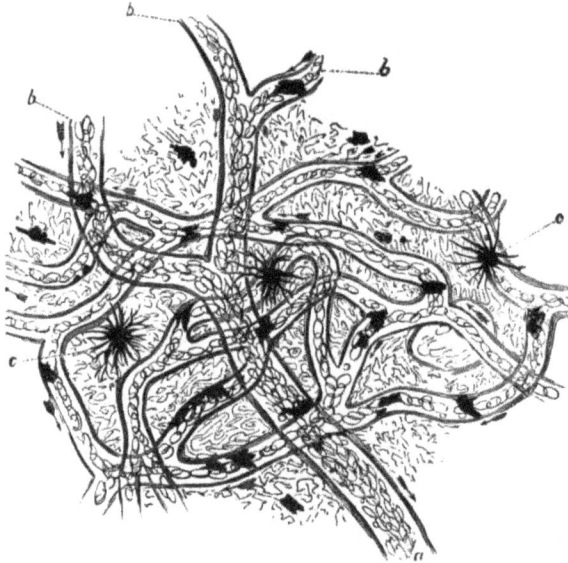

Capillary circulation in the web of a frog's foot, as seen under the compound microscope. *a, b*, small veins; *c*, pigment cells in the skin; *d*, capillaries in which the oval corpuscles are seen to follow one another in single series.

[1] See Hough and Sedgwick, *The Human Mechanism*, page 136.

pound microscope, a network of blood vessels will be seen. In some of the larger vessels the corpuscles are moving rapidly and in spurts; these are *arteries*. The arteries lead into smaller vessels hardly greater in diameter than the width of a single corpuscle. This network of *capillaries* may be followed into larger *veins* in which the blood moves regularly. This illustrates the condition in any tissue of man where the arteries break up into capillaries, and these in turn unite to form veins.

Structure of the Arteries. — A distinct difference in structure exists between the arteries and the veins in the human body. The arteries, because of the greater strain received from the blood which is pumped from the heart, have thicker muscular walls, and in addition are very elastic.

Cause of the Pulse. — The *pulse,* which can easily be detected by pressing the large artery in the wrist or the small one in front of and above the external ear, is caused by the gushing of blood through the arteries after each pulsation of the heart. As the large arteries pass away from the heart, the diameter of each individual artery becomes smaller. At the very end of their course, these arteries are so small as to be almost microscopic in size and are very numerous. There are so many that if they were placed together, side by side, their united diameter would be much greater than the diameter of the large artery (*aorta*) which passes blood from the left side of the heart. This fact is of very great importance, for the force of the blood as it gushes through the arteries becomes very much less when it reaches the smaller vessels. This gushing movement is quite lost when the capillaries are reached, first, because there is so much more space for the blood to fill, and second, because there is considerable friction caused by the very tiny diameter of the capillaries.

Capillaries. — The capillaries form a network of minute tubes everywhere in the body, but especially near the surface and in the lungs. It is through their walls that the food and oxygen pass to the tissues, and carbon dioxide is given up to the plasma. They form the connection that completes the system of circulation of blood in the body.

Function and Structure of the Veins. — If the arteries are supply pipes which convey fluid food to the tissues, then the veins may be likened to drain pipes which carry away waste material from the tissues. Extremely numerous in the extremities and in the muscles and among other tissues of the body, they, like the branches of a tree, become larger and unite with each other as they approach the heart.

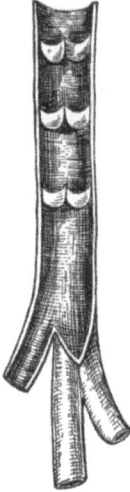

If the wall of a vein is carefully examined, it will be found to be neither so thick nor so tough as an artery wall. When empty, a vein collapses; the wall of an artery holds its shape. If you hold your hand downward for a little time and then examine it, you will find that the veins, which are relatively much nearer the surface than are the arteries, appear to be very much knotted. This appearance is due to the presence of tiny valves within. These valves open in the direction of the blood current, but would close if the direction of the blood flow should be reversed (as in case a deep cut severed a vein). As the pressure of blood in the veins is much less than in the arteries, the valves thus aid in keeping the flow of blood in the veins toward the heart. The higher pressure in arteries and the suction in the veins (caused by the enlargement of the chest cavity in breathing) are the chief factors which cause a steady flow of blood through the veins in the body.

Valves in a vein. Notice the thin walls of the vein.

Lymph Vessels. — The lymph is collected from the various tissues of the body by means of a number of very thin-walled tubes, which are at first very tiny, but after repeated connection with other tubes ultimately unite to form large ducts. These lymph ducts are provided, like the veins, with valves. The pressure of the blood within the blood vessels forces continually more plasma into the lymph; thus a slow current is maintained. On its course the lymph passes through many collections of gland cells, the *lymph glands*. In these glands some impurities appear to be removed and colorless corpuscles made. The lymph ultimately passes into a large tube, the *thoracic duct*, which flows upward near the ventral side of the spinal column, and empties into the large subclavian vein in the left side of the neck. Another smaller lymph duct enters the right subclavian vein.

The Lacteals. — We have already found that part of the digested food (chiefly carbohydrates, proteins, salts, and water) is absorbed

directly into the blood through the walls of the villi and carried to
the liver. Fat, however, is passed into the spaces in the central
part of the villi, and from there into other spaces between the
tissues, known as the *lacteals*.
The lacteals carry the fats into
the blood by way of the thoracic
duct. The lacteals and lymph
vessels have in part the same
course. It will be thus seen
that lymph at different parts of
its course would have a very
different composition.

**The Nervous Control of the
Heart and Blood Vessels.** — Although the muscles of the heart
contract and relax without our being able to stop them or force them
to go faster, yet in cases of sudden
fright, or after a sudden blow, the
heart may stop beating for a short
interval. This shows that the heart
is under the control of the nervous
system. Two sets of nerve fibers,

The lymph vessels; the dark spots are
lymph glands: *lac*, lacteals; *rc*, thoracic duct.

both of which are connected with the central nervous system, pass to
the heart. One set of fibers accelerates, the other slows or inhibits, the
heart beat. The arteries and veins are also under the control of the
sympathetic nervous system. This allows of a change in the diameter
of the blood vessels. Thus, blushing is due to a sudden rush of blood to
the surface of the body caused by an expansion of the blood vessels at
the surface. The blood vessels of the body are always full of blood. This
results from an automatic regulation of the diameter of the blood tubes by
a part of the nervous system called the *vasomotor nerves*. These nerves
act upon the muscles in the walls of the blood vessels. In this way, each
vessel adapts itself to the amount of blood in it at a given time. After
a hearty meal, a large supply of blood is needed in the walls of the stomach
and intestines. At this time, the arteries going to this region are dilated
so as to receive an extra supply. When the brain performs hard work,
blood is supplied in the same manner to that region. Hence, one should
not study or do mental work immediately after a hearty meal, for blood

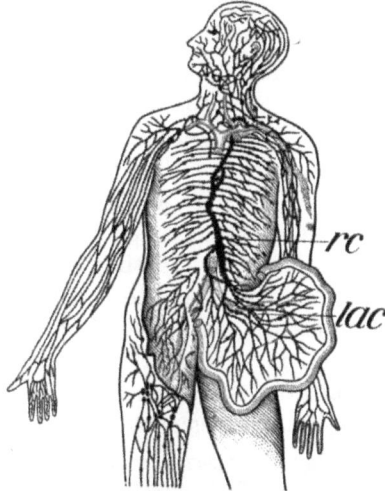

will be drawn away to the brain, leaving the digestive tract with an insufficient supply. Indigestion may follow as a result.

The Effect of Exercise on the Circulation. — It is a fact familiar to all that the heart beats more violently and quickly when we are doing hard work than when we are resting.   Count your own pulse when sitting quietly, and then again after some brisk exercise in the gymnasium.   Exercise in moderation is of undoubted value, because it sends the increased amount of blood to such parts of the body where increased oxidation has been taking place as the result of the exercise.   The best forms of exercise are those which give as many muscles as possible work — walking, out-of-door sports, any exercise that is not violent.   Exercise should not be attempted immediately after eating, as this causes a withdrawal of blood from the digestive tract to the muscles of the body.   Neither should exercise be continued after becoming tired, as poisons are then formed in the muscles, which cause the feeling we call *fatigue*. Remember that extra work given to the heart by extreme exercise may injure it, causing possible trouble with the valves.

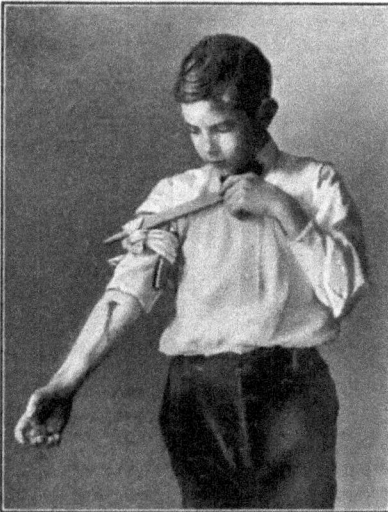

Stopping flow of blood from an artery by applying a tight bandage (ligature) between the cut and the heart.

Treatment of Cuts and Bruises. — Blood which oozes slowly from a cut will usually stop flowing by the natural means of the formation of a clot. A cut or bruise should, however, be washed in a weak solution of carbolic acid or some other antiseptic in order to prevent bacteria from obtaining a foothold on the exposed flesh. If blood, issuing from a wound, gushes in distinct pulsations, then we know that an artery has been severed. To prevent the flow of blood, a tight bandage known as a *tourniquet* must be tied between the cut and the heart.

A handkerchief with a knot placed over the artery may stop bleeding if the cut is on one of the limbs. If this does not serve, then insert a stick in the handkerchief and twist it so as to make the pressure around the limb still greater. Thus we may close the artery until the doctor is called, who may sew up the injured blood vessel.

The Effect of Alcohol upon the Blood. — It has recently been discovered that alcohol has an extremely injurious effect upon the colorless corpuscles of the blood, lowering their ability to fight disease germs to a marked degree. This is well seen in a comparison of deaths from certain infectious diseases in drinkers and abstainers, the percentage of mortality being much greater in the former.

Dr. T. Alexander MacNichol, in a recent address, said : —

" Massart and Bordet, Metchnikoff and Sims Woodhead, have proved that alcohol, even in very dilute solution, prevents the white blood corpuscles from attacking invading germs, thus depriving the system of the coöperation of these important defenders, and reducing the powers of resisting disease. The experiments of Richardson, Harley, Kales, and others have demonstrated the fact that one to five per cent of alcohol in the blood of the living human body in a notable degree alters the appearance of the corpuscular elements, reduces the oxygen bearing elements, and prevents their reoxygenation."

Alcohol weakens Resistance to Disease. — In acute illnesses, grippe, fevers, blood poisoning, etc., substances formed in the blood termed " antibodies " antagonize the action of bacteria, facilitating their destruction by the white blood cells and neutralizing their poisonous influence. In a person with good "resistance" this protective machinery, which we do not yet thoroughly understand, works with beautiful precision, and the patient " gets well." Experiments by scientific experts have demonstrated that alcohol restrains the formation of these marvelous antibodies. Alcohol puts to sleep the sentinels that guard your body from disease.

The Effect of Alcohol on the Circulation. — Alcoholic drinks affect the very delicate adjustment of the nervous centers controlling the blood vessels and heart. Even very dilute alcohol acts upon the muscles of the tiny blood vessels; consequently, more blood is allowed to enter them, and, as the small vessels are usually near the surface of the body, the habitual redness seen in the face of hard drinkers is the ultimate result.

" The first effect of diluted alcohol is to make the heart beat faster. This fills the small vessels near the surface. A feeling of warmth is produced which causes the drinker to feel that he was warmed by the drink. This feeling, however, soon passes away, and is succeeded by one of chilliness. The body temperature, at first raised by the rather rapid oxidation of the alcohol, is soon lowered by the increased radiation from the surface.

" The immediate stimulation to the heart's action soon passes away and, like other muscles, the muscles of the heart lose power and contract with less force after having been excited by alcohol." — MACY, *Physiology*.

Alcohol, when brought to act directly on heart muscle, lessens the force of the beat. It may even cause changes in the tissues, which eventually result in the breaking of the walls of a blood vessel or the plugging of a vessel with a blood clot. This condition may cause the disease known as *apoplexy*.

Effects of Tobacco upon the Circulation. — " The frequent use of cigars or cigarettes by the young seriously affects the quality of the blood. The red blood corpuscles are not fully developed and charged with their normal supply of life-giving oxygen. This causes paleness of the skin, often noticed in the face of the young smoker. Palpitation of the heart is also a common result, followed by permanent weakness, so that the whole system is enfeebled, and mental vigor is impaired as well as physical strength." — MACY, *Physiology*.

XXII. RESPIRATION AND EXCRETION

Problems.—*A study of respiration to find out:*—

(a) *What changes in blood and air take place within the lungs.*

(b) *The mechanics of respiration.*

A study of ventilation to discover:—

(a) *The reason for ventilation.*

(b) *The best method of ventilation.*

A study of the organs of excretion.

LABORATORY SUGGESTIONS

Demonstration. — Comparison of lungs of frog with those of bird or mammal.

Experiment. — The changes of blood within the lungs.

Experiment. — Changes taking place in air in the lungs.

Experiment. — The use of the ribs in respiration.

Demonstration experiment. — What causes the filling of air sacs of the lungs?

Demonstration experiment. — What are the best methods of ventilating a room?

Demonstration. — Best methods of dusting and cleaning.

Demonstration. — Beef or sheep's kidney to show areas.

Necessity for Respiration. — We have seen that plants and animals need oxygen in order that the life processes may go on. Food is oxidized to release energy, just as coal is burned to give heat to run an engine. As a draft of air is required to make fire under the boiler, so, in the human body, oxygen must be given so that food in tissues may be oxidized to release energy used in work. This oxidation takes place in the cells of the body, be they part of a muscle, a gland, or the brain. *Blood, in its circulation to all parts of the body, is the medium which conveys the oxygen to that place in the body where it will be used.*

329

The Organs of Respiration in Man. — We have alluded to the fact that the lungs are the organs which give oxygen to the blood and take from it carbon dioxide. The course of the air passing to the lungs in man is much the same as in the frog. Air passes through the nose, and into the windpipe. This cartilaginous tube, the top of which may easily be felt as the Adam's apple of the throat, divides into two *bronchi*. The bronchi within the lungs break up into a great number of smaller tubes, the *bronchial tubes*, which divide somewhat like the small branches of a tree. The bronchial tubes, indeed all the air passages, are lined with ciliated cells. The cilia of these cells are constantly in motion, beating with a quick stroke toward the outer end of the tube, that is, toward the mouth. Hence any foreign material will be raised from the throat first by the action of the cilia and then by coughing or "clearing the throat." The bronchi end in very minute air sacs, little pouches having elastic walls, into which air is taken when we inspire, or take a deep breath. In the walls of these pouches are numerous capillaries, the ends of arteries which pass from the heart into the lung. *It is through the very thin walls of the air sacs that an interchange of gases takes place which results in the blood giving up part of its load of carbon dioxide, and taking up oxygen in its place.* This exchange appears to be aided by the presence of an enzyme in the lung tissues. This is another example of the various kinds of work done by the enzymes of the body.

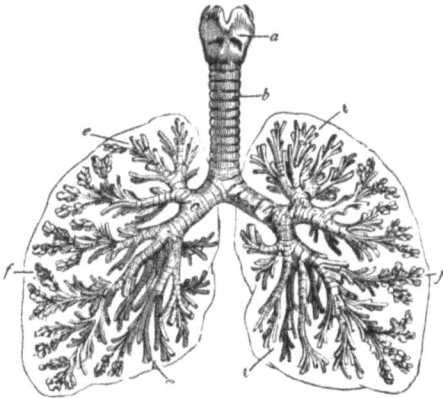

Air passages in the human lungs. *a*, larynx; *b*, trachea (or windpipe); *c, d*, bronchi; *e*, bronchial tubes; *f*, cluster of air cells.

Changes in the Blood within the Lungs. — Blood, after leaving the lungs, is much brighter red than just before entering them. The change in color is due to a taking up of oxygen by the *hæmo-*

globin of the red corpuscles. Changes taking place in blood are, obviously the reverse of those which take place in air in the lungs. Every hundred cubic centimeters of blood going into the lungs contains 8 to 12 c.c. of oxygen, 45 to 50 c.c. of carbon dioxide, and 1 to 2 c.c. of nitrogen. The same amount of blood passing out of the lungs contains 20 c.c. of oxygen, 38 c.c. of carbon dioxide, and 1 to 2 c.c. of nitrogen. The water, of which about half a pint is given off daily, is mostly lost from the blood.

Changes in Air in the Lungs. — Air is much warmer after leaving the lungs than before it enters them. Breathe on the bulb of a thermometer to prove this. Expired air con- tains a considerable amount

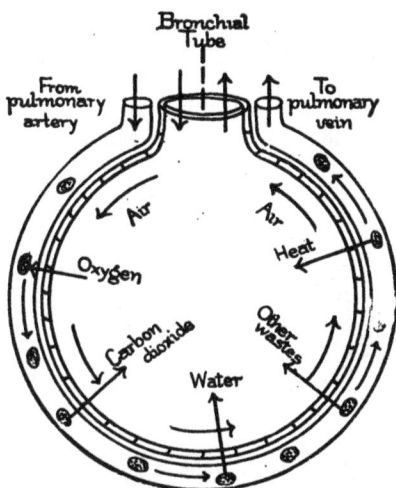

Diagram to show what the blood loses and gains in one of the air sacs of the lungs.

of moisture, as may be proved by breathing on a cold polished surface. This it has taken up in the air sacs of the lungs. The presence of carbon dioxide in expired air may easily be detected by the limewater test. Air such as we breathe out of doors con- tains, by volume : —

Nitrogen	76.95
Oxygen	20.61
Carbon dioxide	.03
Argon	1.00
Water vapor (average)	1.40

Air expired from the lungs contains : —

Nitrogen	76.95
Oxygen	15.67
Carbon dioxide	4.38
Water vapor	2
Argon	1

In other words, there is a loss between 4 and 5 per cent oxygen, and nearly a corresponding gain in carbon dioxide, in expired air. There are also some other organic substances present.

Cell Respiration. — It has been shown, in the case of very simple animals, such as the *amœba*, that when oxidation takes place in a cell, work results from this oxidation. The oxygen taken into the lungs is not used there, but is carried by the blood to such parts of the body as need oxygen to oxidize food materials in the cells. Since work is done in the cells of the body, food and oxygen are therefore required. The quantity of oxygen used by the body is nearly dependent on the amount of work performed. Oxygen is constantly taken from the blood by tissues in a state of rest and is used up when the body is at work. This is suggested by the fact that in a given time a man, when working, gives off more oxygen (in carbon dioxide) than he takes in during that time.

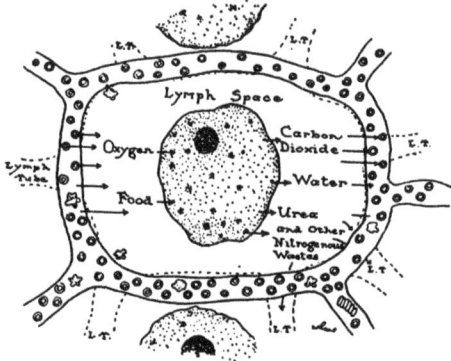

The respiration of cells.

While work is being done certain wastes are formed in the cell. Carbon dioxide is given off when carbon is burned. But when proteins are burned, another waste product containing nitrogen is formed. This must be passed off from the cells, as it is a poison. Here again the lymph and blood, the common carriers, take the waste material to points where it may be *excreted* or passed out of the body.

The Mechanics of Respiration. The Pleura. — The lungs are covered with a thin elastic membrane, the *pleura*. This forms a bag in which the lungs are hung. Between the walls of the bag and the lungs is a space filled with lymph. By this means the lungs are prevented from rubbing against the walls of the chest.

Breathing. — In every full breath there are two distinct movements, inspiration (taking air in) and expiration (forcing air out). In man an inspiration is produced by the contraction of the muscles between the ribs, together with the contraction of the diaphragm, the muscular wall just below the heart and lungs; this results in pulling down the dia-

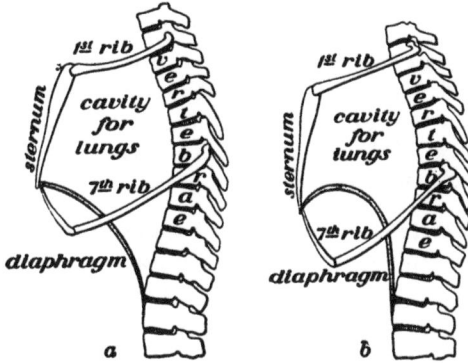

The chest cavity (*a*) at the time of a full breath; (*b*), after an expiration. Explain how the cavity for lungs is made larger.

phragm and pulling upward and outward of the ribs, thus making the space within the chest cavity larger. The lungs, which lie within this cavity, are filled by the air rushing into the larger space thus made. That this cavity is larger than it was at first may be demonstrated by a glance at the accompanying figure. An expiration is simpler than an inspiration, for it requires no muscular effort; the muscles relax, the breastbone and ribs sink into place, while the diaphragm returns to its original position.

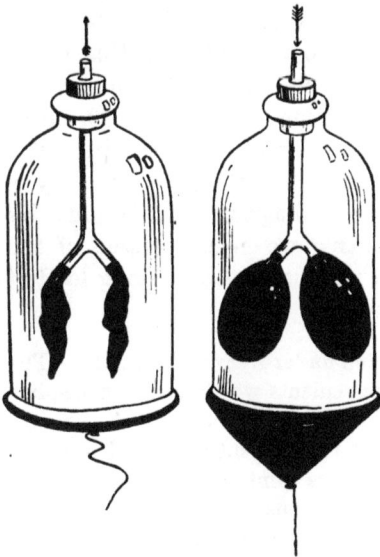

Apparatus to show the mechanics of breathing.

A piece of apparatus which illustrates to a degree the mechanics of breathing may be made as follows: Attach a string to the middle of a piece of sheet rubber. Tie the rubber over the large end of a bell jar. Pass a glass Y-tube through a rubber stopper. Fasten two small toy balloons to the branches of the tube. Close the small end of the jar with the stopper. Adjust the tube so that the balloons shall hang free in the jar. If now the rubber sheet is pulled down by means of the string, the air pressure in the jar is reduced and the toy balloons within expand, owing to the air pressure down the tube. When the rubber is allowed to go back to its former position, the balloons collapse.

Rate of Breathing and Amount of Air Breathed. — During quiet breathing, the rate of inspiration is from fifteen to eighteen times per minute; this rate largely depends on the amount of physical work performed. About 30 cubic inches of air are taken in and expelled during the ordinary quiet respiration. The air so breathed is called *tidal air*. In a "long" breath, we take in about 100 cubic inches in addition to the tidal air. This is called *complemental air*. By means of a forced expiration, it is possible to expel from 75 to 100 cubic inches more than tidal air; this air is called *reserve air*. What remains in the lungs, amounting to about 100 cubic inches, is called the *residual air*. The value of deep breathing is seen by a glance at the diagram. It is only by this means that we clear the lungs of the reserve air with its accompanying load of carbon dioxide.

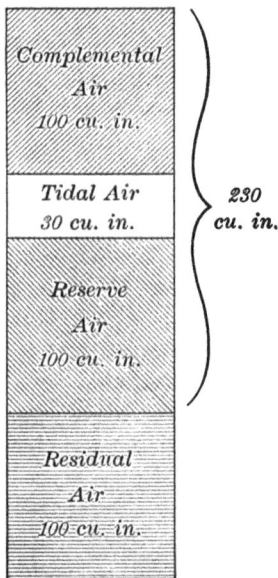

Diagram showing the relative amounts of tidal, complemental, reserve, and residual air. The brace shows the average lung capacity for the adult man.

Respiration under Nervous Control. — The muscular movements which cause an inspiration are partly under the control of the will, but in part the movement is beyond our control. The nerve centers which govern inspiration are part of the sympathetic nervous system. Anything of an irritating nature in the trachea or larynx will cause a sudden expiration or cough. When a boy runs, the quickened respiration is due to the fact that oxygen is used up rapidly and a larger quantity of carbon dioxide is formed. The carbon dioxide in the blood stimulates the nervous center which has control of respiration to greater activity, and quickened inspiration follows.

Need of Ventilation. — During the course of a day the lungs lose to the surrounding air nearly two pounds of carbon dioxide. This means that about three fifths of a cubic foot is given off by each person during an hour. When we are confined for some time in a room, it becomes necessary to get rid of this carbon dioxide. This can be done only by means of proper ventilation. A considerable amount of moisture is given off from the body, and this moisture in a crowded room is responsible for much of the discomfort. The air becomes humid and uncomfortable. It has been found that by keeping the air in motion in such a room (as through the use of electric fans) much of this discomfort is obviated.

The presence of impurities in the air of a room may easily be determined by its odor. The odor of a poorly ventilated room is due to organic impurities given off with the carbon dioxide. This, fortunately, gives us an index of the amount of waste material in the air. Among the factors which take oxygen from the air in a closed room and produce carbon dioxide are burning gas or oil lamps and stoves, and the presence of a number of people.

Proper Ventilation. — Ventilation consists in the removal of air that has been used, and the introduction of a fresh supply to take its place. Heated air rises, carrying with it much of the carbon dioxide and other impurities. A good method

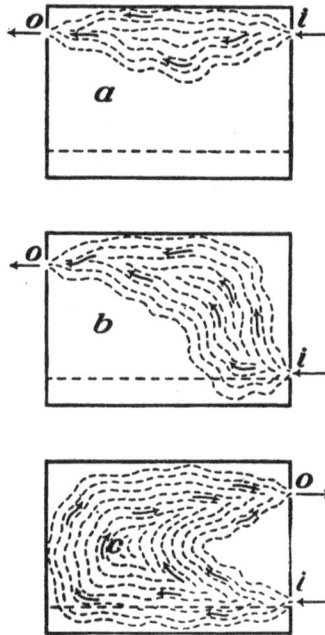

Three ways of ventilating a room. *i*, inlet for air; *o*, outlet for air. Which is the best method of ventilation? Explain.

of ventilation for the home is to place a board two or three inches high between the lower sash and the frame of a window or to have the window open an inch or so at the top and the bottom. An open fireplace in a room aids in ventilation because of the constant draft up the flue.

Sweeping and Dusting. — It is very easy to demonstrate the amount of dust in the air by following the course of a beam of light in a darkened room. We have already proved that spores of mold and yeast exist in the air. That bacteria are also present can be proved by exposing a sterilized gelatin plate to the air in a schoolroom for a few moments.[1]

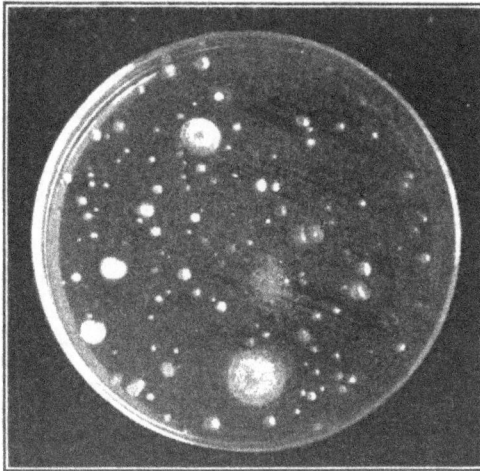

Plate culture exposed for five minutes in a school hall where pupils were passing to recitations. Each spot is a colony of bacteria or mold.

Many of the bacteria present in the air are active in causing diseases of the respiratory tract, such as diphtheria, membranous croup, and tuberculosis. Other diseases, as colds, bronchitis (inflammation of the bronchial tubes), and pneumonia (inflammation of the tiny air sacs of the lungs), are also caused by bacteria.

Dust, with its load of bacteria, will settle on any horizontal surface in a room not used for three or four hours. Dusting and sweeping should always be done with a damp cloth or broom, otherwise the bacteria are simply stirred up and sent into the air

---

[1] Expose two sterilized dishes containing culture media; one in a room being swept with a damp broom, and the other in a room which is being swept in the usual manner. Note the formation of colonies of bacteria in each dish. In which dish does the more abundant growth take place?

again. The proper watering of streets before they are swept is
also an important factor in health. Much dust is composed largely
of dried excreta of animals. Soft-coal smoke does its share to
add to the impurities of the air, while sewer gas and illuminating
gas are frequently found in sufficient quantities to poison people.
Pure air is, as can be seen, almost an impossibility in a great city.

How to get Fresh Air. — As we know, green plants give off in
the sunlight considerable more oxygen than they use, and they
use up carbon dioxide. The air in the country is naturally purer
than in the city, as smoke and bacteria are not so prevalent there,
and the plants in abundance give off oxygen.   In the city the
night air is purer than day air,
because the factories have stopped
work, the dust has settled, and
fewer people are on the streets.
The old myth of " night air "
being injurious has long since been
exploded, and thousands of people
of delicate health, especially those
who have weak throat or lungs,
are regaining health by sleeping
out of doors or with the windows
wide open. The only essential in
sleeping out of doors or in a room

·A sleeping porch, an ideal way to
get fresh air at night.

with a low temperature is that the body be kept warm and the
head be protected from strong drafts by a nightcap or hood.
Proper ventilation at *all* times is one of the greatest factors in
good health.

Change of Air. — Persons in poor health, especially those having
tuberculosis, are often cured by a change of air. This is not always
so much due to the composition of the air as to change of occupation, rest, and good food. Mountain air is dry, and relatively
free from dust and bacteria, and often helps a person having tuberculosis. Air at the seaside is beneficial for some forms of disease,
especially hay fever and bone tuberculosis. Many sanitariums
have been established for this latter disease near the ocean, and
thousands of lives are being annually saved in this way.

HUNTER, CIV. BI. — **22**

Ventilation of Sleeping Rooms. — Sleeping in close rooms is the cause of much illness. Beds ought to be placed so that a constant supply of fresh air is given without a direct draft. This may often be managed with the use of screens. Bedroom windows should be thrown open in the morning to allow free entrance of the sun and air, bedclothes should be washed frequently, and sheets

Unfavorable sleeping conditions. Explain why unfavorable.

and pillow covers often changed. Bedroom furniture should be simple, and but little drapery allowed in the room.

Hygienic Habits of Breathing. — Every one ought to accustom himself upon going into the open air to inspire slowly and deeply to the full capacity of the lungs. A slow expiration should follow. Take care to force the air out. Breathe through the nose, thus warming the air you inspire before it enters the lungs and chills the blood. Repeat this exercise several times every day. You will thus prevent certain of the air sacs which are not often used from becoming hardened and permanently closed.

Relation of Proper Exercise to Health. — We are all aware that exercise in moderation has a beneficial effect upon the human organism. The pale face, drooping shoulders, and narrow chest of the boy or girl who takes no regular exercise is too well known. Exercise, besides giving direct use of the muscles, increases the work of the heart and lungs, causing deeper breathing and giving the heart muscles increased work; it liberates heat and carbon dioxide from the tissues where the work is taking place, thus increasing the respiration of the tissues themselves, and aids mechanically in the removal of wastes from tissues. It is well known that exercise, when taken some little time after eating, has a very beneficial effect upon digestion. Exercise and especially games are of immense importance to the nervous system as a means of rest. The increasing number of playgrounds in this country is due to this acknowledged need of exercise, especially for growing children.

Proper exercise should be moderate and varied. Walking in itself is a valuable means of exercising certain muscles, so is bicycling, but neither is ideal as the *only* form to be used. *Vary* your exercise so as to bring different muscles into play, take exercise that will allow free breathing out of doors if possible, and the natural fatigue which follows will lead you to take the rest and sleep that every normal body requires.

Exercise should always be limited by fatigue, which brings with it fatigue poisons. This is nature's signal when to rest. If one's use of diet and air is proper, the fatigue point will be much further off than otherwise. One should learn to *relax* when not in activity. The habit produces rest, even between exertions very close together, and enables one to continue to repeat those exertions for a much longer time than otherwise. The habit of lying down when tired is a good one.

The Relation of Tight Clothing to Correct Breathing. — It is impossible to breathe correctly unless the clothing is worn loosely over the chest and abdomen. Tight corsets and tight belts prevent the walls of the chest and the abdomen from pushing outward and interfere with the drawing of air into the lungs. They may also result in permanent distortion of parts of the skeleton directly under the pressure. Other organs of the body cavity, as the stomach and intestines, may be forced downward, out of place, and in consequence cannot perform their work properly.

Suffocation and Artificial Respiration. — Suffocation results from the shutting off of the supply of oxygen from the lungs. It may be brought about by an obstruction in the windpipe, by a lack of oxygen in the air, by inhaling some other gas in quantity, or by drowning. A severe electric shock may paralyze the nervous centers which control respiration, thus causing a kind of suffocation. In the above cases, death often may be prevented by prompt recourse to artificial respiration. To accomplish this, place the patient on his back with the head lower than the body; grasp the arms near the elbows and draw them *upward* and *outward* until they are stretched above the head, on a line with the body. By this means the chest cavity is enlarged and an inspiration produced. To produce an expiration, carry the arms downward, and press them against the chest, thus forcing the air out of the lungs. This exercise, regularly repeated every few seconds, if necessary for hours, has been the source of saving many lives.

Common Diseases of the Nose and Throat. — Catarrh is a disease to which people with sensitive mucous membrane of the nose and throat are subject. It is indicated by the constant secretion of mucus from these membranes. Frequent spraying of the nose and throat with some mild antiseptic solutions is found helpful. Chronic catarrh should be attended to by a physician. Often we find children breathing entirely through the mouth, the nose being seemingly stopped up. When this goes on for some time the nose and throat should be examined by a physician for *adenoids*, or growths of soft masses of tissue which fill up the nose cavity, thus causing a shortage of the air supply for the body. Many a child, backward at school, thin and irritable, has been changed to a healthy, normal, bright scholar by the removal of adenoids. Sometimes the tonsils at the back of the mouth cavity may become enlarged, thus shutting off the air supply and causing the same trouble as we see in a case of adenoids. The simple removal of the obstacle by a doctor soon cures this condition. (See page 395.)

Organs of Excretion. — All the life processes which take place in a living thing result ultimately, in addition to giving off of carbon dioxide, in the formation of organic wastes within the body. The retention of these wastes which contain nitrogen, is harmful to animals. In man, the skin and kidneys remove this waste from the body, hence they are called the organs of excretion.

The Human Kidney. — The human kidney is about four inches long, two and one half inches wide, and one inch in thickness. Its color is dark red. If the structure of the medulla and cortex (see figure above) is examined under the compound microscope, you will find these regions to be composed of a vast number of tiny branched and twisted tubules. The outer

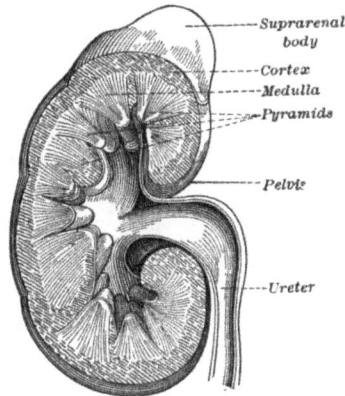

Longitudinal section through a kidney.

end of each of these tubules opens into the *pelvis*, the space within the kidney; the inner end, in the cortex, forms a tiny closed sac. In each sac, the outer wall of the tube has grown inward and

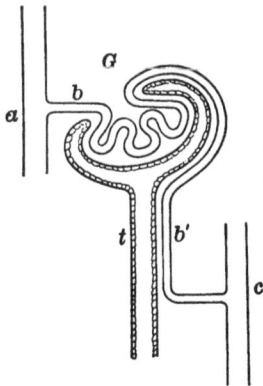

Diagram of kidney circulation, showing a glomerulus and tubule: *a*, artery bringing blood to part; *b*, capillary bringing blood to glomerulus; *b'*, vessel continuing with blood to vein; *c*, vein; *t*, tubule; *G*, glomerulus.

carried with it a very tiny artery. This artery breaks up into a mass of capillaries. These capillaries, in turn, unite to form a small vein as they leave the little sac. Each of these sacs with its contained blood vessels is called a *glomerulus*.

Wastes given off by the Blood in the Kidney. — In the glomerulus the blood loses by osmosis, through the very thin walls of the capillaries, first, a considerable amount of water (amounting to nearly three pints daily); second, a nitrogenous waste material known as urea; third, salts and other waste organic substances, uric acid among them.

These waste products, together with the water containing them, are known as *urine*. The total amount of nitrogenous waste leaving the body each day is about twenty grams. It

is passed through the *ureter* to the *urinary bladder;* from this reservoir
it is passed out of the body, through a tube called the *urethra*. After
the blood has passed through the glomeruli of the kidneys it is purer
than in any other place in the body, because, before coming there, it
lost a large part of its burden of carbon dioxide in the lungs. After
leaving the kidney it has lost much of its nitrogenous waste. So dependent is the body upon the excretion of its poisonous material that,
in cases where the kidneys do not do their work properly, death may
ensue within a few hours.

Structure and Use of Sweat Glands. — If you examine the
palm of your hand with a lens, you will notice the surface is thrown

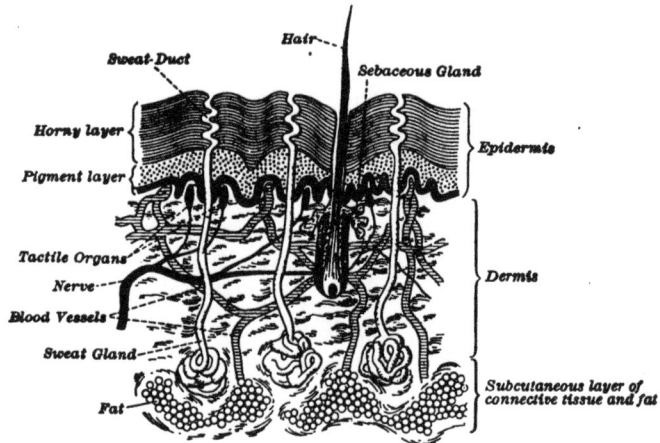

Diagram of a section of the skin. (Highly magnified.)

into little ridges. In these ridges may be found a large number of
very tiny pits; these are the pores or openings of the sweat-
secreting glands. From each opening a little tube penetrates deep
within the epidermis; there, coiling around upon itself several
times, it forms the sweat gland. Close around this coiled tube are
found many capillaries. From the blood in these capillaries, cells
lining the wall of the gland take water, and with it a little carbon
dioxide, urea, and some salts (common salt among others). This
forms the excretion known as *sweat*. The combined secretions
from these glands amount normally to a little over a pint during

twenty-four hours. At all times, a small amount of sweat is given off, but this is evaporated or is absorbed by the underwear; as this passes off unnoticed, it is called *insensible perspiration*. In hot weather or after hard manual labor the amount of perspiration is greatly increased.

Regulation of Heat of the Body. — The bodily temperature of a person engaged in manual labor will be found to be but little higher than the temperature of the same person at rest. We know from our previous experiments that heat is released. Muscles, nearly one half the weight of the body, release about five sixths of their energy as heat. At all times they are giving up some heat. How is it that the bodily temperature does not differ greatly at such times? The temperature of the body is largely regulated by means of the activity of the sweat glands. The blood carries much of the heat, liberated in the various parts of the body by the oxidation of food, to the surface of the body, where it is lost in the evaporation of sweat. In hot weather the blood vessels of the skin are dilated; in cold weather they are made smaller by the action of the nervous system. The blood thus loses water in the skin, the water evaporates, and we are cooled off. *The object of increased perspiration, then, is to remove heat from the body.* With a large amount of blood present in the skin, perspiration is increased; with a small amount, it is diminished. Hence, we have in the skin an automatic regulator of bodily temperature.

Sweat Glands under Nervous Control. — The sweat glands, like the other glands in the body, are under the control of the sympathetic nervous system. Frequently the nerves dilate the blood vessels of the skin, thus helping the sweat glands to secrete, by giving them more blood.

" Thus regulation is carried out by the nervous system determining, on the one hand, the *loss* by governing the supply of blood to the skin and the action of the sweat glands; and on the other, the *production* by diminishing or increasing the oxidation of the tissues." — FOSTER AND SHORE, *Physiology*.

Colds and Fevers. — The regulation of blood passing through the blood vessels is under control of the nervous system. If this mechanism is interfered with in any way, the sweat glands may not do their work, perspiration may be stopped, and the heat from oxidation held within the body. The body temperature goes up, and a fever results.

If the blood vessels in the skin are suddenly cooled when full of blood, they contract and send the blood elsewhere. As a result a congestion or cold may follow. Colds are, in reality, a congestion of membranes lining certain parts of the body, as the nose, throat, windpipe, or lungs.

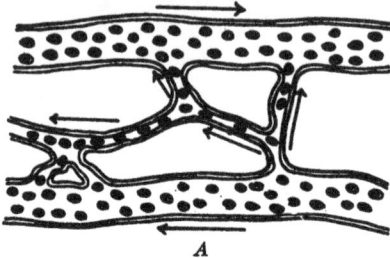

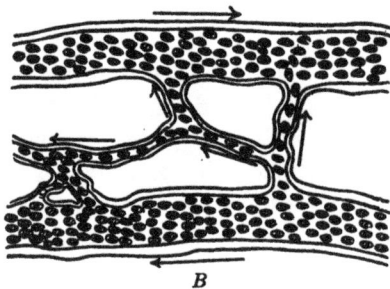

A, blood vessels in skin normal; B, when congested.

When suffering from a cold, it is therefore important not to chill the skin, as a full blood supply should be kept in it and so kept from the seat of the congestion. For this reason hot baths (which call the blood to the skin), the avoiding of drafts (which chill the skin), and warm clothing are useful factors in the care of colds.

Hygiene of the Skin. — The skin is of importance both as an organ of excretion and as a regulator of bodily temperature. The skin of the entire body should be bathed frequently so that this function of excretion may be properly performed. Pride in one's own appearance forbids a dirty skin. For those who can stand it, a cold sponge bath is best. Soap should be used daily on parts exposed to dirt. Exercise in the open air is important to all who desire a good complexion. The body should be kept at an even temperature by the use of proper underclothing. Wool, a poor conductor of heat, should be used in winter, and cotton, which allows of a free escape of heat, in summer.

Cuts, Bruises, and Burns. — In case the skin is badly broken, it is necessary to prevent the entrance and growth of bacteria. This may be done by washing the wound with weak antiseptic solutions such as 3 per cent *carbolic acid*, 3 per cent *lysol, or peroxide of hydrogen* (full strength). These solutions should be applied immediately. A burn or scald should be covered at once with a paste of baking soda, with olive oil, or with a mixture of limewater and linseed oil. These tend to lessen the pain by keeping out the air and reducing the inflammation.

Summary of Changes in Blood within the Body. — We have already seen that red corpuscles in the lungs lose part of their load of carbon dioxide that they have taken from the tissues, replacing it with oxygen. This is accompanied by a change of color from purple (in blood which is poor in oxygen) to that of bright red (in richly oxygenated blood). Other changes take place in other parts of the body. In the walls of the food tube, especially in the small intestine, the blood receives its load of fluid food. In the muscles and other working tissues the blood gives up food and oxygen, receiving carbon dioxide and organic waste in return. In the liver, the blood gives up its sugar, and the worn-out red corpuscles which break down are removed (as they are in the spleen) from the circulation. In glands, it gives up materials used by the gland cells in their manufacture of secretions. In the kidneys, it loses water and nitrogenous wastes (*urea*). In the skin, it also loses some waste materials, salts, and water.

" **The Effect of Alcohol on Body Heat.** — It is usually believed that 'taking a drink' when cold makes one warmer. But such is not the case. In reality alcohol lowers the temperature of the body by dilating the blood vessels of the skin. It does this by means of its influence on the nervous system. It is, therefore, a mistake to drink alcoholic beverages when one is extremely cold, because by means of this more bodily heat is allowed to escape.

" Because alcohol is quickly oxidized, and because heat is produced in the process, it was long believed to be of value in maintaining the heat of the body. A different view now prevails as the result of much observation and experiment. Physiologists show by careful experiments that though the temperature of the body rises during digestion of food, it is lowered for some hours when alcohol is taken. The flush which is felt upon the skin after a drink of wine or spirits is due in part to an increase of heat in the body, but also to the paralyzing effect of the alcohol upon the capillary walls, allowing them to dilate, and so permitting more of the warm blood of the interior of the body to reach the surface. There it is cooled by radiation, and the general temperature is lowered." — MACY, *Physiology*.

Effect of Alcohol on Respiration. — Alcohol tends to congest the membrane of the throat and lungs. It does this by paralyzing the nerves which take care of the tiny blood vessels in the walls of the air tubes and air sacs. The capillaries become full of blood, the air spaces are lessened, and breathing is interfered with. The use of alcohol is believed by many physicians to predispose a person to tuberculosis. Certainly this disease attacks drinkers more readily than those who do not drink. Alcohol interferes with the respiration of the cells because it is oxidized very quickly within the body as it is quickly absorbed and sent to the cells. So rapid is this oxidation that it interferes with the oxidation of other substances. Using alcohol has been likened to burning kerosene in a stove; the operation is a dangerous one.

Effects of Tobacco on Respiration. — Tobacco smoke contains the same kind of poisons as the tobacco, with other irritating substances added. It is extremely irritating to the throat; it often causes a cough, and renders it more liable to inflammation. If the smoke is inhaled more deeply, the vaporized nicotine is still more readily absorbed and may thus produce greater irritation in the bronchi and lungs. Cigarettes are worse than other forms of tobacco, for they contain the same poisons with others in addition.

Effect of Alcohol on the Kidneys. — It is said that alcohol is one of the greatest causes of disease in the kidneys. The forms of disease known as "fatty degeneration of the kidney" and "Bright's disease" are both frequently due to this cause. The kidneys are the most important organs for the removal of nitrogenous waste.

Alcohol unites more easily with oxygen than most other food

materials, hence it takes away oxygen that would otherwise be used in oxidizing these foods. Imperfect oxidation of foods causes the development and retention of poisons in the blood which it becomes the work of the kidneys to remove. If the kidneys become overworked, disease will occur. Such disease is likely to make itself felt as rheumatism or gout, both of which are believed to be due to waste products (poisons) in the blood.

Poisons produced by Alcohol. — When too little oxygen enters the draft of the stove, the wood is burned imperfectly, and there are clouds of smoke and irritating gases. So, if oxygen unites with the alcohol and too little reaches the cells, instead of carbon dioxide, water, and urea being formed, there are other products, some of which are exceedingly poisonous and which the kidneys handle with difficulty. The poisons retained in the circulation never fail to produce their poisonous effects, as shown by headaches, clouded brain, pain, and weakness of the body. The word " intoxication " means " in a state of poisoning." These poisons gradually accumulate as the alcohol takes oxygen from the cells. The worst effects come last, when the brain is too benumbed to judge fairly of their harm.

REFERENCE BOOKS

ELEMENTARY

Hunter, *Laboratory Problems in Civic Biology.* American Book Company.
Davison, *Human Body and Health.* American Book Company.
Gulick, *Hygiene Series, Emergencies, Good Health.* Ginn and Company.
Hough and Sedgwick, *The Human Mechanism.* Ginn and Company.
Macy, *General Physiology.* American Book Company.
Ritchie, *Human Physiology.* World Book Company.

XXIII. BODY CONTROL AND HABIT FORMATION

Problems. — How is body control maintained?

(a) What is the mechanism of direction and control?

(b) What is the method of direction and control?

(c) What are habits? How are they formed and how broken?

(d) What are the organs of sense? What are their uses?

(e) How does alcohol affect the nervous system?

LABORATORY SUGGESTIONS

Demonstration. — Sensory motor reactions.

Demonstration. — Nervous system. Models and frog dissections.

Demonstration. — Neurones under compound microscope (optional).

Demonstration. — Reflex acts are unconscious acts: show how conscious acts may become habitual.

Home exercise in habit forming.

The senses. — *Home exercises.* — (1) To determine areas most sensitive to touch. (2) To determine or map out hot and cold spots on an area on the wrist. (3) To determine functions of different areas on tongue.

Demonstration. — Show how eye defects are tested.

Laboratory summary. — The effects of alcohol on the nervous system.

The Body a Self-directed Machine. — Throughout the preceding chapters the body has been likened to an engine, which, while burning its fuel, food, has done work. If we were to carry our comparison further, however, the simile ceases. For the engineer runs the engine, while the bodily machine is self-directive.

Moreover, most of the acts we perform during a day's work are results of the automatic working of this bodily machine. The heart pumps; the blood circulates its load of food, oxygen, and wastes; the movements of breathing are performed; the thousand and one complicated acts that go on every day within the body are *seemingly* undirected.

Automatic Activity. — In addition to this, numbers of other of our daily acts are not thought about. If we are well-regulated

348

body machines, we get up in the morning, automatically wash, clean our teeth, dress, go to the toilet, get our breakfast, walk to school, even perform such complicated processes as that of writing, without *thinking* about or *directing* the machine. In these respects we have become creatures of habit. Certain acts which once we might have learned consciously, have become automatic.

But once at school, if we are really making good in our work in the classroom, we begin a higher control of our bodily functions. Automatic control acts no longer, and sensation is not the

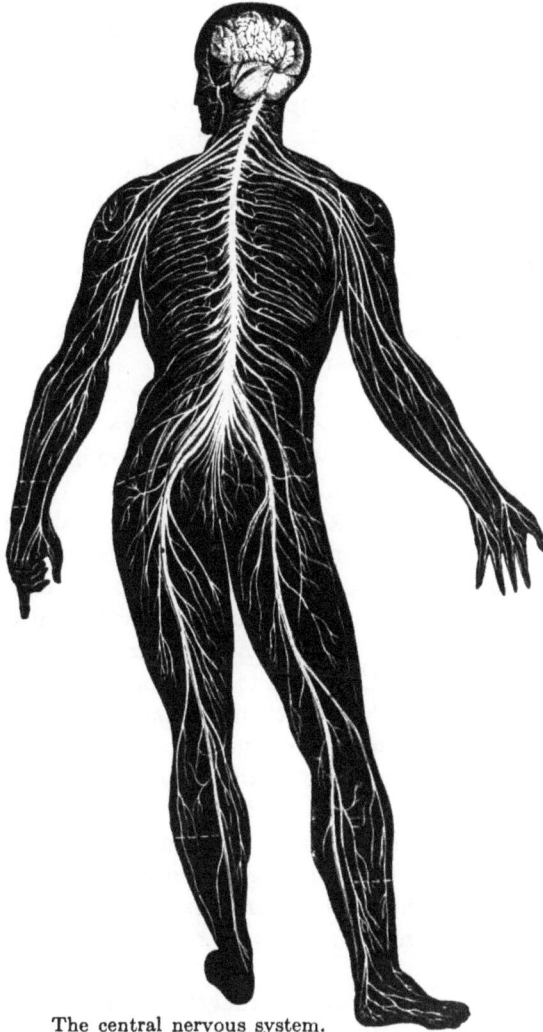

The central nervous system.

only guide — for we now begin to make *conscious choice;* we weigh this matter against another, — in short, we *think*.

Parts of the Nervous System. — This wonderful self-directive apparatus placed within us, which is in part under control of our will, is known as the nervous system. In the vertebrate animals, including man, it consists of two divisions. One includes the brain, spinal cord, the cranial and spinal nerves, which together make up the *cerebro-spinal nervous system*. The other division is called the *sympathetic nervous system* and has to do with those bodily functions which are beyond our control. Every group of cells in the body that has work to do (excepting the floating cells of the blood) is directly influenced by these nerves. Our bodily comfort is dependent upon their directive work. The organs which put us in touch with our surroundings are naturally at the *surface* of the body. Small collections of nerve cells, called *ganglia*, are found in all parts of the body. These nerve centers are connected, to a greater or less degree, with the surface of the body by the nerves, which serve as pathways between the end organs of touch, sight, taste, etc., and the centers in the brain or spinal cord. Thus sensation is obtained.

Sensations and Reactions. — We have already seen that simpler forms of life perform certain acts because certain outside forces acting upon them cause them to *react* to the stimulus from without. The one-celled animal responds to the presence of food, to heat, to oxygen, to other conditions in its surroundings. An earthworm is repelled by light, is attracted by food. All animals, including man, are put in touch with their surroundings by what we call the organs of sensation. The senses of man, besides those we commonly know as those of sight, hearing, taste, smell, and touch, are those of temperature, pressure, and pain. It is obvious that such organs, if they are to be of use to an animal, must be at the outside of the body. Thus we find eyes and ears in the head, and taste cells in the mouth, while other cells in the nose perceive odors, and still others in the skin are sensitive to heat or cold, pressure or pain.

But this is not all. Strangely enough, we do not see with our eyes or taste with our taste cells. These organs receive the sensations, and by means of a complicated system of greatly elongated cell structures, the message is sent inward, relayed by other elongated cells until the sensory message reaches an inner station, in the central nervous system. We see and hear and smell in our brain. Let us next examine the structure of the nerve cells or *neurons* part of which serve as pathways for these messages.

Neurones. — A nerve cell, like other cells in the body, is a mass of protoplasm containing a nucleus. But the body of the nerve cell is usually rather irregular in shape, and distinguished from most other cells by possessing several delicate, branched protoplasmic projections called *dendrites*. One of these processes, the axon, is much longer than the others and ends in a muscle or organ of sensation. The axon forms the pathway over which nervous impulses travel to and from the nerve centers.

A nerve consists of a bundle of such tiny axons, bound together by connective tissue. As a nerve ganglia is a center of activity in the nervous system, so a cell body is a center of activity which may send an impulse over this thin strand of protoplasm (the axon) prolonged many hundreds of thousands of times the length of the cell. Some neurones in the human body, although visible only under the compound microscope, give rise to axons several feet in length.

Because some bundles of axons originate in organs that receive sensations and send those sensations to the central nervous system, they are called *sensory nerves*. Other axons originate in the central nervous system and pass outward as nerves producing movement of muscles. These are called *motor nerves*.

Diagram of a neuron or nerve unit.

The Brain of Man. — In man, the central nervous system consists of a brain and spinal cord inclosed in a bony case. From the brain, twelve pairs of nerves are given off; thirty-one pairs more leave the spinal cord. The brain has three divisions. The *cerebrum* makes up the largest part. In this respect it differs from the cerebrum of the frog and other vertebrates. It is divided into two lobes, the *hemispheres*, which are connected with each other by a broad band of nerve fibers. The outer surface of the cerebrum is thrown into folds or *convolutions* which give a large surface, the cell bodies of the neurons being found in this part of the cerebrum. Holding the cell bodies and fibers in place is a kind of connective tissue. The inner part (white in color) is composed largely of fibers which pass to other parts of the brain and down into the spinal cord. Under the cerebrum, and dorsal to it, lies the little brain, or *cerebellum*. The two sides of the cerebellum are connected by a band of nerve fibers which run around into the lower hindbrain or *medulla*. This band of fibers is called the *pons*. The medulla is, in structure, part of the spinal cord, and is made up largely of fibers running longitudinally.

The Sympathetic Nervous System. — Connected with the central nervous system is that part of the nervous apparatus that controls the muscles of the digestive tract and blood vessels, the secretions of gland cells, and all functions which have to do with life processes in the body. This is called the sympathetic nervous system.

Functions of the Parts of the Central Nervous System of the Frog. — From careful study of living frogs, birds, and some mammals we have learned much of what we know of the functions of the parts of the central nervous system in man.

It has been found that if the entire brain of a frog is destroyed and separated from the spinal cord, "the frog will continue to live, but with a very peculiarly modified activity." It does not appear to breathe, nor does it swallow. It will not move or croak, but if acid is placed upon the skin so as to irritate it, the legs make movements to push away and to clean off the irritating substance. The spinal cord is thus shown to be a center for defensive movements. If the cerebrum is separated from the rest of the nervous system, the frog seems to act a little differently from the normal animal. It jumps when touched, and swims when placed in water. It will croak when stroked, or swallow if food be placed in its mouth. But it manifests no hunger or fear, and is in every sense a machine which will perform certain actions after certain stimulations. Its movements are automatic. If now we watch the movements of a frog which has the brain uninjured in any way, we find that it acts *spontaneously*. It tries to escape when caught. It feels hungry and seeks food. It is capable of voluntary action. It acts like a normal individual.

Functions of the Cerebrum. — In general, the functions of the different parts of the brain in man agree with those functions we have already observed in the frog. The cerebrum has to do with conscious activity; that is, thought. It presides over what we call our thoughts, our will, and our sensations. A large part of the area of the outer layer of the cerebrum seems to be given over to some one of the different functions of speech, hearing, sight, touch, movements of bodily parts. The movement of the

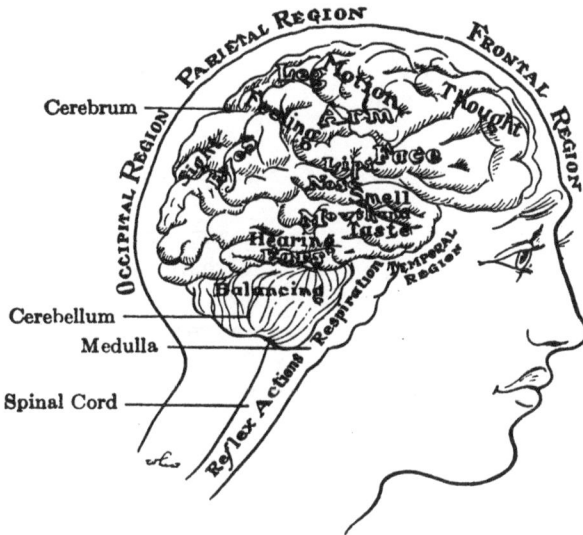

Diagram to show the parts of the brain and action of the different parts of the brain.

smallest part of the body appears to have its definite localized center in the cerebrum. Experiments have been performed on monkeys, and these, together with observations made on persons who had lost the power of movement of certain parts of the body, and who, after death, were found to have had diseases localized in certain parts of the cerebrum, have given to us our knowledge on this subject.

Reflex Actions; their Meaning. — If through disease or for other reasons the cerebrum does not function, no will power is

HUNTER, CIV. BI. — 23

exerted, nor are intelligent acts performed. All acts performed in such a state are known as *reflex actions*. The involuntary brushing of a fly from the face, or the attempt to move away from the source of annoyance when tickled with a feather, are examples of reflexes. In a reflex act, a person does not think before acting. The nervous impulse comes from the outside to cells that are not in the cerebrum. The message is short-circuited back to the surface by motor nerves, without ever having reached the thinking centers. The nerve cells which take charge of such acts are located in the cerebellum or spinal cord.

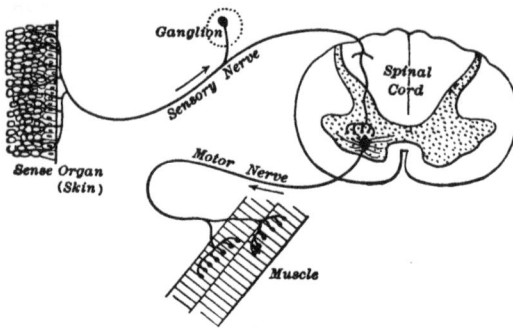

Diagram of the nerve path of a simple reflex action.

Automatic Acts. — Some acts, however, are learned by conscious thought, as writing, walking, running, or swimming. Later in life, however, these activities become automatic. The actual performance of the action is then taken up by the cerebellum, medulla, and spinal ganglia. Thus the thinking portion of the brain is relieved of part of its work.

Bundles of Habits. — It is surprising how little real thinking we do during a day, for most of our acts are habitual. Habit takes care of our dressing, our bathing, our care of the body organs, our methods of eating; even our movements in walking and the kind of hand we write are matters of habit forming. We are bundles of habits, be they good ones or bad ones.

Habit Formation. — The training of the different areas in the cerebrum to do their work well is the object of education. When we learned to write, we exerted conscious effort in order to make the letters. Now the act of forming the letters is done without thought. By training, the act has become automatic. In the beginning, a process may take much thought and many trials before we are able to complete it. After a little practice, the same process may become almost automatic. We have formed a habit. Habits are really acquired reflex actions. They are the result of nature's method of training. The conscious part of the brain has trained the cerebellum or spinal cord to do certain things that, at first, were taken charge of by the cerebrum.

Importance of Forming Right Habits. — Among the habits early to be acquired are the habits of studying properly, of concentrating the mind, of learning self-control, and, above all, of contentment. Get the most out of the world about you. Remember that the immediate effect in the study of some subjects in school may not be great, but the cultivation of correct methods of thinking may be of the greatest importance later in life. The man or woman who has learned how to concentrate on a problem, how to weigh all sides with an unbiased mind, and then to decide on what they believe to be best and right are the efficient and happy ones of their generation.

" The hell to be endured hereafter, of which theology tells, is no worse than the hell we make for ourselves in this world by habitually fashioning our characters in the wrong way. Could the young but realize how soon they will become mere walking bundles of habits, they would give more heed to their conduct while in the plastic state. We are spinning our own fates, good or evil, and never to be undone. Every smallest stroke of virtue or of vice leaves its never-so-little scar. The drunken Rip Van Winkle, in Jefferson's play, excuses himself for every fresh dereliction by saying, ' I won't count this time! ' Well! he may not count it, and a kind Heaven may not count it; but it is being counted none the less. Down among his nerve cells and fibers the molecules are counting it, registering and storing it up to be used against him when the next temptation comes. Nothing we ever do is, in strict scientific literalness, wiped out. Of course this has its good side as well as its bad one. As we become permanent drunkards by so many separate drinks, so we become saints in the moral, and authorities in the practical and scientific, spheres by so many separate acts and hours of work. Let no youth have any anxiety about the upshot of his education, whatever the line of it may be. If he keep faithfully busy each hour of the working day, he may safely leave the final result to itself. He can with perfect certainty count on waking up some fine morning, to find himself one of the competent ones of his generation, in whatever pursuit he may have singled out." — JAMES, *Psychology.*

Some Rules for Forming Good Habits. — Professor Horne gives several rules for making good or breaking bad habits. They are : " First, *act on every opportunity.* Second, *make a strong start.* Third, *allow no exception.* Fourth, *for the bad habit establish a good one.* Fifth, summoning all the man within, *use effort of will.*" Why not try these out in forming some good habit? You will find them effective.

Necessity of Food, Fresh Air, and Rest. — The nerve cells, like all other cells in the body, are continually wasting away and being rebuilt. Oxidation of food material is more rapid when we do mental work. The cells of the brain, like muscle cells, are not only capable of fatigue, but show this in changes of form and of contents. *Food* brought to them in the blood, plenty of *fresh air*, especially when engaged in active brain work, and *rest* at proper times, are essential in keeping the nervous system in condition. One of the best methods of resting the brain cells is a change of occupation. Tennis, golf, baseball, and other outdoor sports combine muscular exercise with brain activity of a different sort from that of business or school work. But change of occupation will not rest exhausted neurones. For this, sleep is necessary. Especially is sleep an important factor in the health of the nervous system of growing children.

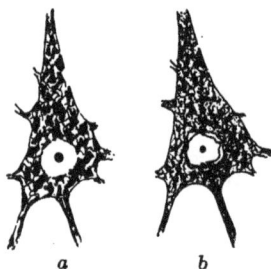

The effect of fatigue on nerve cells. *a,* healthy brain cell; *b,* fatigued brain cell.

Necessity of Sleep. — Most brain cells attain their growth early in life. Changes occur, however, until some time after the school age. Ten hours of sleep should be allowed for a child, and at least eight hours for an adult. At this time, only, do the brain cells have opportunity to rest and store food and energy for their working period.

Sleep is one way in which all cells in the body, and particularly those of the nervous system, get their rest. The nervous system, by far the most delicate and hardest-worked set of tissues in the body, needs rest more than do other tissues, for its work directing the body only ends with sleep or unconsciousness. The afternoon nap, snatched by the brain worker, gives him renewed energy for his evening's work. It is not hard application to a task that wearies the brain; it is *continuous* work without rest.

THE SENSES

Touch. — In animals having a hard outside covering, such as certain worms, insects, and crustaceans, minute hairs, which are sensitive to touch, are found growing out from the body covering. At the base of these hairs are found neurones which send axons inward to the central nervous system.

Organs of Touch. — In man, the nervous mechanism which governs touch is located in the folds of the dermis or in the skin. Special nerve endings, called the *tactile corpuscles*, are found there, each inclosed in a sheath or capsule of connective tissue. Inside is a complicated nerve ending, and axons pass inward to the central nervous system. The number of tactile corpuscles present in a given area of the skin determines the accuracy and ease with which objects may be known by touch.

If you test the different parts of the body, as the back of the hand, the neck, the skin of the arm, of the back, or the tip of the tongue, with a pair of open

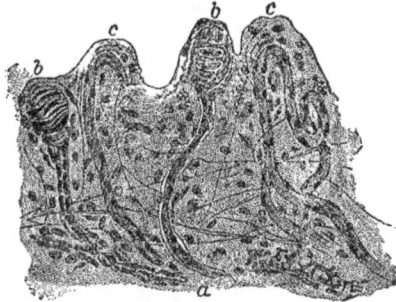

Nerves in the skin: *a*, nerve fiber; *b*, tactile papillæ, containing a tactile corpuscle; *c*, papillæ containing blood vessels. (After Benda.)

dividers, a vast difference in the accuracy with which the two points may be distinguished is noticed. On the tip of the tongue, the two points need only be separated by $\frac{1}{24}$ of an inch to be so distinguished. In the small of the back, a distance of 2 inches may be reached before the dividers feel like two points.

Temperature, Pressure, Pain. — The feeling of temperature, pressure, and pain is determined by different end organs in the skin. Two kinds of nerve fibers exist in the skin, which give distinct sensations of heat and cold. These nerve endings can be located by careful experimentation. There are also areas of nerve endings which are sensitive to pressure, and still others, most numerous of all, sensitive to pain.

Taste Organs. — The surface of the tongue is folded into a number of little projections known as papillæ. These may be more easily found on your own tongue if a drop of vinegar is placed on its broad surface. In the folds, between these projections on the top and back part of the tongue, are located the organs of taste. These organs are called *taste buds*.

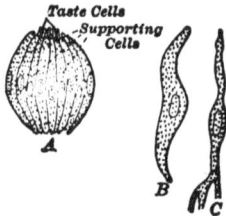

Each taste bud consists of a collection of spindle-shaped neurones, each cell tipped at its outer end with a hairlike projection. These cells send inward fibers to other cells, the fibers from which ultimately reach the brain. The sensory cells are surrounded by a number of projecting cells which are arranged in layers about them. Thus the organ in longitudinal section looks somewhat like an onion cut lengthwise.

A, isolated taste bud, from whose upper free end project the ends of the taste cells; *B*, supporting or protecting cell; *C*, sensory cell.

How we Taste. — Four kinds of substances may be distinguished by the sense of taste. These are sweet, sour, bitter, and salt. Certain taste cells located near the back of the tongue are stimulated only by a bitter taste. Sweet substances are perceived by cells near the tip of the tongue, sour substances along the sides, and salt about equally all over the surface. A substance must be dissolved in fluid in order to be tasted. Many things which we believe we taste are in reality perceived by the sense of smell. Such are spicy sauces and flavors of meats and vegetables. This may easily be proved by holding the nose and chewing, with closed eyes, several different substances, such as an apple, an onion, and a raw potato.

Smell. — The sense of smell is located in the membrane lining the upper part of the nose. Here are found a large number of rod-shaped cells which are connected with the brain by means of the olfactory nerve. In order to perceive odors, it is necessary to have them diffused in the air; hence we sniff so as to draw in more air over the olfactory cells.

The Organ of Hearing. — The organ of hearing is the ear. The outer ear consists of a funnel-like organ composed largely of cartilage which is of use in collecting sound waves. This part of the ear incloses the auditory canal, which is closed at the inner end by a tightly stretched membrane, the *tympanic membrane* or ear drum. The function of the tympanic membrane is to receive sound waves, for all sound is caused by vibrations in the air, these vibrations being transmitted, by the means of a complicated apparatus found in the middle ear, to the real organ of hearing located in the inner ear.

Middle Ear. — The middle ear in man is a cavity inclosed by the temporal bone, and separated from the outer ear by the tympanic membrane. A little tube called the *Eustachian tube* connects the inner ear with the mouth cavity. By allowing air to enter from the mouth, the air pressure is equalized on the ear drum. For this reason, we open the mouth at the time of a heavy concussion and thus prevent the rupture of the delicate tympanic membrane.

Placed directly against the tympanic membrane and connecting it with the inner ear is a chain of three tiny bones, the smallest bones of the body. The outermost is called the *hammer;* the next the *anvil;* the third the *stirrup.* All three bones are so called from their resemblances in shape to the articles for which they are named. These bones are held in place by very small muscles which are delicately adjusted so as to tighten or relax the membranes guarding the middle and inner ear.

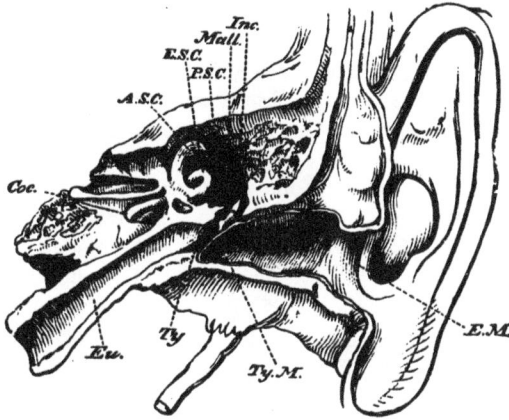

Section of ear: *E.M.*, auditory canal; *Ty.M.*, tympanic membrane; *Eu.*, Eustachian tube; *Ty*, middle ear; *Coc.*, *A.S.C.*, *E.S.C.*, etc., internal ear.

The Inner Ear. — The inner ear is one of the most complicated, as well as one of the most delicate, organs of the body. Deep within the temporal bone there are found two parts, one of which is called, collectively, the *semicircular canal region*, the other the *cochlea*, or organ of hearing.

It has been discovered by experimenting with fish, in which the semicircular canal region forms the chief part of the ear, that this region has to do with the equilibrium or balancing of the body. We gain in part our knowledge of our position and movements in space by means of the *semicircular canals*.

That part of the ear which receives sound waves is known as the *cochlea*, or snail shell, because of its shape. This very complicated organ is lined with sensory cells provided with cilia. The cavity of the cochlea is filled with a fluid. It is believed that somewhat as a stone thrown into water causes ripples to emanate from the spot where it strikes, so sound waves are transmitted by means of the fluid filling the cavity to the sensory cells of the cochlea (collectively known as the *organ of Corti*) and thence to the brain by means of the auditory nerve.

The Character of Sound. — When vibrations which are received by the ear follow each other at regular intervals, the sound is said to be musical. If the vibrations come irregularly, we call the sound a noise. If the vibrations come slowly, the pitch of the sound is low; if they come rapidly, the pitch is high. The ear is able to perceive as low as thirty vibrations per second and as high as almost thirty thousand. The ear can be trained to recognize sounds which are unnoticed in untrained ears.

The Eye. — The eye or organ of vision is an almost spherical body which fits into a socket of bone, the *orbit*. A stalklike structure, the *optic nerve*, connects the eye with the brain. Free movement is obtained by means of six little muscles which are attached to the outer coat, the *eyeball*, and to the bony socket around the eye.

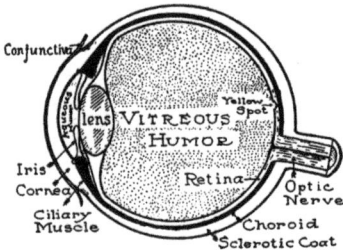

Longitudinal section through the eye.

The wall of the eyeball is made up of three coats. An outer tough white coat, of connective tissue, is called the *sclerotic coat*. Under the sclerotic coat, in front, the eye bulges outward a little. Here the outer coat is continuous with a transparent tough layer called the *cornea*. A second coat, the *choroid*, is supplied with blood vessels and cells which bear pigments. It is a part of this coat which we see through the cornea as the colored part of the eye (the *iris*). In the center of the iris is a small circular hole (the *pupil*). The iris is under the control of muscles, and may be adjusted to varying amounts of light, the hole becoming larger in dim light, and smaller in bright light. The inmost layer of the eye is called the *retina*. This is, perhaps, the most delicate layer in the entire body. Despite the fact that the retina is less than $\frac{1}{10}$ of an inch in thickness, there are several layers of cells in its composition. The optic nerve enters the eye from behind and spreads out to form the surface of the retina. Its finest fibers are ultimately connected with numerous elongated cells which are stimulated by light. The retina is dark purple in color, this color being caused by a layer of cells next to the choroid coat. This accounts for the black appearance of the pupil of the eye, when we look through the pupil into the darkened space within the eyeball. The retina acts as the sensitized plate in the camera, for on it are received the impressions which are transformed and sent to the brain as sensations of sight. The eye, like the camera, has a lens. This lens is formed of transparent, elastic material. It is found directly behind the iris and is attached to the choroid coat by means of delicate ligaments. In front of the lens is a small cavity filled with a watery fluid, the *aqueous humor*, while behind it is the main cavity of the eye, filled with a transparent, almost jellylike, *vitreous humor*. The lens itself is elastic. This circumstance permits of a change of form and, in consequence, a change of focus upon the retina of the lens. By means of this change in form, or *accommodation*, we are able to distinguish between near and distant objects.

Defects in the Eye. — In some eyes, the lens is in focus for near objects, but is not easily focused upon distant objects; such an eye is said to be nearsighted. Other eyes which do not focus clearly on objects near at hand are said to be farsighted. Still another eye defect is astigmatism, which causes images of lines in a certain direction

Y F E V

How far away can you read these letters? Measure the distance.   Twenty feet is a test for the normal eye.

to be indistinct, while images of lines transverse to the former are distinct. Many nervous troubles, especially headaches, may be due to eye strain. We should have our eyes examined from time to time, especially if we are subject to headaches.

The Alcohol Question. — It is agreed by investigators that in large or continued amounts alcohol has a narcotic effect; that it first dulls or paralyzes the nerve centers which control our judgment, and later acts upon the so-called motor centers, those which control our muscular activities.

The reason, then, that a man in the first stages of intoxication talks rapidly and sometimes wittily, is because the centers of judgment are paralyzed. This frees the speech centers from control exercised by our judgment, with the resultant rapid and free flow of speech.

In small amounts alcohol is believed by some physiologists to have always this same narcotic effect, while other physiologists think that alcohol does stimulate the brain centers, especially the higher centers, to increased activity. Some scientific and professional men use alcohol in small amounts for this stimulation and report no seeming harm from the indulgence. Others, and by far the larger number, agree that this stimulation from alcohol is only apparent and that even in the smallest amounts alcohol has a narcotic effect.

The Paralyzing Effects of Alcohol on the Nervous System. — Alcohol has the effect of temporarily paralyzing the nerve centers. The first effect is that of exhilaration. A man may do more work for a time under the stimulation of alcohol. This stimulation, however, is of short duration and is invariably followed by a period of depression and inertia. In this latter state, a man will do less work than before. In larger quantities, alcohol has the effect of completely paralyzing the nerve centers. This is seen in the case of a man " dead drunk." He falls in a stupor because all of the centers governing speech, sight, locomotion, etc., have been temporarily paralyzed. If a man takes a very large amount of alcohol, even the nerve centers governing respiration and circulation may become poisoned, and the victim will die.

Effect on the Organs of Special Sense. — Professor Forel, one of the foremost European experts on the question of the effect of alcohol on the nervous system, says: " Through all parts of nervous activity from the innervation of the muscles and the simplest sensation to the highest activity of the soul the paralyzing effect of alcohol can be demonstrated." Several experimenters of undoubted ability have noted the paralyzing effect of alcohol even in small doses. By the use of delicate instruments of precision, Ridge tested the effect of alcohol on the senses of smell, vision, and muscular sense of weight. He found that two drams of absolute alcohol produced a positive decrease in the sensitiveness of the nerves of feeling, that so small a quantity as one half dram of absolute alcohol diminished the power of vision and the muscular sense of weight. Kraepelin and Kurz by experiment determined that the acuteness of the special senses of sight, hearing, touch, taste, and smell was diminished by an ounce of alcohol, the power of vision being lost to one third of its extent and a similar effect being produced on the other special senses. Other investigators have reached like conclusions. There is no doubt but that alcohol, even in small quantities, renders the organs of sense less sensitive and therefore less accurate.

Effect of Alcohol on the Ability to Resist Disease. — Among certain classes of people the belief exists that alcohol in the form of brandy or some other drink or in patent medicines, malt tonics,

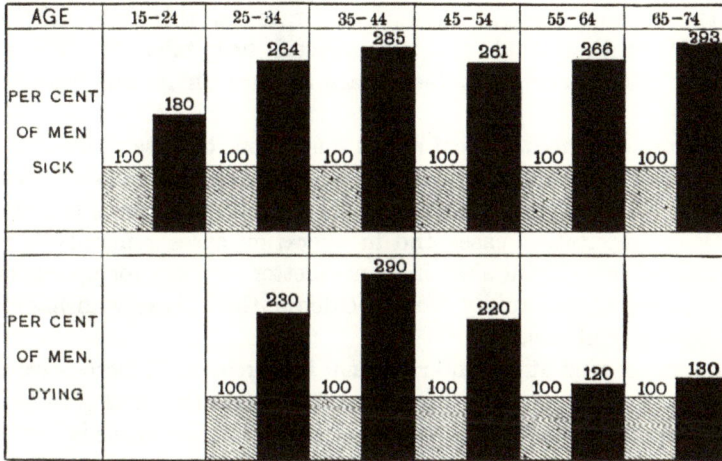

AGE	15-24	25-34	35-44	45-54	55-64	65-74
PER CENT OF MEN SICK	180 / 100	264 / 100	285 / 100	261 / 100	266 / 100	293 / 100
PER CENT OF MEN DYING	/ 100	230 / 100	290 / 100	220 / 100	120 / 100	130 / 100

Table to show a comparison of chances of illness and death in drinkers and non-drinkers. Solid black, drinkers. (From German sources.)

and the like is of great importance in building up the body so as to resist disease or to cure it after disease has attacked it. Nothing is further from the truth. In experiments on a large number of animals, including dogs, rabbits, guinea pigs, fowls, and pigeons, Laitenen, of the University of Helsingsfors, found that alcohol, without exception, made these animals more susceptible to disease than were the controls.

One of the most serious effects of alcohol is the lowered resistance of the body to disease. It has been proved that a much larger proportion of hard drinkers die from infectious or contagious diseases than from special diseased conditions due to the direct action of alcohol on the organs of the body. This lowered resistance is shown in increased liability to contract disease and increased severity of the disease. We have already alluded to the findings of insurance companies with reference to the length of life — the abstainers from alcohol have a much better chance of a longer life and much less likelihood of infection by disease germs.

Use of Alcohol in the Treatment of Disease. — In the London Temperance Hospital alcohol was prescribed seventy-five times in thirty-three years. The death rate in this hospital has been lower than that of most general hospitals. Sir William Collins, after serving nineteen years as surgeon in this hospital, said : —

" In my experience, speaking as a surgeon, the use of alcohol is not essential for successful surgery. . . . At the London Temperance Hospital, where alcohol is very rarely prescribed, the mortality in amputation cases and in operation cases generally is remarkably low. Total abstainers are better subjects for operation, and recover more rapidly from accidents, than those who habitually take stimulants."

In a paper read at the International Congress on Tuberculosis, in New York, 1906, Dr. Crothers remarked that alcohol as a remedy or a preventive medicine in the treatment of tuberculosis is a most dangerous drug, and that all preparations of sirups containing spirits increase, rather than diminish, the disease.

Dr. Kellogg says : " The paralyzing influence of alcohol upon the white cells of the blood — a fact which is attested by all investigators — is alone sufficient to condemn the use of this drug in acute or chronic infections of any sort."

The Effect of Alcohol upon Intellectual Ability. — With regard to the supposed quickening of the mental processes Horsley and Sturge, in their recent book, *Alcohol and the Human Body*, say : " Kraepelin found that the simple reaction period, by which is meant the time occupied in making a mere response to a signal, as, for instance, to the sudden appearance of a flag, was, after the ingestion of a small quantity of alcohol ($\frac{1}{4}$ to $\frac{1}{2}$ ounce), slightly accelerated ; that there was, in fact, a slight shortening of the time, as though the brain were enabled to operate more quickly than before. But he found that after a few minutes, in most cases, a slowing of mental action began, becoming more and more marked, and enduring as long as the alcohol was in active operation in the body, *i.e.* four to five hours. . . . Kraepelin found that it was only more or less automatic work, such as reading aloud, which was quickened by alcohol, though even this was rendered less trustworthy and accurate." Again: "Kraepelin had always shared the popular belief that a small quantity of alcohol (one to two teaspoonfuls) had an accelerating effect on the activity of his mind,

Conditions.	Average number figures memorized
Six non-alcohol days.	1280.66
Twelve alcohol days.	1063.3
Seven non-alcohol days	1830.8
Two alcohol days.	1086

Effect of use of alcohol on memory.

enabling him to perform test operations, as the adding and subtracting and learning of figures more quickly. But when he came to measure with his instruments the exact period and time occupied, he found, to his astonishment, that he had accomplished these mental operations, not more, but less, quickly than before. . . . Numerous further experiments were carried out in order to test this matter, and these proved that *alcohol lengthens the time taken to perform complex mental processes*, while by a singular illusion the person experimented upon imagines that his psychical actions are rendered more rapid."

Attention — that is, the power of the mind to grasp and consider impressions obtained through the senses — is weakened by drink. The ability of the mind to associate or combine ideas, the faculty involved in sound *judgment*, showed that when the persons had taken the amounts of alcohol mentioned, the combinations of ideas or judgments expressed by them were confused, foggy, sentimental, and general. When the persons had taken no alcohol,

	Conditions	Average time in minutes
before breakfast.	10 days WITHOUT alcohol	18 min. 2 sec.
before breakfast.	8 days WITH alcohol	30 min 48 sec.
after breakfast.	42 days WITHOUT alcohol	21 min. 47 sec.
after breakfast.	26 days WITH alcohol	24 min 16 sec.

The effect of alcohol upon ability to do mental work.

their judgments were rational, specific, keen, showing closer observation.

" The words of Professor Helmholtz at the celebration of his seventieth birthday are very interesting in this connection. He spoke of the ideas flashing up from the depths of the unknown soul, that lies at the foundation of every truly creative intellectual production, and closed his account of their origin with these words: ' The smallest quantity of an alcoholic beverage seemed to frighten these ideas away.' " — DR. G. SIMS WOODHEAD, Professor of Pathology, Cambridge University, England.

Professor Von Bunge (*Textbook of Physiological and Pathological Chemistry*) of Switzerland says that: " The stimulating action which alcohol appears to exert on the brain functions is only a paralytic action. The cerebral functions which are first interfered with are *the power of clear judgment and reason.* No man ever became witty by aid of spirituous drinks. The lively gesticulations and useless exertions of intoxicated people are due to paralysis, — the restraining influences, which prevent a sober man from uselessly expending his strength, being removed."

The Drink Habit. — The harmful effects of alcohol (aside from the purely physiological effect upon the tissues and organs of the body) are most terribly seen in the formation of the alcohol habit. The first effect of drinking alcoholic liquors is that of exhilaration. After the feeling of exhilaration is gone, for this is a temporary state, the subject feels depressed and less able to work than before he took the drink. To overcome this feeling, he takes another drink. The result is that before long he finds a habit formed from which he cannot escape. With body and mind weakened, he attempts to break off the habit. But meanwhile his will, too, has suffered from overindulgence. He has become a victim of the drink habit!

" The capital argument against alcohol, that which must eventually condemn its use, is this, that *it takes away all the reserved control, the power of mastership, and therefore offends against the splendid pride in himself or herself, which is fundamental in every man or woman worth anything.*" — DR. JOHN JOHNSON, quoting Walt Whitman.

Self-indulgence, be it in gratification of such a simple desire as that for candy or the more harmful indulgence in tobacco or alcoholic beverages, is dangerous — not only in its immediate effects on the tissues and organs, but in its more far-reaching effects on habit formation. Each one of us is a bundle of appetites. If we gratify appetites of the wrong kind, we are surely laying the foundation for the habit of excess. Self-denial is a good thing for each of us to practice at one time or another, if for no other purpose than to be ready to fight temptation when it comes.

The Economic Effect of Alcoholic Poisoning. — In the struggle for existence, it is evident that the man whose intellect is the quickest and keenest, whose judgment is most sound, is the man who is most likely to succeed. The paralyzing effect of alcohol upon the nerve centers must place the drinker at a disadvantage. In a hundred ways, the drinker sooner or later feels the handicap that the habit of drink has imposed upon him. Many corporations, notably several of our greatest railroads (the Pennsylvania and the New York Central Railroad among them), refuse to employ any but abstainers in positions of trust. Few persons know the number of railway accidents due to the uncertain eye of some engineer who mistook his signal, or the hazy inactivity of the brain of some train dispatcher who, because of drink, forgot to send the telegram that was to hold the train from wreck. In business and in the professions, the story is the same. The abstainer wins out over the drinking man.

Effect of Alcohol on Ability to do Work. — In *Physiological Aspects of the Liquor Problem*, Professor Hodge, formerly of Clark University, describes many of his own experiments showing the effect of alcohol on animals. He trained four selected puppies to recover a ball thrown across a gymnasium. To two of the dogs he gave food mixed with doses of alcohol, while the others were fed normally. The ball was thrown 100 feet as rapidly as recovered. This was repeated 100 times each day for fourteen successive days. Out of 1400 times the dogs to which alcohol had been given brought back the ball only 478 times, while the others secured it 922 times.

Dr. Parkes experimented with two gangs of men, selected to be as nearly similar as possible, in mowing. He found that with one gang abstaining from alcoholic drinks and the other not, the abstaining gang could accomplish more. On transposing the gangs, the same results were repeatedly obtained. Similar results were obtained by Professor Aschaffenburg of Heidelberg University, who found experimentally that men " were able to do 15 per cent less work after taking alcohol."

Recently many experiments along the same lines have been made. In typewriting, in typesetting, in bricklaying, or in the highest type of mental work the result is the same. The quality and quantity of work done on days when alcohol is taken is less than on days when no alcohol is taken.

The Relation of Alcohol to Efficiency. — We have already seen that work is neither so well done nor is as much accomplished by drinkers as by non-drinkers.

A Massachusetts shoe manufacturer told a recent writer on temperance that in one year his firm lost over $5000 in shoes spoiled by drinking men, and that he had himself traced these spoiled shoes to the workmen who, through their use of alcoholic liquors, had thus rendered themselves incapable. This is a serious handicap to our modern factory system, and explains why so many factory towns and cities are strongly favoring a policy of " No license " in opposition to the saloons.

" It is believed that the largest number of accidents in shops and mills takes place on Monday, because the alcohol that is drunk on Sunday takes away the skill and attentive care of the workman. To prove the truth of this opinion, the accidents of the building trades in Zurich were studied during a period of six years, with the result shown by this table " : —

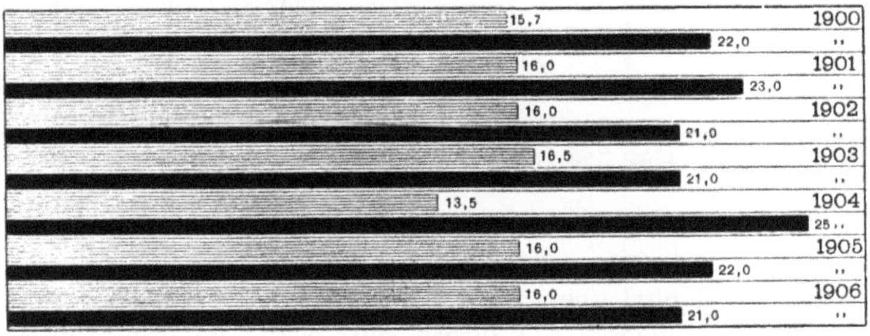

15,7		1900
	22,0	''
16,0		1901
	23,0	''
16,0		1902
	21,0	''
16,5		1903
	21,0	''
13,5		1904
	25,,	''
16,0		1905
	22,0	''
16,0		1906
	21,0	''

(From Tolman, *Hygiene for the Worker.*)
Shaded, non-alcoholic; black, alcoholic, accidents.

Another relation to efficiency is shown by the following chart. During the week the curve of working efficiency is highest on Friday and lowest on Monday. The number of accidents were also least on Friday and greatest on Monday. Lastly the assaults were fewest in number on Friday and greatest on Sunday and Monday The moral is plain. Workingmen are apt to spend their week's wages freely on Saturday. Much of this goes into drink, and as a result comes crime on Sunday because of the deadened moral and

HUNTER, CIV. BI. — 24

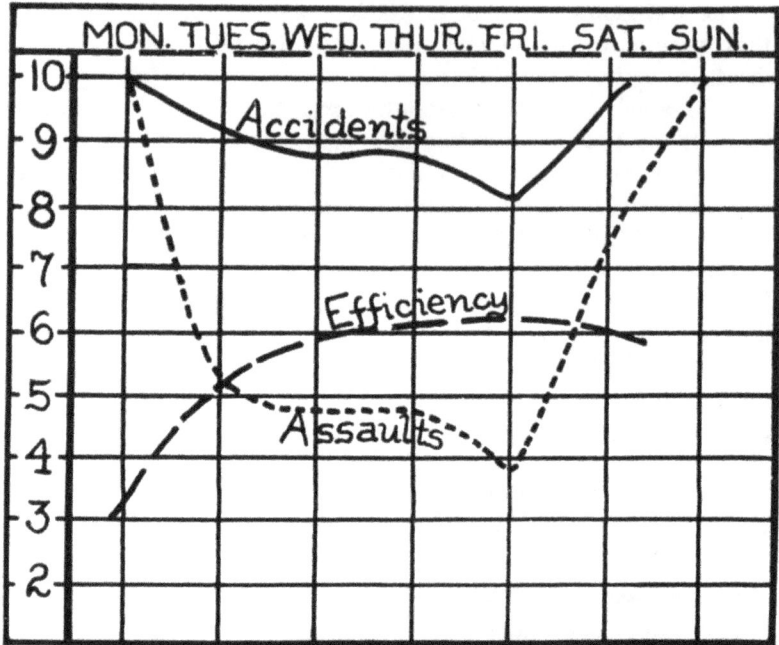

Notice that the curve of efficiency is lowest on Monday and that crimes and accidents are most frequent on Sunday and Monday. Account for this.

mental condition of the drinker, and loss of efficiency on Monday, because of the poisonous effects of the drug.

Effect of Alcohol upon Duration of Life. — Still more serious is the relation of alcohol as a direct cause of disease (see table).

It is as yet quite impossible, in the United States at least, to tell just how many deaths are brought about, directly or indirectly, by alcohol. Especially is this true in trying to determine the number of cases of deaths from disease promoted by alcohol. In Switzerland provision is made for learning these facts, and the records of that country throw some light on the subject.

Dr. Rudolph Pfister made a study of the records of the city of Basle for the years 1892–1906, finding the percentage of deaths in which alcohol had been reported by the attending physician as one cause of death. He found that 18.1 per cent of all deaths of men between 40 and 50 years of age were caused, in part at least, by alcohol, and this at what should be the most active period in a man's life, the time when he is most needed by his family and community. Taking all ages between 20 and 80, he found that alcohol was one cause of death in one man in every ten who died.

Another study was made by a certain doctor in Sweden, from records of 1082 deaths occurring in his own practice and the local hospital. No case was counted as alcoholic of which there was the slightest doubt. Of deaths of adult men, 18 in every 100 were due, directly or indirectly, to alcoholism. In middle life, between the ages of 40 and 50, 29 ; and between 50 and 60 years of age, 25.6 out of every 100 deaths had alcohol as one cause, thus agreeing

15721	17418	
ALCOHOLISM + ALCOHOLIC LIVER CIRRHOSIS		33,139
22,211		
TYPHOID		
2214		
SMALLPOX		

with other statistics we have been quoting. — From the *Metropolitan*, Vol. XXV, Number 11.

The Relation of Alcohol to Crime. — A recent study of more than 2500 habitual users of alcohol showed that over 66 per cent had committed crime. Usually the crimes had been done in saloons or as a result of quarrels after drinking. Of another lot of 23,581 criminals questioned, 20,070 said that alcohol had led them to commit crime.

The Relation of Alcohol to Pauperism. — We have already spoken of the Jukes family. These and many other families of a similar sort are more or less directly a burden upon the state. Alcohol is in part at least responsible for the condition of such families. Alcohol weakens the efficiency and moral courage, and thus leads to begging, pauperism, petty stealing or worse, and ul-

COUNTRY	PERCENTAGE.
	10 20 30 40 50 60 70 80 90 100
BELGIUM	
ENGLAND	
FRANCE	
GERMANY	
UNITED STATES	
HOLLAND	

The proportion of crime due to alcohol is shown in black.

timately to life in some public institution. In Massachusetts, of 3230 inmates of such institutions, 66 per cent were alcoholics.

The Relation of Alcohol to Heredity. — Perhaps the gravest side of the alcohol question lies here. If each one of us had only himself to think of, the question of alcohol might not be so serious. But drinkers may hand down to their unfortunate children tendencies toward drink as well as nervous diseases of various sorts; an alcoholic parent may beget children who are epileptic, neurotic, or even insane.

In the State of New York there are at the present time some 30,000 insane persons in public and private hospitals. It is believed that about one fifth of them, or 6000 patients, owe their insanity to alcohol used either by themselves or by their parents. In the asylums of the United States there are 150,000 insane people. Taking the same proportions as before, there are 30,000 persons in this country whom alcohol has made or has helped to make insane. This is the most terrible side of the alcohol problem.

REFERENCE READING

ELEMENTARY

Hunter, *Laboratory Problems in Civic Biology*. American Book Company.
Overton, *General Hygiene*. American Book Company.
The Gulick Hygiene Series, *Emergencies, Good Health, The Body at Work, Control of Body and Mind*. Ginn and Company.
Ritchie, *Human Physiology*. World Book Company.
Hough and Sedgwick, *The Human Mechanism*. Ginn and Company.

XXIV. MAN'S IMPROVEMENT OF HIS ENVIRONMENT

Problems.—How may we improve our home conditions of living?

How may we help improve our conditions at school?

How does the city care for the improvement of our environ ment?

(a) In inspection of buildings, etc.

(b) In inspection of food supplies.

(c) In inspection of milk.

(d) In care of water supplies.

(e) In disposal of wastes.

(f) In care of public health.

LABORATORY SUGGESTIONS

Home exercise. — How to ventilate my bedroom.

Demonstration. — Effect of use of duster and damp cloth upon bacteria in schoolroom.

Home exercise. — Luncheon dietaries.

Home exercise. — Sanitary map of my own block.

Demonstration. — The bacterial content of milk of various grades and from different sources.

Demonstration. — Bacterial content of distilled water, rain water, tap water, dilute sewage.

Laboratory exercise. — Study of board of health tables to plot curves of mortality from certain diseases during certain times of year.

The Purpose of this Chapter. — In the preceding chapters we have traced the lives of both plants and animals within their own environment. We have seen that man, as well as plants and other animals, needs a favorable environment in order to live in comfort and health. It will be the purpose of the following pages first to show how we as individuals may better our home environment, and secondly, to see how we may aid the civic authorities in the betterment of conditions in the city in which we live.

Home Conditions. — The Bedroom. — We spend about one third of our total time in our bedroom. This room, therefore, deserves more than passing attention. First of all, it should have good ventilation. Two windows make an ideal condition, especially if the windows receive some sun. Such a condition as this is manifestly impossible in a crowded city, where too often the apartment bedrooms open upon narrow and ill-ventilated courts. Until comparatively recent time, tenement houses were built so that the bedrooms had practically no light or air; now, thanks to good tenement-house laws, wide airshafts and larger windows are required by statute.

How I should ventilate my bedroom.

Care of the Bedroom. — Since sunlight cannot always be obtained for a bedroom, we must so care for and furnish the room that it will be difficult for germs to grow there. Bedroom furniture should be light and easy to clean, the bedstead of iron, the floors painted or of hardwood. No hangings should be allowed at the windows to collect dust, nor should carpets be allowed for the same reason. Rugs on the floor may easily be removed when cleaning is done. The furniture and woodwork should be wiped with a damp cloth every day. Why a *damp* cloth? In certain tenements in New York City, tuberculosis is believed to have been spread by people occupying rooms in which a previous tenant has had tuberculosis. A new tenant should insist on a thorough cleaning of the bedrooms and removal of old wall paper before occupancy.

Sunlight Important. — In choosing a house in the country we would take a location in which the sunlight was abundant. A shaded location might be too damp for health. Sunlight should enter at least some of the rooms. In choosing an apartment we should have this matter in mind, for, as we know, germs cannot long exist in sunlight.

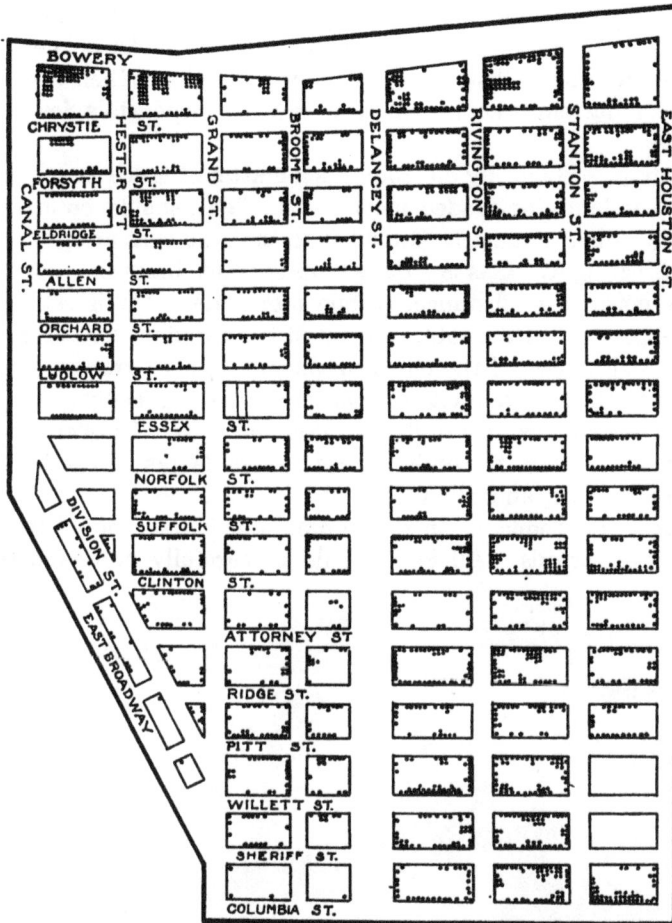

This map shows how cases of tuberculosis are found recurring in the same locality and in the same houses year after year. Each black dot is one case of tuberculosis.

Heating. — Houses in the country are often heated by open fires, stoves or hot-air furnaces, all of which make use of heated currents of air to warm the rooms. But in the city apartments, usually pipes conduct steam or hot water from a central plant to our rooms. The difficulty with this system is that it does not give us fresh air, but warms over the stale air in a room. Steam causes our rooms to be too warm part of the time, and not warm enough part of the time. Thus we become overheated and then take cold by becoming chilled. Steam heat is thus responsible for much sickness.

Lighting. — Lighting our rooms is a matter of much importance. A student lamp, or shaded incandescent light, should be used for reading. Shades must be provided so that the eyes are protected from direct light. Gas is a dangerous servant, because it contains a very poisonous substance, carbon monoxide. " It is estimated that 14 per cent of the total product of the gas plant leaks into the streets and houses of the cities supplied." This forms an unseen menace to the health in cities. Gas pipes, and especially gas cocks, should be watched carefully for escaping gas. Rubber tubing should not be used to conduct gas to movable gas lamps, because it becomes worn and allows gas to escape.

Insects and Foods. — In the summer our houses should be provided with screens. All food should be carefully protected from

During the summer all food should be protected from flies. Why?

flies. Dirty dishes, scraps of food, and such garbage should be quickly cleaned up and disposed of after a meal. Insect powder (pyrethrum) will help keep out "croton bugs" and other undesirable household pests, but cleanliness will do far more. Most kitchen pests, as the roach, simply stay with us because they find dirt and food abundant.

Use of Ice. — Food should be properly cared for at all times, but especially during the summer. Iceboxes are a necessity, especially where children live, in order to keep milk fresh. A dirty icebox is almost as bad as none at all, because food will decay or take on unpleasant odors from other foods.

Disposal of Wastes. — In city houses the disposal of human wastes is provided for by a city system of sewers. The wastes from the kitchen, the garbage, should be disposed of each day. The garbage pail should be frequently sterilized by rinsing it with boiling water. Plenty of lye or soap should be used. Remember that flies frequent the uncovered garbage pail, and that they may next walk on your food. Collection and disposal of garbage is the work of the municipality.

The wrong and the right kind of garbage cans.

School Surroundings. — How to Improve Them. — From five to six hours a day for forty weeks is spent by the average boy or girl in the schoolroom. It is part of our environment and should therefore be considered as worthy of our care. Not only should a schoolroom be attractive, but it should be clean and sanitary. City schools, because of their locations, of the sometimes poor janitorial service, and especially because of the selfishness and carelessness of children who use them, may be very dirty and unsanitary. Dirt and dust breed and carry bacteria. Plate cultures show greatly increased numbers of bacteria to be in the air when pupils are moving about, for then dust, bearing bacteria, is stirred up and circulated through the air. Sweeping and dusting with dry brooms or feather dusters only stirs up the dust, leaving it to settle in some other place with its load of bacteria. Professor Hodge tells of an experience in a school in Worcester, Mass. A health brigade was formed among the children, whose duty was to clean the rooms every morning by wiping all exposed surface with a damp

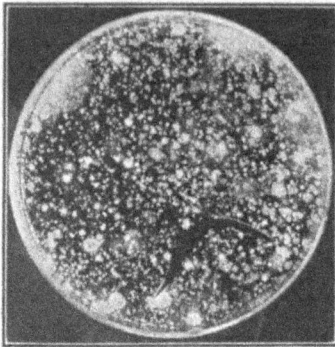 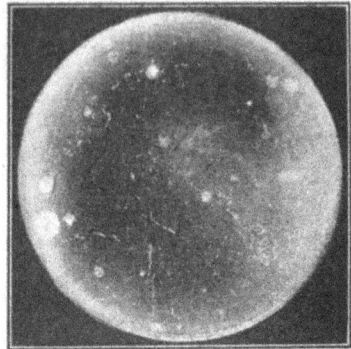

A   B

The culture (A) was exposed to the air of a dirty street in the crowded part of Manhattan. (B) was exposed to the air of a well-cleaned and watered street in the uptown residence portion. Which culture has the more colonies of bacteria? How do you account for this?

cloth. In a school of 425 pupils not a single case of contagious diseases appeared during the entire year. Why not try this in your own school?

Unselfishness the Motto. — Pupils should be unselfish in the care of a school building. Papers and scraps dropped by some careless boy or girl make unpleasant the surroundings for hundreds of others. Chalk thrown by some mischievous boy and then tramped underfoot may irritate the lungs of a hundred innocent schoolmates. Colds or worse diseases may be spread through the filthy habits of some boys who spit in the halls or on the stairways.

Lunch Time and Lunches. — If you bring your own lunch to school, it should be clean, tasty, and well balanced as a ration. In most large schools well-managed lunch rooms are part of the school equipment, and balanced lunches can be obtained at low cost. Do not make a lunch entirely from cold food, if hot can be obtained. Do not eat only sweets. Ice cream is a good food, if taken with something else, but be sure of your ice cream. "Hokey pokey" cream, tested in a New York school laboratory, showed the presence of many more colonies of bacteria than *good* milk would show. Above all, be sure the food you buy is clean. Stands on the street, exposed to dust and germs, often sell food far from fit for human consumption.

A sensible lunch box, sanitary and compact.

If you eat your lunch on the street near your school, remember not to scatter refuse. Paper, bits of lunch, and the like scattered on the streets around your school show lack of school spirit and lack of civic pride. Let us learn above all other things to be good citizens.

Inspection of Factories, Public Buildings, etc. — It is the duty of a city to inspect the condition of all public buildings and especially of factories. Inspection should include, first, the supervision of the work undertaken. Certain trades where grit, dirt, or poison fumes are given off are dangerous to human health, hence care for the workers becomes a necessity. Factories should also be inspected as to cleanliness, the amount of air space per person employed, ventilation, toilet facilities, and proper fire protection. Tenement inspection should be thorough and should aim to provide safe and sanitary homes.

Dust exhausts on grinding wheels protect lungs of the workmen.

Inspection of Food Supplies. — In a city certain regulations for the care of public supplies are necessary. Foods, both fresh and preserved, must be inspected and rendered safe for the thousands of people who are to use them. All raw foods exposed on stands should be covered so as to prevent insects or dust laden with bacteria from coming in contact with them. Meats must be inspected for diseases, such as tuberculosis in beef, or trichinosis in pork. Cold storage plants must be inspected to prevent the keeping of food until it becomes unfit for use. Inspection of sanitary conditions of factories where products are canned, or bakeries where foods are prepared, must be part of the work of a city in caring for its citizens.

Care of Raw Foods. — Each one of us may coöperate with the city government by remembering that fruits and vegetables can be carriers of disease, especially if they are sold from exposed stalls or carts and handled by the passers-by. All vegetables, fruits, or raw foods should be carefully washed before using. Spoiled or overripe fruit, as well as meat which is decayed, is swarming with bacteria and should not be used.

An interesting exercise would be the inspection of conditions in your own home block. Make a map showing the houses on the block. Locate all stores, saloons, factories, etc. Notice any cases of contagious disease, marking this fact on the map. Mark all heaps of refuse in the street, all uncovered garbage pails, any street stands that sell uncovered fruit, and any stores with an excessive number of flies.

In addition to food inspection, two very important supplies must be rendered safe by a city for its citizens. These are milk and water.

Care in Production of Milk. — Milk when drawn from a healthy cow should

Clean cows in clean barns with clean milkers and clean milk pails means clean milk in the city.

be free from bacteria. But immediately on reaching the air it may receive bacteria from the air, from the hands of the person who milks the cows, from the pail, or from the cow herself. Cows should, therefore, be milked in surroundings that are sanitary, the milkers should wear clean garments, put on over their ordinary clothes at milking time, while pails and all utensils used should be kept clean. Especially the surface exposed on the udder from which the milk is drawn should be cleansed before milking.

Most large cities now send inspectors to the farms from which milk is supplied. Farms that do not accept certain standards of cleanliness are not allowed to have their milk become part of the city supply.

Tuberculosis and Milk. — It is recognized that in some European countries from 30 to 40 per cent of all cattle have tuberculosis. Many dairy herds in this country are also infected. It is also known that the tubercle bacillus of cattle and man are much alike in form and action and that the germ from cattle would cause tuberculosis in man. Fortunately, the tuberculosis germ does not *grow* in milk, so that even if milk from tubercular cattle should get into our supply, it would be diluted with the milk of healthy cattle. In order to protect our milk supply from these germs it would be necessary to kill all tubercular cattle (almost an impossibility) or to pasteurize our milk so as to kill the germs in it.

Other Disease Germs in Milk. — We have already shown how typhoid may be spread through milk. Usually such outbreaks may be traced to a single case of typhoid, often a person who is a " typhoid carrier," *i.e.* one who may not suffer from the effects of the disease, but who carries the germs in his body, spreading them by contact. A recent epidemic of typhoid in New York City was traced to a single typhoid carrier on a farm far from the city. Sometimes the milk cans may be washed in contaminated water or the cows may even get the germs on their udders by wading in a polluted stream. Diphtheria, scarlet fever, and Asiatic cholera are also undoubtedly spread through milk supplies. Milk also plays a very important part in the high death rate from diarrheal diseases among young children in warm weather. Why?

Grades of Milk in a City Supply. — Milk which comes to a city

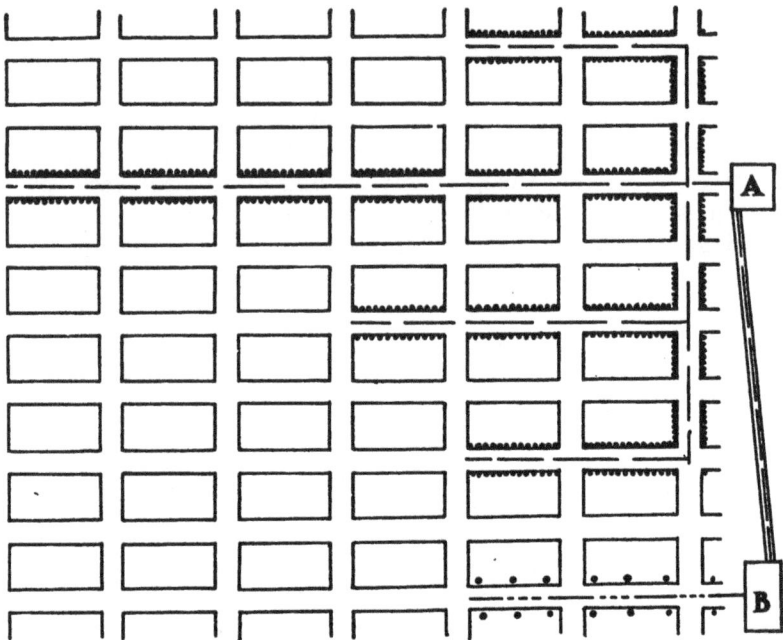

A diagram to show how typhoid may be spread in a city through an infected milk supply. The black spots in the blocks mean cases of typhoid. *A*, a farm where typhoid exists; the dashes in the streets represent the milk route. *B* is a second farm which sends part of its milk to *A*; the milk cans from *B* are washed at farm *A* and sent back to *B*. A few cases of typhoid appear along *B*'s milk route. How do you account for that?

may be roughly placed in three different classes. The best milk, coming from farms where the highest sanitary standards exist, where the cows are all tubercular tested, where modern appliances for handling and cooling the milk exist, is known as certified or, in New York City, grade A milk. Most of the milk sold, however, is not so pure nor is so much care taken in handling it. Such milk, known in New York as grade B milk, is pasteurized before delivery, and is sold only in bottles. A still lower grade of milk (dipped milk) is sold direct from cans. It is evident that such milk, often exposed to dust and other dirt, is unfit for any purpose except for cooking. It should under no circumstances be used for children. A regulation recently made by the New York City

Department of Health states that milk sold "loose" in restaurants, lunch-rooms, soda fountains, and hotels must be pasteurized.

Care of a City Milk Supply. — Besides caring for milk in its production on the farm, proper transportation facilities must be provided. Much of the milk used in New York City is forty-eight hours old before it reaches the consumer. During shipment it must be kept in refrigerator cars, and during transit to customers it should be iced. Why? All but the highest grade milk should be pasteurized. Why? Milk should be bottled by machinery if possible so as to insure no personal contact; it should be kept in clean, cool places; and no milk should be sold by dipping from cans. Why is this a method of dispensing impure milk?

Care of Milk in the Home. — Finally, milk at home should receive the best of care. It should be kept on ice and in covered bottles, because it readily takes up the odors of other foods. If we are not certain of its purity or keeping qualities, it should be pasteurized at home. Why?

Water Supplies. — One of the greatest assets to the health of a large city is pure water. By pure water we mean water free from all *organic* impurities, including germs. Water from springs and deep driven wells is the safest water, that from large reservoirs next best, while water that has drainage in it, river water for example, is very unsafe.

The waters from deep wells or springs if properly

New York City is spending $350,000,000 to have a pure and abundant water supply. This is the tunnel which will bring the water from the Catskill Mountains to New York City.

protected will contain no bacteria. Water taken from protected streams into which no sewage flows will have but few bacteria, and these will be destroyed if exposed to the action of the sun and the constant aëration (mixing with oxygen) which the surface water receives in a large lake or reservoir. But water taken from a river

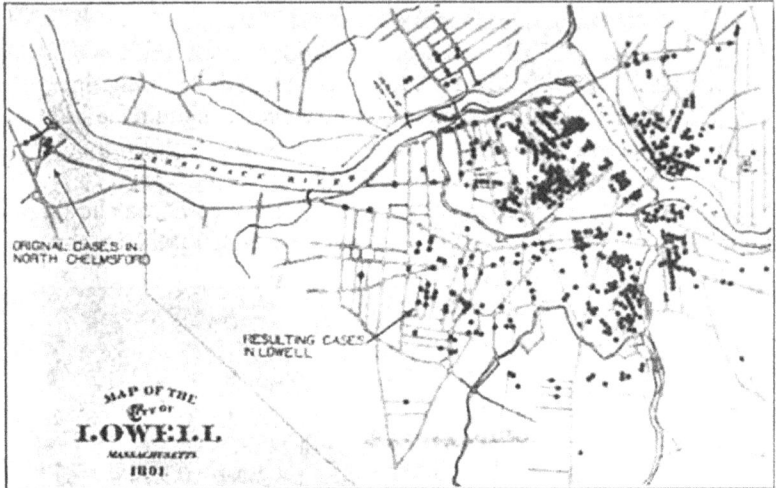

The city of Lowell in 1891 took its water *without filtering*, *i.e.* from the Merrimack River at the point shown on the map.
Typhoid fever broke out in North Chelmsford and about two weeks later cases began to appear in Lowell until a great epidemic occurred. Explain this outbreak. Each black dot is a case of typhoid.

into which the sewage of other towns and cities flows must be filtered before it is fit for use.

Typhoid fever germs live in the food tube, hence the excreta of a typhoid patient will contain large numbers of germs. In a city with a system of sewage such germs might eventually pass from the sewers into a river. Many cities take their water supply directly from rivers, sometimes not far below another large town. Such cities must take many germs into their water supply. Many cities, as Cleveland and Buffalo, take their water from lakes into which their sewage flows. Others, as Albany, Pittsburgh, and Philadelphia, take their drinking water directly from rivers into which

Filter beds at Albany, N. Y.

sewage from cities above them on the river has flowed. Filtering such water by means of passing the water through settling basins and sand filters removes about 98 per cent of the germs. The result of drinking unfiltered and filtered water in certain large cities is shown graphically at right. In cities which drain their sewage into rivers and lakes, the question of sewage disposal is a large one, and many cities now have means of disposing of their sewage in some manner as to render it harmless to their neighbors.

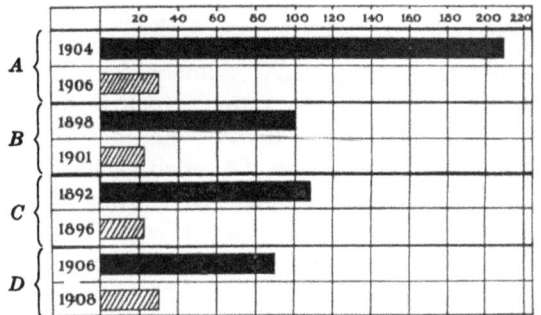

Cases of typhoid per 100,000 inhabitants before filtering water supply (solid) and after (shaded) in A, Watertown, N. Y.; B, Albany, N. Y.; C, Lawrence, Mass.; D, Cincinnati, Ohio. What is the effect of filtering the water supply?

Railroads are often responsible for carrying typhoid and spreading it. It is said that a recent outbreak of typhoid in Scranton, Pa., was due to the fact that the excreta from a typhoid patient traveling in a sleeping car was washed by rain into a reservoir near which the train was passing. Railroads are thus seen to be great open sewers. A sanitary car toilet is the only remedy.

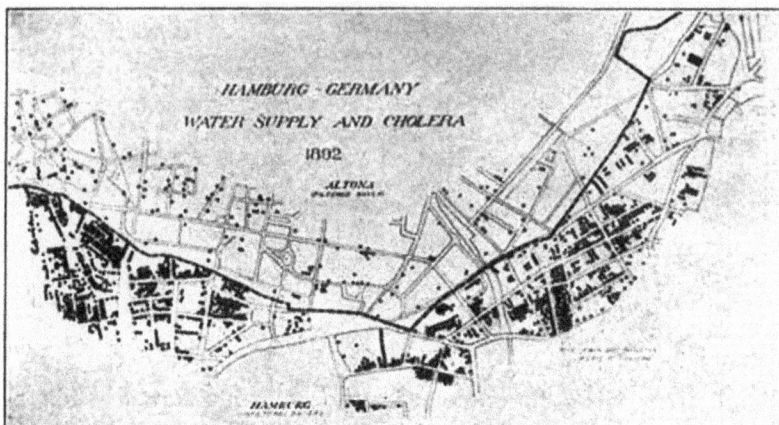

This chart shows that during a cholera epidemic in 1892 there were hundreds of cases of cholera in Hamburg, which used unfiltered water from the Elbe, but in adjoining Altona, where filtered water was used, the cases were very few.

Sewage Disposal. — Sewage disposal is an important sanitary problem for any city. Some cities, like New York, pour their sewage directly into rivers which flow into the ocean. Consequently much of the liquid which bathes the shores of Manhattan Island is dilute sewage. Other cities, like Buffalo or Cleveland, send their sewage into the lakes from which they obtain their supply of drinking water. Still other cities which are on rivers are forced to dispose of their sewage in various ways. Some have a system of

Stone filter beds in a sewage disposal plant.

filter beds in which the solid wastes are acted upon by the bacteria of decay, so that they can be collected and used as fertilizer. Others precipitate or condense the solid materials in the sewage and then dispose of it. Another method is to flow the sewage over large areas of land, later using this land for the cultivation of crops. This method is used by many small European cities.

The Work of the Department of Street Cleaning. — In any city a menace to the health of its citizens exists in the refuse and garbage. The city streets, when dirty, contain countless millions of germs which have come from decaying material, or from people ill with disease. In most large cities a department of street cleaning not only cares for the removal of dust from the streets, but also has the removal of garbage, ashes, and other waste as a part of its work. The disposal of solid wastes is a tremen-

Collecting ashes.

dous task. In Manhattan the dry wastes are estimated to be 1,000,000 tons a year in addition to about 175,000 tons of garbage. Prior to 1895 in the city of New York garbage was not separated from ashes ; now the law requires that garbage be placed in separate receptacles from ashes. Do you see why? The street-cleaning department should be aided by every citizen ; rules for the separation of garbage, papers, and ashes should be kept. Garbage and ash cans should be *covered*. The practice of upsetting ash or garbage cans is one which no young citizen should allow in his neighborhood, for sanitary reasons. The best results in summer street cleaning are obtained by washing or flushing the streets, for thus the dirt containing germs is prevented from getting into the air. The garbage is removed in carts, and part of it is burned in huge furnaces. The animal and plant refuse is cooked in great tanks ; from this material the fats are extracted, and the solid matter is sold for fertilizer. Ashes are used for filling marsh land. Thus the removal of waste matter may pay for itself in a large city.

An Experiment in Civic Hygiene. — During the summer of 1913 an interesting experiment on the relation of flies and filth to disease was carried on in New York City by the Bureau of Public Health and Hygiene of the New York Association for improving the condition of the poor. Two adjoining blocks were chosen in a thickly populated part of the Bronx near a number of stables which were the sources of great numbers of flies. In one block all houses were screened, garbage pails were furnished with covers, refuse was removed and the surroundings made as sanitary as possible. In the adjoining block conditions were left unchanged. During the summer as flies began to breed in the manure heaps near the stables all manure was disinfected. Thus the breeding of flies was checked. The campaign of education was

The upper picture shows the stables where millions of flies were bred; the lower picture, the disinfection of manure so as to prevent the breeding of flies.

continued during the summer by means of moving pictures, nurses, boy scouts, and school children who became interested.

At the end of the summer it was found that there had been a considerable decrease in the number of cases of fly-carried diseases and a still greater decrease in the total days of sickness (especially of children) in the screened and sanitary block. The table and pictures speak for themselves. If such a small experiment shows results like this, then what might a general cleanup of a city show?

Public Hygiene. — Although it is absolutely necessary for each individual to obey the laws of health if he or she wishes to keep well, it has also become necessary, especially in large cities, to have general supervision over the health of people living in a community. This is done by means of a department or board of health. It is the function of this department to care for public health. In addition to such a body in cities, supervision over the health of its citizens is also exercised by state boards of health. But as yet the government of the United States has not established a Bureau of Health, important as such a bureau would be.

The Functions of a City Board of Health. — The administration

In the upper picture a little girl can be seen dumping garbage from the fire escape. She was a foreigner and knew no better. The picture below shows the result of such garbage disposal.

of the Board of Health in New York City includes a number of divisions, each of which has a different work to do. Each is in itself important, and, working together, the entire machine provides ways and means for making the great city a safe and sanitary place in which to live. Let us take up the work of each division

of the health board in order to find out how we may coöperate with them.

The Division of Infectious Diseases. — Infectious diseases are chiefly spread through *personal contact.* It is the duty of a government to prevent a person having such a disease from spreading it broadcast among his neighbors. This can be done by *quarantine* or *isolation* of the person having the disease. So the board of health at once isolates any case of disease which may be communi-

DISEASES	FILTHY AREA	CLEANED-UP AREA
TOTAL SICKNESS	165	110
NON COMMUNICABLE	40	36
COMMUNICABLE	125	74
POSSIBLY FLY-BORNE	65	22

Comparison of cases of illness during the summer of 1913 in two city blocks, one clean and the other dirty. What are your conclusions?

cated from one person to another. No one save the doctor or nurse should enter the room of the person quarantined. After the disease has run its course, the clothing, bedding, etc., in the sick room is fumigated. This is usually done by the board of health. Formaldehyde in the form of candles for burning or in a liquid form is a good disinfectant. In disinfecting the room should be tightly closed to prevent the escape of the gas used, as the object of the disinfection is to kill all the disease germs left in the room. In some cases of infectious disease, as scarlet fever, it is found best to isolate the patients in a hospital used for that purpose. Examples of the most infectious diseases are measles, scarlet fever, whooping cough, and diphtheria.

Immunity. — In the prevention of germ diseases we must fight the germ by attacking the parasites directly with poisons that will kill them (such poisons are called *germicides* or *disinfectants*), and we must strive to make the persons coming in contact with the disease unlikely to take it. This insusceptibility or *immunity* may be either natural or acquired. Natural immunity seems to be in the constitution of a person, and may be inherited. Immunity may be acquired by means of such treatment as the antitoxin treatment for diphtheria. This treatment, as the name denotes, is a method of neutralizing the poison (toxin) caused by the bacteria in the system. It was discovered a few years ago by a German, Von Behring, that the serum of the blood of an animal immune to diphtheria is capable of neutralizing the poison produced by the diphtheria-causing bacteria. Horses are rendered immune by giving them the diphtheria toxin in gradually increasing doses.

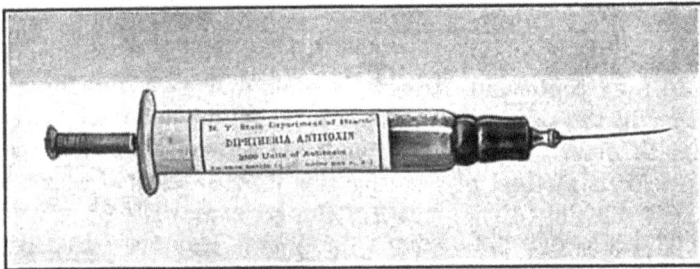

Antitoxin for diphtheria prepared by the New York Board of Health.

The serum of the blood of these horses is then used to inoculate the patient suffering from or exposed to diphtheria, and thus the disease is checked or prevented altogether by the antitoxin injected into the blood. The laboratories of the board of health prepare this antitoxin and supply it fresh for public use.

It has been found from experience in hospitals that deaths from diphtheria are largely preventable by *early use* of antitoxin. When antitoxin was used on the first day of the disease no deaths took place. If not used until the second day, 5 deaths occurred in every hundred cases, on the third day 11 deaths, on the 4th day 19 deaths, and on the 5th day 20 deaths out of every hundred cases. It is therefore advisable, in a suspected case of diphtheria, to have antitoxin used at once to prevent serious results.

Vaccination. — Smallpox was once the most feared disease in this country; 95 per cent of all people suffered from it. As late as 1898, over 50,000 persons lost their lives annually in Russia from this disease. It is probably not caused by bacteria, but by a tiny animal parasite. Smallpox has been brought under absolute control by vaccination, — the inoculation of man with the substance (called *virus*) which causes cowpox in a cow. Cowpox is like a mild form of smallpox, and the introduction of this virus gives complete immunity to smallpox for several years after vaccination. This immunity is caused by the formation of a germicidal substance in the blood, due to the introduction of the virus. Another function of the board of health is the preparation and distribution of vaccine (material containing the virus of cowpox).

Rabies (Hydrophobia). — This disease, which is believed to be caused by a protozoan parasite, is communicated from one dog to another in the saliva by biting. In a similar manner it is transferred to man. The great French bacteriologist, Louis Pasteur, discovered a method of treating this disease so that when taken early at the time of the entry of the germ into the body of man, the disease can be prevented. In some large cities (among them New York) the board of health has established a laboratory where free treatment is given to all persons bitten by dogs suspected of having rabies.

Vaccination against Typhoid. — Typhoid fever has within the past five years received a new check from vaccination which has been introduced into our army and which is being used with good effect by the health departments of several large cities.

The following figures show the differences between number of cases and mortality in the army in 1898 during the war with Spain and in 1911 during the concentration of certain of our troops at San Antonio, Texas.

1898 — 2nd Division, 7th Army Corps, Jacksonville, Florida. June–October, 1898

Mean strength, 10,759.
Cases of typhoid certain and probable, 2693.
Death from typhoid, 258.
Death from all diseases, 281.

Manœuver Division, San Antonio, Texas. March 10–July 11, 1911.

Mean strength, 12,801.
Cases of typhoid, 1.
Death from typhoid, 0.
Deaths all diseases, 11.

During this period there were 49 cases of typhoid and 19 deaths in the near-by city of San Antonio. But in camp, *where vaccination*

	2ND DIV. 7TH ARMY CORPS JACKSONVILLE. FLA.-JUNE-OCT. 1898	MANŒUVER, DIV.-SAN ANTONIO TEXAS. MAR.10–JULY 11, 1911.
MEAN STRENGTH	10,759	12,801
CASES OF TYPHOID	2693	ONE
TYPHOID DEATHS	258	NONE
DEATHS ALL DISEASES	281	11

Comparison of cases of and death from typhoid in 1898 and 1911. What have we learned about combating typhoid since 1898?

for typhoid was required, all were practically immune. In the army at large, since typhoid vaccination has been practiced, 1908–1909, the death rate from typhoid has dropped from 2.9 per 1000 to .03 per 1000, a wonderful record when we remember that during the Spanish-American War 86 per cent of the deaths in the army were from typhoid fever.

How the Board of Health fights Tuberculosis. — Tuberculosis, which a few years ago killed fully one seventh of the people who died from disease in this country, now kills less than one tenth. This decrease has been largely brought about because of the treatment of the disease. Since it has been proved that tuberculosis if taken early enough is curable, by quiet living, good food, and *plenty* of fresh air and light, we find that numerous sanitaria have come into existence which are supported by private or public means. At these sanitaria the patients *live* out of doors, especially sleep in the air, while they have plenty of nourishing food and little exercise. The department of health of New York City main-

tains a sanitarium at Otisville in the Catskill Mountains. Here
people who are unable to provide means for getting away from the

The best cures for tuberculosis are rest, plenty of fresh out-of-door air, and
wholesome food.

city are cared for at the city's expense and a large percentage of
them are cured. In this way and by tenement house laws which
require proper air shafts and window ventilation in dwellings, by
laws against spitting in public places, and in other ways, the boards
of health in our towns and cities are waging war on tuberculosis.

A sanitarium for tuberculosis. Notice the outdoor sleeping rooms.

Ex-President Roosevelt said, in one of his latest messages to Congress : —

" There are about 3,000,000 people seriously ill in the United States, of whom 500,000 are consumptives. *More than half of this illness is preventable.* If we count the value of each life lost at only $1700 and reckon the average earning lost by illness at $700 a year for grown men, we find that the economic gain from mitigation of preventable disease in the United States would exceed $1,500,000,000 a year. This gain can be had through medical investigation and practice, school and factory hygiene, restriction of labor by women and children, the education of the people in both public and private hygiene, and through improving the efficiency of our health service, municipal, state, and national."

Work of the Division of School and Infant Hygiene. — Besides the work of the division of infectious disease, the division of sanitation, which regulates the general sanitary conditions of houses and their surroundings and the division of inspection, which looks after the purity and conditions of sale and delivery of milk and foods, there is another department which most vitally concerns school children. This is the division of school and infant hygiene. The work of this department is that of the care of the children of the city. During the year 1912, 279,776 visits were made to the homes of school children of the city of New York by inspectors and nurses. Besides this, thousands of children in school were cared for and aided by the city.

Adenoids. — Many children suffer needlessly from adenoids, — growths in the back of the nose or mouth which prevent sufficient oxygen being admitted to the lungs. A child suffering from these growths is known as a " mouth breather " because the mouth is opened in crder to get more air. The result to the child may be a handicap of deafness, chronic running of the nose, nervousness, and lack of power to think. His body cells are starving for oxygen. A very simple operation removes this growth. Coöperation on the part of the children and parents with the doctors or nurses of the board of health will do much in removing this handicap from many young lives.

Eyestrain. — Another handicap to a boy or girl is eyestrain,

Twenty-two per cent of the school children of Massachusetts were recently found to have defects in vision. Tests for defective eyesight may be made at school easily by competent doctors, and if the child or parent takes the advice given to correct this by procuring proper glasses, a handicap on future success will be removed.

Decayed Teeth. — Decayed teeth are another handicap, cared for by this division. Free dental clinics have been established in many cities, and if children will do their share, the chances of their success in later life will be greatly aided. Boys and girls, if handicapped with poor eyes or teeth, do not have a fair chance in life's competition. In a certain school in New York City there were 236 pupils marked " C " in their school work. These children were examined, and 126 were found to have bad teeth, 54 defective vision, and 56 other defects, as poor hearing, adenoids, enlarged tonsils, etc. Of these children 185 were treated for these various difficulties, and 51 did not take treatment. During the following year's work 176 of these pupils *improved* from " C " to " B " or " A ", while 60 did not improve. If defects *are* such a handicap in school, then what would be the chances of success in life outside.

In conclusion : this department of school hygiene deserves the earnest aid of every young citizen, girl or boy. If each of us would honestly help by maintaining quarantine in the case of contagious disease, by observing the rules of the health department in fumigation, by acting upon advice given in case of eyestrain, bad teeth, or adenoids, and most of all by observing the rules of personal hygiene as laid down in this book, the city in which we live would, a generation hence, contain stronger, more prosperous, and more efficient citizens than it does to-day.

REFERENCE BOOKS

ELEMENTARY

Hunter, *Laboratory Problems in Civic Biology*. American Book Company.
Davison, *The Human Body and Health*. American Book Company.
Gulick Hygiene Series, *Town and City*. Ginn and Company.
Hough and Sedgwick, *The Human Mechanism*, Part II. Ginn and Company.
Overton, *General Hygiene*. American Book Company.

Richards, *Sanitation in Daily Life*. Whitcomb and Barrows.
Richmond and Wallach, *Good Citizenship*. American Book Company.
Ritchie, *Primer of Sanitation*. World Book Company.
Sharpe, *Laboratory Manual of Biology*, pages 320–334. American Book Company.

ADVANCED

Allen, *Civics and Health*. Ginn and Company.
Chapin, *Municipal Sanitation in the United States*. Snow and Farnham.
Chapin, *Sources and Modes of Infection*. Wiley and Sons.
Conn, *Practical Dairy Bacteriology*. Orange Judd Company.
Hough and Sedgwick, *The Human Mechanism*. Part II. Ginn and Company.
Hutchinson, *Preventable Diseases*. The Houghton, Mifflin Company.
Morse, *The Collection and Disposal of Municipal Waste*. Municipal Journal and
    Engineer.
Overlock, *The Working People, Their Health and How to Protect It*. Mass. Health
    Book Publishing Co.
Price, *Handbook of Sanitation*. Wiley and Sons.
Tolman, *Hygiene for the Worker*. American Book Company.

REPORTS, ETC.

American Health Magazine.
Annual Report of Department of Health, City of New York (and other cities).
Bulletins and Publications of Committee of One Hundred on National Health.
School Hygiene, American School Hygiene Association.
Grinnell, *Our Army versus a Bacillus*. National Geographic Magazine.

XXV. SOME GREAT NAMES IN BIOLOGY

If we were to attempt to group the names associated with the study of biology, we would find that in a general way they were connected either with discoveries of a purely scientific nature or with the benefiting of man's condition by the *application* of the purely scientific discoveries. The first group are necessary in a science in order that the second group may apply their work. It was necessary for men like Charles Darwin or Gregor Mendel to prove their theories before men like Luther Burbank or any of the men now working in the Department of Agriculture could benefit mankind by growing new varieties of plants. The discovery of scientific truths must be achieved before the men of modern medicine can apply these great truths to the cure or prevention of disease. Since we are most interested in discoveries which touch directly upon human life, the men of whom this chapter treats will be those who, directly or indirectly, have benefited mankind.

The Discoverers of Living Matter. — The names of a number of men living at different periods are associated with our first knowledge of cells. About the middle of the seventeenth century microscopes came into use. Through their use plant cells were first described and pictured as hollow boxes or " cells." But it was not until 1838 that two German friends, Schleiden and Schwann by name, working on plants and animals, discovered that both of these forms of life contained a jellylike substance that later came to be called *protoplasm*. Another German named Max Schultz in 1861 gave the name protoplasm to *all living matter*, and a little later still Professor Huxley, a famous Englishman, friend and champion of Charles Darwin, called attention to the physical and chemical qualities of protoplasm so that it came to be known as the chemical and physical basis of life.

Life comes from Life. — Another group of men, after years of patient experimentation, worked out the fact that *life comes from other life.*   In ancient times it was thought that life arose *spontaneously;* for example, that fish or frogs arose out of the mud of the river bottoms, and that insects came from the dew or rotting meat.   It was believed that bacteria arose spontaneously in water, even as late as 1876, when Professor Tyndall proved by experiment the contrary to be true.

As early as 1651 William Harvey, the court physician of Charles I of England, showed that all life came from the egg. It was much later, however, that the part played by the sperm and egg cell in fertilization was carefully worked out. It is to Harvey, too, that we owe the beginnings of our knowledge of the circulation of the blood.   He showed that

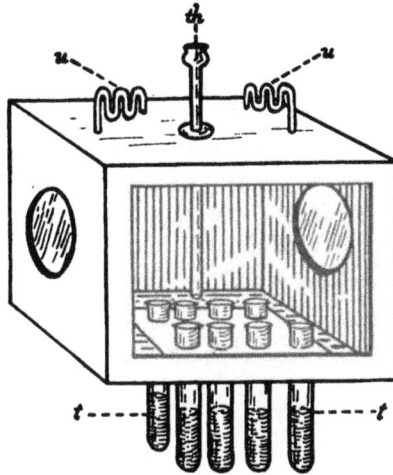

Prof. Tyndall's experiment to show that if air containing germs is kept from organic substances, such substances will not decay.   The box is sterilized; likewise the tubes (*t*) containing nutrients. Air is allowed to enter by the tubes (*u*), which are so made that dust is prevented from entering.   A thermometer (*th*) records the temperature.   The substances in the tubes do not decay, no matter how favorable the temperature.

blood moved through tubes in the body and that the heart pumped it.   He might be called the father of modern physiology as well as the father of embryology.   A long list of names might be added to that of Harvey to show how gradually our knowledge of the working of the human body has been added to.   At the present time we are far from knowing all the functions of the various parts of the human engine, as is shown by the number of investigators in physiology at the present time.   Present-day problems have much to do with the care of the human mechanism and with its surroundings.   The solution of these problems will come from applying the sciences of hygiene, preventive medicine, and sanitation.

In the preceding chapters of this book we have learned something about our bodies and their care. We have found that man is able within limitations to control his environment so as to make it better to live in. All of the scientific facts that have been of use to man in the control of disease have been found out by men who have devoted their lives in the hope that their experiments and their sacrifices of time, energy, and sometimes life itself might make for the betterment of the human race. Such men were Harvey, Jenner, Lister, Koch, and Pasteur.

Edward Jenner and Vaccination. — The civilized world owes much to Edward Jenner, the discoverer of vaccination against smallpox. Born in Berkeley, a little town of Gloucestershire, England, in 1749, as a boy he showed a strong liking for natural history. He studied medicine and also gave much time to the working out of biological problems. As early as 1775 he began to associate the disease called cowpox with that of smallpox, and gradually the idea of inoculation against this terrible scourge, which killed or disfigured hundreds of thousands every year in England alone, was worked out and applied. He believed that if the two diseases were similar, a person inoculated with the mild disease (cowpox) would after a slight attack of this disease be immune against the more deadly and loathsome smallpox. It was not until 1796 that he was able to prove his theory, as at first few people would submit to vaccination. War at this time was being waged between France and England, so that the former country, usually so quick to appreciate the value of scientific discoveries, was slow to give this method a trial. In spite of much opposition, how-

Edward Jenner, the discoverer of vaccination.

ever, by the year 1802, vaccination was practiced in most of the civilized countries of the world.   At the present time the death rate in Great Britain, the home of vaccination, is less than .3 to every 1,000,000 living persons.   This shows that the disease is practically wiped out in England.   An interesting comparison with these figures might be made from the history of the disease in parts of Russia where vaccination is not practiced.   There, thousands of deaths from smallpox occur annually.   During the winter of 1913–1914 an epidemic of smallpox with more than 250 cases broke out in the city of Niagara Falls.   This epidemic appears to be due to a campaign conducted by people who do not believe in vaccination.   In cities and towns near by, where vaccination was practiced, no cases of smallpox occurred.   Naturally if opposition to vaccination is found nowadays, Jenner had a much harder battle to fight in his day.   He also had many failures, due to the imperfect methods of his time.   The full worth of his discovery was not fully appreciated until long after his death, which occurred in 1823.

Louis Pasteur. — The one man who, in biological science, did more than any other to directly benefit mankind was Louis Pasteur. Born in 1822, in the mountains near the border of northeastern France, he spent the early part of his life as a normal boy, fond of fishing and not very partial to study.   He inherited from his father, however, a fine character and grim determination, so that when he became interested in scientific pursuits he settled down to work with enthusiasm and energy.

At the age of twenty-five he became well known throughout France as a physicist.   Shortly

Louis Pasteur.

after this he became interested in the tiny plants we call bacteria, and it was in the field of bacteriology that he became most

famous. First as professor at Strassburg and at Lille, later as director of scientific studies in the École Normale at Paris, he showed his interest in the application of his discoveries to human welfare.

In 1857 Pasteur showed that fermentation was due to the presence of bacteria, it having been thought up to this time that it was a purely chemical process. This discovery led to very practical ends, for France was a great wine-producing country, and with a knowledge of the cause of fermentation many of the diseases which spoiled wine were checked.

In 1865–1868 Pasteur turned his attention to a silkworm disease which threatened to wipe out the silk industry of France and Italy. He found that this disease was caused by bacteria. After a careful study of the case he made certain recommendations which, when carried out, resulted in the complete overthrow of the disease and the saving of millions of dollars to the poor people of France and Italy.

The greatest service to mankind came later in his life when he applied certain of his discoveries to the treatment of disease. First experimenting upon chickens and later with cattle, he proved that by making a virus (poison) from the germs which caused certain diseases he could reduce this virus to any desired strength. He then inoculated the animals with the virus of reduced strength, giving the inoculated animals a mild attack of the disease, and found that this made them *immune* from future attacks. This discovery, first applied to chicken cholera, laid the foundation for all future work in the uses of serums, vaccines, and antitoxins.

Pasteur was perhaps the best known through his study of rabies. The great Pasteur Institute, founded by popular subscriptions from all over the world, has successfully treated over 22,000 cases of rabies with a death rate of less than 1 per cent. But more than that it has been the place where Roux, a fellow worker with Pasteur, discovered the antitoxin for diphtheria which has resulted in the saving of thousands of human lives. Here also have been established the principles of inoculation against bubonic plague, lockjaw, and other germ diseases.

Pasteur died in 1895 at the age of seventy-three, " the most perfect man in the realm of science," a man beloved by his countrymen and honored by the entire world.

Robert Koch. — Another name associated with the battle against disease germs is that of Robert Koch. Born in Klausthal, Hanover, in 1843, he later became a practicing physician, and about 1880 was called to Berlin to become a member of the sanitary commission and professor in the school of medicine. In 1881 he discovered the germ that causes tuberculosis and two years later the germ that causes Asiatic cholera. His later work has been directed toward the discovery of a cure for tuberculosis and other germ diseases. As yet, however, no certain cure seems to have been found.

Lister and Antiseptic Treatment of Wounds. — A third great benefactor of mankind was Sir Joseph Lister, an Englishman who was born in 1827.

Robert Koch.

As a professor of surgery he first applied antiseptics in the operating room. By means of the use of carbolic acid or other antiseptics on the surface of wounds, on instruments, and on the hands and clothing of the operating surgeons, disease germs were prevented from taking a foothold in the wounds. Thus blood poisoning was prevented. This single discovery has done more to prevent death after operations than any other of recent time.

Modern Workers on the Blood. — At the present time several names stand out among investigators on the blood. Paul Ehrlich, a German born in 1854, is justly famous for his work on the blood and its relation to immunity from certain diseases. His able research work has given the world a much better understanding of the problem of acquired immunity.

Another name associated with the blood is that of Elias Metchnikoff, a Russian. He was born in 1845. Metchnikoff first advanced the belief that the colorless blood corpuscles, or *phagocytes*, did service as the sanitary police of the body. He has found that there are several different kinds of colorless corpuscles, each having somewhat different work to do. Much of the modern work done by physiologists on the blood are directly founded on the discoveries of Metchnikoff.

Heredity and Evolution. Charles Darwin. — There is still another important line of investigation in biology that we have not mentioned. This is the doctrine of evolution and the allied discoveries along the line of heredity. The development or evolution of plants and animals from simpler forms to the many and present complex forms of life have a practical bearing on the betterment of plants and animals, including man himself. The one name indelibly associated with the word evolution is that of Charles Darwin.

Charles Darwin, the grand old man of biology.

Charles Darwin was born on February 12, 1809, a son of well-to-do parents, in the pretty English village of Shrewsbury. As a boy he was very fond of out-of-door life, was a collector of birds' eggs, stamps, coins, shells, and minerals. He was an ardent fisherman, and as a young man became an expert shot. His studies, those of the English classical school, were not altogether to his liking. It is not strange, perhaps, that he was thought a very ordinary boy, because his interest in the out-of-doors led him to neglect his studies. Later he was sent to Edinburgh University to study medicine. Here the dull lectures, coupled with his intense dislike for operations, made him determine never to become a physician. But all this time he showed his intense interest in natural history and took frequent part in the discussions at the meetings of one of the student zoological societies.

In 1828 his father sent him to Cambridge to study for the ministry. His three years at the university were wasted so far as preparation for the ministry were concerned, but they were invaluable in shaping his future. He made the acquaintance of one or two professors who were naturalists like himself, and in their company he spent many happy hours in roaming over the countryside collecting beetles and other insects. In 1831 an event occurred which changed his career and made Darwin one of the world's greatest naturalists. He received word through one of his professional friends that the position of naturalist on her Majesty's ship *Beagle* was open for a trip around the world. Darwin applied for the position, was accepted, and shortly after started on an eventful five years' trip around the world. He returned to England a famous naturalist and spent the remainder of his long and busy life producing books which have done more than those of any other writer to account in a satisfactory way for the changes of form and habits of plants and animals on the earth. His theories established a foundation upon which plant and animal breeders were able to work.

His wonderful discovery of the doctrine of evolution was due not only to his information and experimental evidence, but also to an iron determination and undaunted energy. In spite of almost constant illness brought about by eyestrain, he accomplished more than most well men have done. His life should mean to us not so much the association of his name with the *Origin of Species* or *Plants and Animals under Domestication*, two of his most famous books, but rather that of a patient, courteous, and brave gentleman who struggled with true English pluck against the odds of disease and the attacks of hostile critics. He gave to the world the proofs of the theory on which we to-day base the progress of the world. Darwin lived long enough to see many of his critics turn about and come over to his beliefs. He died on the 19th of April, 1882, at seventy-four years of age.

Associated with Darwin's name we must place two other co-workers on heredity and evolution, Alfred Russel Wallace, an Englishman who independently and at about the same time reached many of the conclusions that Darwin came to, and August Weissman, a German. The latter showed that the protoplasm of the germ cells (eggs and sperms) is directly handed down from generation to generation, they being different from the other body cells from the very beginning. In 1883 a German named Boveri discovered that the chromosomes of the egg and the sperm cell were at the time of fertilization just half in number of the other cells (see page 252) so that a *fertilized* egg was really a *whole cell* made up of *two half cells*, one from each parent. The chromosomes within the nucleus, we remember, are believed to be the bearers of the hereditary qualities handed down from parent to child. This discovery shows us some of the mechanics of heredity.

Applications to Plant and Animal Breeding. — Turning to the practical applications of the scientific work on the method of heredity, the name of Gregor Mendel, an Austrian monk, stands out most prominently. Mendel lived from 1822 until 1884. His work, of which we already have learned something (see page 258), remained undiscovered until a few years ago. The application of his methods to plant and animal raising are of the utmost importance because the breeder is able to separate the qualities he desires and breed for those qualities only. Another name we have mentioned with reference to plant breeding is Hugo de Vries, the Dutchman who recently showed that in some cases plants arise as new species by sudden and great variations known as *mutations*. And lastly, in our own California, Luther Burbank, by careful hybridizing, is making lasting fame with his new and useful hybrid plants.

REFERENCES

Conn, *Biology*. Silver, Burdett & Co.
Darwin, *Life and Letters of Charles Darwin*. Appletons.
Galton, *Hereditary Genius*. London (1892).
Thompson, *Heredity*. John Murray, London England.
Wasmann, *Problem of Evolution*. Kegan Paul, Trench, Trübner and Co., London, E. C.

APPENDIX

A SUGGESTED OUTLINE FOR BIOLOGY BEGINNING IN THE FALL

LIST OF TOPICS

FIRST TERM

First week. WHY STUDY BIOLOGY? Relation to human health, hygiene. Relations existing between plants and animals. Relation of bacteria to man. Uses of plants and animals. Conservation of plants and animals. Relation to life of citizen in the city. Plants and animals in relation to their environment. What is the environment; light, heat, water, soil, food, etc. What plants take out of the environment. What animals take out of the environment. Dependence of plants and animals upon the factors of the environment. *Laboratory:* Study of a plant or an animal in the school or at home to determine what it takes from its environment.

Second week. SOME RELATIONS EXISTING BETWEEN PLANTS (GREEN) AND ANIMALS. Field trip planned to show that insects feed upon plants; make their homes upon plants. That flowers are pollinated by insects. Insects lay eggs upon certain food plants. Green plants make food for animals. Other relations. (Time allotment. One day trip, collecting, etc.; two days' discussion of trip in all its relations.) Make a careful study of the locality you wish to visit, have a plan that the pupils know about beforehand. Review and hygiene of pupil's environment, 2 days.

Third week. STUDY OF A FLOWER, PARTS ESSENTIAL TO POLLINATION NAMED. Adaptations for insect pollination worked out in laboratory. Study of bee or butterfly as an insect carrier of pollen. Names of parts of insect learned. Elementary knowledge of groups of insects seen on field trip. Bees, butterflies, grasshoppers, beetles, possibly flies and bugs. Drawing of a flower, parts labeled. Drawing of an insect, outline only, parts labeled. Careful study of some fall flower fitted for insect pollination with an insect as pollinating agent. Some examples of cross-pollination explained. Practical value of cross-pollination.

Fourth week. LIVING PLANTS AND ANIMALS COMPARED. Parts of plants, functions; organs, tissues, cells. Demonstration cells of onion or elodea. How cells form others. What living matter can do. Reproduction. Growth of pollen tube, fertilization. Development of ovule into seed. Fruits, how formed. Uses, to man.

Fifth week. WHAT MAKES A SEED GROW. Bean seed, a baby plant, and food supply. Food, what is it? Organic nutrients, tests for starch, protein, oil. Show their presence in seeds.

407

Sixth week. NEED FOR FOODS. Germination of bean due to (a) presence of foods, (b) outside factors. What is done with the food. Release of energy. Examples of engine, plants, human body. Oxidation in body. Proof by experiment. Test for presence of CO_2. Oxidation in growing plant, experiment. Respiration a general need for both plants and animals.

Seventh week. NEED FOR DIGESTION. The corn grain. Parts, growth, food supply outside body of plant, how does it get inside. Digestion, need for. Test for grape sugar. Enzymes, their function. Action of diastase on starch.

Eighth week. WHAT PLANTS TAKE FROM THE SOIL, HOW THEY DO THIS. Use of root. Influence of gravity and water. Why? Absorption a function Root hairs. Demonstration. Pocket gardens, optional home work, but each pupil must work on root hairs from actual specimen. How root absorbs. Osmosis; what substances will osmose. Experiments to demonstrate this.

Ninth week. COMPOSITION OF SOIL. What root hairs take out of soil. Plant needs mineral matter to make living matter. Why? Nitrogen necessary. Sources of nitrogen, the nitrogen-fixing bacteria. Relation of this to man. Rotation of crops.

Tenth week. HOW GREEN PLANTS MAKE FOOD. Passage of liquids up stem. Demonstration. Structure of a green leaf. Cellular structure demonstrated. Microscopic demonstration of cells, stoma, air spaces, chlorophyll bodies. Evaporation of water from green leaf, regulation of transpiration.

Eleventh week. *Midterm Examinations.* Sun a source of energy. Effect of light on green plants. Experimental proof. Starch made in green leaf. Light and air necessary for starch making. Proof. Protein making in leaf. By-products in starch making. Proof. Respiration.

Twelfth week. THE CIRCULATION AND DISTRIBUTION OF FOOD IN GREEN PLANTS. Uses of bark, wood, what part of stem does food pass down. Willow twig experiment. Summary of functions of living matter in plant. Forestry lecture. Economic uses of green plants. Reports.

Thirteenth week. PLANTS WITHOUT CHLOROPHYLL IN THEIR RELATION TO MAN. Saprophytic fungi. Molds. Growth on bread or other substances. Conditions most favorable for growth. Favorite foods. Methods of prevention. Economic importance.

Fourteenth week. YEASTS IN THEIR RELATION TO MAN. Experiments to show fermentation is caused by yeasts. Experiments to show conditions necessary for fermentation. The part played by yeasts in bread making, in wine making, in other industries. Structure of yeast demonstrated. Summary.

Fifteenth week. EXPERIMENTS TO SHOW WHERE BACTERIA MAY BE FOUND AND CONDITIONS NECESSARY TO GROWTH BEGUN. Have cultures collected and placed in a warm room during the holidays. Suggested experiments are exposure to air of quiet room and room with persons moving, dust of floor, knife blade, etc.

Sixteenth, seventeenth, and eighteenth weeks. THE MONTH OF JANUARY SHOULD BE DEVOTED TO THE STUDY OF BACTERIA IN THEIR GENERAL RELATIONS TO MAN. Economically, both directly and indirectly. Especial emphasis placed on the nature and necessity of decay. Bacteria in relation to disease should also be emphasized. The experiments to be performed and the topics expected to be covered follow.

CONDITIONS FAVORABLE AND UNFAVORABLE FOR GROWTH OF BACTERIA. (Use bouillon cultures.) Effect of intense heat, sterile bouillon exposed to air, effect of boiling, effect of cold, effect of antiseptics (corrosive sublimate, carbolic acid, boric acid, formalin, etc.), effect of large amounts of sugar and salt and the relation of this to preserving, etc. Bring out practical application of principles demonstrated. Discuss sterilization in medicine and surgery, cold storage, canning, sterilization, *e.g.* laundries, etc., use of antiseptics, preserving by means of salt and sugar. Microscopic demonstration of bacteria. Methods of reproduction. Importance in causing organic decay, fixation of nitrogen, various useful forms in cheese making, butter ripening, etc. · Harmfulness of bacteria as disease producers. Specific diseases discussed: tuberculosis, typhoid, infective colds, blood poisoning, etc. Vaccination. Antitoxins begun — continued after knowledge of human body is gained. Work of Lister and Pasteur.

Nineteenth and twentieth weeks. REVIEW AND EXAMINATIONS.

SECOND TERM

First week. THE BALANCED AQUARIUM. Carbon and nitrogen cycles. Balanced aquarium and hay infusion compared.

Second week. ONE PROTOZOAN, DEMONSTRATION TO SHOW CHANGES IN SHAPE, RESPONSE TO STIMULI, SUMMARY OF VITAL PROCESSES IN CELL. Food getting, digestion, assimilation, oxidation, excretion, growth, reproduction. Internal structure of protozoan. Protozoa as cause of disease.

Third week. GENERAL SURVEY OF ANIMAL KINGDOM. Survey introduced by museum trip if possible. Protozoa, worm, insect, fish, mammal. Distinction between vertebrate and invertebrate. Character of mammalia. Division of labor emphasized. Man's place in nature.

Fourth week. STUDY OF THE FROG. Relation to habitat, adaptations for locomotion, food getting, respiration, comparison of frog and fish on latter point. Osmotic exchange of gases emphasized. Cell respiration.

Fifth week. METAMORPHOSIS OF FROG. Fertilization, cell division, and differentiation emphasized. Touch on plant and animal breeding. Function of chromosomes as bearers of heredity. Comparison of bird's egg and mammal embryo.

Sixth week. FACTORS IN BREEDING. 1. Variation. 2. Selection. 3. Heredity fixes variation. 4. Hybridizing. 5. Control of environment. Eugenics in relation to (a) crime, (b) disease, (c) genius. Continuity of germ plasm. Work of Darwin, Mendel, De Vries, Burbank.

Seventh week. A BRIEF STUDY OF THE GROSS STRUCTURE OF THE HUMAN BODY. Skin, muscles, bones. Removal of lime from bone by HCl to show other substances and need for lime. Effect of posture, spinal curvature, fractures, sprains.

Eighth week. NEED FOR FOOD. Nutritive value of food. Use of charts to show foods rich in carbohydrates, fats, proteins, minerals, water, refuse. The relation of age, sex, work, and environment to the food requirements. What is a cheap food. Price list of common foods at present time. Efforts of government to secure a cheap food supply for the people. Digestibility of foods.

Ninth week. How THE FUEL VALUE OF FOOD HAS BEEN DETERMINED. Meaning of calorie. The 100-caloric portion, its use in determining a daily or weekly dietary. Standard dietary as determined by Atwater. Comparison of standards of Chittenden and Voit with those of Atwater.

Tenth week. STUDY OF PUPIL'S DIETARY. Planning ideal meals. Individual dietaries for one day required from each pupil. Discussions and corrections. The family dietary. Relation to cost.

Eleventh week. DIGESTION. The digestive system in the frog and in man compared. Drawings of each. Glands and enzymes. Internal secretions and their importance. Demonstration of glandular tissues. Experiment to show digestion of starch in mouth.

Twelfth week. DIGESTION CONTINUED. Digestion of white of egg by gastric juice. Digestion of starch with pancreatic fluid. Functions of pancreatic juice. Microscopic examination of emulsion. Reasons for digestion. Part played by osmosis. Demonstration of osmosis. Non-osmosis of non-digested foods, comparison between osmosable qualities of starch and grape sugar.

Thirteenth week. ABSORPTION. Where and how foods are absorbed. The structure of a villus explained. Course taken by foods after absorption. Function of liver. Blood making the result of absorption. Composition of blood, red and colorless corpuscles, plasma, blood plates, antibodies. Microscopic drawing of corpuscles of frog's and man's blood.

Fourteenth week. CIRCULATION OF BLOOD. The heart and lungs of frog demonstrated. Heart of man a force pump, explain with use of force pump. Demonstration of beef's heart. Circulation and changes of blood in various parts of body. Work of cells with reference to blood made clear. Capillary circulation (demonstration of circulation in tadpole's tail or web of frog's foot).

Fifteenth week. RESPIRATION AND EXCRETION. Necessity for taking of oxygen to cells and removal of wastes from cells. Part played by blood and lymph. Mechanics of breathing (use of experiments). Changes of air and blood in lungs (experiments). Best methods of ventilation (experiments). Elimination of wastes from blood by lungs, skin, and kidneys. Cell respiration.

Sixteenth week. HYGIENE OF ORGANS OF EXCRETION, especially care of skin. The general structure and functions of the central nervous system. Sensory and motor nerves. Reflexes, instincts, habits. Habit formation, importance of right habits. Rules for habit formation. Habit-forming drugs and other agents. Lecture.

Seventeenth, eighteenth, nineteenth weeks. CIVIC HYGIENE AND SANITATION. Hygiene of special senses, eye and ear. A well citizen an efficient citizen. Public health is purchasable. Improvement of environment a means of obtaining this. Civic hygiene and sanitation. Cleaning up neighborhood, inquiry into home and street conditions. Fighting the fly. Conditions of milk and water supply. Relation of above to disease. Work of Board of Health, etc. Review and Examinations.

SUGGESTED SYLLABUS FOR COURSE BEGINNING FEBRUARY 1 AND ENDING THE FOLLOWING JANUARY

First Term

First week. WHY STUDY BIOLOGY? Relation to human health, hygiene. Relations existing between plants and animals. Relation of bacteria to man. Uses of plants and animals. Conservation of plants and animals. Relation to life of citizen in this city. Needs of plants and animals: (1) food, (2) water, (3) air, (4) proper temperature. Study of a single plant or animal in relation to its environment. Problems of city government: (a) storage, preservation and distribution of foods, (b) water supply, (c) overcrowded tenements, (d) street cleaning, (e) clean schools. Biological problems in city government.

Second week. INTERRELATIONS BETWEEN PLANTS AND ANIMALS. Plants furnish food, clothing, shelter, and medicine. Animals use food, shelter. Man's use of plants as above. Man's use of animals as above. Plant and animal industries. Use of balanced aquarium as illustrative material.

Third week. DESTRUCTION OF FOOD AND OTHER THINGS BY MOLD. Home experiment. Conditions favorable to growth of mold. Food, moisture, temperature. Destruction of commodities by mold: food, leather, clothing.

Fourth week, fifth week. DESTRUCTION OF FOODS BY BACTERIA. Experiment. To show where bacteria are found. Soil, dust, water, milk, hands, mouth. Use and harm of decay. Relation to agriculture. Experiment. Conditions favorable and unfavorable to growth of bacteria: boiling, cold, sugar, salt. Bacteria in relation to disease briefly mentioned. Bacteria in industries.

Sixth week. USE OF STORED FOOD BY YOUNG GREEN PLANT: (a) for energy, (b) for construction of tissue. Experiment. Structure of bean seed. Draw to show outer coat, cotyledon, hypocotyl, and plumule. Test for starch and sugar (grape). Test for oil, protein, water, mineral matter. Use of all nutrients to seedling.

Seventh week. OTHER NEEDS OF YOUNG PLANTS. Home experiments to show (a) temperature, (b) amount of water most favorable to germination. Experiment. To show need of oxygen. To show that germinating seeds give off carbon dioxide. Proof of presence of carbon dioxide in breath. The needs of a young plant compared with those of a boy or girl.

Eighth week. DIGESTION IN SEEDLING. Structure of corn grain. Experiment. To show that starch is digested in a growing seedling (corn). Experiment. To show that diastase digests starch. Discussion of experiments.

Ninth week. WHAT PLANTS TAKE FROM THE SOIL AND HOW THEY DO THIS. Use of roots. Proof that it holds plant in position, takes in water and mineral matter, and in some cases stores food. Influence of gravity and water. Labeled drawing of root hair. Root hair as a *cell* emphasized. Osmosis demonstrated.

Tenth week. COMPOSITION OF THE SOIL. Demonstration of presence of mineral and organic substances in the soil. What root hairs take from the soil. Mineral matter necessary and why. Importance and sources of nitrogen. Soil exhaustion and its prevention. Nitrogen-fixing bacteria. Review bacteria of decay. Rotation of crops.

Eleventh week. UPWARD COURSE OF MATERIALS IN THE STEM. Demonstration of pea seedlings with eosin to show above. Demonstration of evaporation of water from a leaf. Action of stomata in control of transpiration. Cellular structure of leaf. Demonstration of elodea to show cell.

Twelfth week. SUN A SOURCE OF ENERGY. Heliotropism. Demonstration. Necessity of sunlight for starch manufacture. Necessity of air for starch manufacture. By-products in starch making. Oil manufacture in leaf. Protein manufacture in plant. Respiration.

Thirteenth week. REPRODUCTION. Necessity for (a) perpetuation, (b) regeneration. Study of a typical flower to show sepals, petals, stamens, pistil. Functions of each part. Cross and longitudinal sections of ovary shown and drawn. Emphasis on essential organs. Pollination, self and cross. (NOTE. At least one field trip must be planned for the month of May. This trip will take up the following topics: The relations between flowers and insects. The food and shelter relation between plants and animals. Recognition of 5 to 10 common trees. Need of conservation of forests. An extra trip could well be taken to give child a little knowledge and love for spring flowers and awakening nature.)

Fourteenth week. STUDY OF THE BEE OR BUTTERFLY WITH REFERENCE TO ADAPTATIONS FOR INSECT POLLINATION. Study of an irregular flower to show adaptations for insect visitors. Fertilization begun. Growth of pollen tubes.

Fifteenth week. FERTILIZATION COMPLETED. Use of chart to show part played by egg and sperm cell. Ultimate result the formation of embryo and its growth under favorable conditions into young plant. Relation of flower and fruit, pea, or bean used for this purpose. Development of fleshy fruit. Apple used for this purpose.

Sixteenth week. MATURING OF PARTS AND STORING OF FOOD IN SEED AND FRUIT. The devices for scattering the seeds and relation to future plants. Résumé of processes of nutrition to show how materials found in fruit and seed are obtained by the plant.

Seventeenth week. PLANT BREEDING. Factors: (a) selective planting, (b) cross-pollination, (c) hybridizing. Heredity and variation begun. Darwin and Burbank mentioned.

Eighteenth and nineteenth weeks. THE NATURAL RESOURCES OF MAN: SOIL, WATER, PLANTS, ANIMALS. The relation of plant life to the above factors of the environment. The relation of insects to plants (forage and other crops) and the relation of birds to insects. Need for conservation of the helpful factors in the environment of plants. Attention called to some native birds as insect and wood destroyers.

Twentieth week. REVIEW AND EXAMINATIONS.

SECOND TERM

First week. THE BALANCED AQUARIUM. Study of conditions producing this. The rôle of green plants, the rôle of animals. What causes the balance. How the balance may be upset. The nitrogen cycle. What it means in the world outside the aquarium. Symbiosis as opposed to parasitism. Examples.

Second week. STUDY OF THE PARAMECIUM. Study of a hay infusion to show how environment reacts upon animals. Relation to environment. Study of cell under microscope to show reactions. Structure of cell. Response to stimuli, function of cilia, gullet, nucleus, contractile vacuoles, food vacuoles, asexual reproduction. Drawings to show how locomotion is performed, general structure. Copy chart for fine structure.

Third week. A BIRD'S-EYE VIEW OF THE ANIMAL KINGDOM. One day. Development of a multicellular organism. (Use models.) One day. Physiological division of labor. Tissues, organs. Functions common to all animals. Illustrative material. Optional trip to museum for use of illustrative material to illustrate the principal characteristics of (a) a simple metazoan, sponge, or hydrazoan, (b) a segmented worm, (c) a crustacean (Decapod), (d) an insect, (e) a mollusk and echinoderm, (f) vertebrates. (Differences between vertebrates and invertebrates.) The characteristics of the vertebrates. Distinguish between fishes, amphibia, reptiles, birds, mammals. Two days for discussion. Man's place in the animal series, elementary discussion of what evolution means.

Fourth week. THE ECONOMIC IMPORTANCE OF ANIMALS. Uses of animals: (1) As food. Directly: fish, shellfish, birds, domesticated mammals. (2) Indirectly as food: protozoa, crustacea. (3) They destroy harmful animals and plants. Snakes — birds; birds — insects; birds — weed seeds; herbivorous animals — weeds. (4) Furnish clothing, etc. Pearl buttons, etc. (5) Animal industries, silkworm culture, etc. (6) Domesticated animals.

Animals do harm: (1) To gardens. (2) To crops. (3) To stored food; examples, rats, insects, etc. (4) To forest and shade trees. (5) To human life. Disease: parasitism and its results, — examples, from worms, etc.; disease carriers fly, etc. Preventive measures. Methods of extermination.

References to Toothaker's *Commercial Raw Materials*. Use one day for laboratory work from references.

Fifth week. THE STUDY OF A WATER-BREATHING VERTEBRATE. Two days. The fish, adaptations in body, fins, for food getting, for breathing. Structure of gills shown. Laboratory demonstration to show how water gets to the gills. Drawings. Outline of fish, gills. Required trip to aquarium. Object, to see fish in environment. One day. Home work at market. Why are some fish more expensive than others. Economic importance of fish. Relation of habits of (a) food getting, (b) spawning to catching and extermination of fish. Two days. Means of preventing overfishing, stocking, fishing laws, artificial fertilization of eggs, methods. Development of fish egg. Comparison with that of frog and bird.

Sixth week. THE FACTORS UNDERLYING PLANT AND ANIMAL BREEDING. Study of pupils in class to show heredity and variation. Conclusion. Animals tend to vary and to be like their ancestors. Heredity, rôle of sex cells, chromosomes. Principles of plant breeding. Selective planting, hybridizing, work of Darwin, Mendel, De Vries, and Burbank. Methods and results. Animal breeding, examples given, results. Improvement of man: (1) by control of environment, (a) example of clean-up campaign, 1913; (2) by control of individual, personal hygiene, and control of heredity. Eugenics. Examples from Davenport, Goddard, etc.

Seventh week. THE HUMAN MACHINE. Skin, bones and muscles, function of each. Examples and demonstration with skeleton. Organs of body cavity; show manikin. Work done by cells in body.

Eighth week. STUDY OF FOODS to determine: (a) nutritive value. Exercise with food charts to determine foods rich in water, starch, sugar, fats, proteins, mineral salts, refuse. One day. (b) Nutritive value of foods as related to work, age, sex, environment, cost, and digestibility. Foods compared to determine what is really a cheap food.

Ninth week. HOW THE FUEL VALUE OF FOOD HAS BEEN DETERMINED. The dietaries of Atwater, Chittenden, and Voit. The 100-calorie portion table and its use.

Tenth week. THE APPLICATION OF THE 100-CALORIE PORTION TO THE MAKING OF THE DAILY DIETARIES. Luncheon dietaries. A balanced dietary for pupil for one day. Family dietaries. Relation to cost. Reasons for this.

Eleventh week. FOOD ADULTERATIONS. Tests. Drugs and the alcohol question.

Twelfth week. DIGESTION. The alimentary canal of frog and of man compared. Drawings. (One day.) The work of glands. Work of salivary gland. Enzymes, internal secretions. Experiments to show (a) digestion of starch by saliva, (b) digestion of proteins by gastric or pancreatic juice, (c) emulsification of fats in the presence of an alkaline medium. Functions of other digestive glands. Movements of stomach and intestine discussed and explained.

Thirteenth week. ABSORPTION. How it takes place, where it takes place. Passage of foods into blood, function of liver, glycogen.

Fourteenth week. THE BLOOD AND ITS CIRCULATION. Composition and functions of plasma, red corpuscles, colorless corpuscles, blood plates, antibodies. The lymph and work of tissues. The blood and its method of distribution. Heart a force pump. Demonstration. Arteries, capillaries (demonstration), veins. Hygiene of exercise.

Fifteenth week. WHAT RESPIRATION DOES FOR THE BODY. The apparatus used. Changes of blood within lungs, changes of air within lungs. Demonstration. Cell respiration. The mechanics of respiration. Demonstration. Ventilation, need for, explain proper ventilation. Demonstration. Hygiene of fresh air and proper breathing. Dusting, sweeping, etc.

Sixteenth week. EXCRETION, ORGANS OF. Skin and kidneys, regulation of body heat. Colds and fevers. Proper care of skin, hygiene. Summary of blood changes in body. Explanation of same.

Seventeenth week. BODY CONTROL AND HABIT FORMATION. Nervous system, nerve control. The neuron theory, brain psychology explained in brief. Habits and habit formation. Hygiene of sense organs.

Eighteenth and nineteenth weeks. CIVIC HYGIENE AND SANITATION. THE IMPROVEMENT OF ONE'S ENVIRONMENT. Civic conditions discussed. Water, milk, food supplies. Relation to disease. How safeguarded. How help improve conditions in city.

Twentieth week. REVIEW AND EXAMINATIONS.

HYGIENE OUTLINE

(This outline may be introduced with Plant Biology, or, better, may come as application of the work in Second-term Biology.)

THE ENVIRONMENT. Changes for betterment under control. How a city boy may improve his environment : by proper clothing, proper food and preparation of food, by care in home life; by sanitary conditions in neighborhood and in home.

REVIEW OF ACTIVITIES OF CELL. Irritability, food taking, assimilation, oxidation, excretion, reproduction. Similarity of functions of plant and animal cells. All cells perform these functions. Some cells perform functions especially well, e.g. contracting muscle cells. All cells need food and oxygen. Some must have this carried to them. A system of tubes carries blood which carries food and oxygen. Food must be prepared to get into the blood. Digestive system : mouth, teeth, stomach, intestines, glands, and digestive juices. Uses of above in preparing food to pass into the blood. Absorption of food into the blood. How oxygen gets to the cells. Nose, throat, windpipe, lungs; blood goes to lungs and carries away oxygen. Excretion. Cells give up wastes to blood and these wastes taken out of blood by kidneys and other glands and passed out of body. Sweat, urine, carbon dioxide.

CERTAIN KINDS OF WORK PERFORMED BY CERTAIN KINDS OF CELLS. Advantage of this. Cells of movement. Muscles, tissues. Bones as levers necessary for some movements. This especially true for legs and arms. Skeleton also necessary for protection of internal organs and support of body. Making of special things in the body, e.g. digestive juices given to certain cells called gland cells. Working together or coördination of different organs provided for by nervous system. This is composed of cells which are highly irritable or sensitive. Collections of these nerve cells give us the power of feeling or sensation and of thinking.

DIETETICS. Diet influenced by age, weight, occupation, temperature or climate, cheapness of food, digestibility.

NUTRIENTS. List of nutrients found in seeds and fruits, also other common foods. Need of nutrients for human body. Nitrogenous foods, examples. A mixed diet best.

DIGESTION AND INDIGESTION. What is digestion? Where does it take place? *Causes of indigestion.* Eating too rapidly and not chewing food. Eating foods hard to digest. Overeating. Eating between meals. Hard exercise immediately before or after eating.

CONSTIPATION. A condition in which the bowels do not move at least once every day. Dangers of constipation. Poisonous materials may be absorbed, causing lack of inclination to work, headache. Importance of regular habits of emptying the bowels. Each one must try to get at the cause of constipation in his own case. *Causes of constipation.* Lack of exercise, improper food, not drinking enough water, lack of laxative food, as fruits; lack of sleep, lack of regular habits. *Remedies.* Avoid use of drugs. Half hour before breakfast a glass of hot water, exercise of abdominal muscles, laxative foods, form habit of moving bowels after breakfast.

HYGIENE OF CIRCULATION AND ABSORPTION. How digested foods get to the cells. Absorption. Definition. The passing of the digested food into the blood. How accomplished. Blood vessels. In walls of stomach and food tube. Membrane of cells separating food from blood. Food passes by osmosis through the membrane and by osmosis through the thin walls of the blood vessels.

CIRCULATION OF FOODS. Blood contains foods, oxygen, and waste materials. Heart pumps the blood, blood vessels subdivide until very small and thin, so food, etc., passes from them to cells. Hygiene of the heart.

TRANSPIRATION AND EXCRETION. Skin, function in excretion. Bathing. Care of skin. Hot baths. Bathe at least twice a week. Cold baths, how taken. Bathtub not a necessity. Effect of latter on educating skin to react. Relation to catching cold.

CARE OF SCALP AND NAILS. Scalp should be washed weekly. If dandruff present, wash often enough to keep clean. Baldness often results from dandruff. Finger nails cut even with end of fingers and cleaned daily with scrub brush.

HYGIENE OF RESPIRATION. Definition of respiration. Object of respiration. (Connection between circulation and respiration.) Necessity of oxygen. Organs of respiration. Lungs most important. Deep breath, function. Ventilation, reasons for. Mouth breathing. Results. Lessened mental power, nasal catarrh, colds easily caught.

PLANTS HARMFUL to MAN. Poison ivy and mushrooms. Treatment. Poisoning. Send for physician. Cause vomiting by (1) finger, (2) mustard and water. (NOTE. An unconscious person should not be given anything by the mouth unless he can swallow.) Relation of yeasts and bacteria to man. Fermentation a cause of indigestion. Relation to candy, sirups, sour stomach, formation of gas causes pain.

BACTERIA OF MOUTH AND ALIMENTARY CANAL. Entrance of bacteria by mouth and nose. Nose: "cold in the head," grippe, catarrh. Mouth: decay of teeth, tonsillitis, diphtheria. Germs pass from one person to another, no one originates germs in himself. Precautions against receiving and transferring germs. Common drinking cups, towels, coins, lead pencils, moistening fingers to turn pages in book or to count roll of bills. Tuberculosis germs. Entrance by mouth, lungs favorite place, may be any part of body. Dust of air, sweeping streets, watering a necessity. Spitting in streets and in public buildings. Germs of typhoid fever. Entrance: water, milk, fresh uncooked vegetables, oysters. Thrive in small intestines. Preventable. Typhoid epidemics, methods of prevention of typhoid. Conditions favorable for growth of specific disease germs. Work of Boards of Health.

Home sanitary conditions, sunlight, air, curtains and blinds, open windows. Live out of doors as much as possible. Cleanliness. Bare walls well scrubbed better than carpets and rugs. Lace curtains, iron bedsteads, one thickness of paper on walls. Open plumbing, dry cellars, all garbage promptly removed.

This outline is largely the work of Dr. L. J. Mason and Dr. C. H. Morse of the department of biology of the De Witt Clinton High School.

WEIGHTS, MEASURES, AND TEMPERATURES

As the metric system of weights and measures and the Centigrade measurement of temperatures are employed in scientific work, the following tables showing the English equivalents of those in most frequent use are given for the convenience of those not already familiar with these standards. The values given are approximate only, but will answer for all practical purposes.

WEIGHT

Kilogram	kg.	2¼ pounds
Gram . .	gm.	15½ grains avoirdupois. ⅟₂₈ of an ounce avoirdupois.

CAPACITY

Liter . .	l.	61 cubic inches, or a little more than 1 quart, U. S. measure.
Cubic centimeter .	cc.	⅟₁₆ of a cubic inch.

MEASURES OF LENGTH

METRIC		ENGLISH EQUIVALENTS
Kilometer	km.	⅔ of a mile.
Meter . .	m.	39 inches.
Decimeter	dm.	4 inches.
Centimeter	cm.	⅜ of an inch.
Millimeter	mm.	⅟₂₅ of an inch.

The next table gives the Fahrenheit equivalent for every tenth degree Centigrade from absolute zero to the boiling point of water. To find the corresponding F. for any degree C., multiply the given C. temperature by nine, divide by five, and add thirty-two. Conversely, to change F. to C. equivalent, subtract thirty-two, multiply by five, and divide by nine.

CENT.	FAHR.	CENT.	FAHR.	CENT.	FAHR.	CENT.	FAHR.
100 . . .	212	50 . . .	122	0 . . .	32	− 50 . . .	− 58
90 . . .	194	40 . . .	104	− 10 . . .	14	− 100 . . .	− 148
80 . . .	176	30 . . .	86	− 20 . . .	− 4		
70 . . .	158	20 . . .	68	− 30 . . .	− 22	Absolute zero	
60 . . .	140	10 . . .	50	− 40 . . .	− 40	− 273 . . .	− 459

The following articles comprise a simple equipment for a laboratory class of ten. The equipment for larger classes is proportionately less in price. The following articles may be obtained from any reliable dealer in laboratory supplies, such as the Bausch and Lomb Optical Company of Rochester, N.Y., or the Kny-Scheerer Company, 404, 410 West 27th Street, New York City: —

1 balance, Harvard trip style, with weights on carrier.
1 bell jar, about 365 mm. high by 165 mm. in diameter.
10 wide mouth (salt mouth) bottles, with corks to fit.
10 25 c.c. dropping bottles for iodine, etc.
25 250 c.c. glass-stoppered bottles for stock solutions.
100 test tubes, assorted sizes, principally 6" × $\frac{3}{4}$".
50 test tubes on base (excellent for demonstrations).
2 graduated cylinders, one to 100 c.c., one to 500 c.c.
1 package filter paper 300 mm. in diameter.
10 flasks, Erlenmeyer form, 500 c.c. capacity.
2 glass funnels, one 50, one 150 mm. in diameter.
30 Petri dishes, 100 mm. in diameter, 10 mm. in depth.
10 feet glass tubing, soft, sizes 2, 3, 4, 5, 6, assorted.
1 aquarium jar, 10 liters capacity.
2 specimen jars, glass tops, of about 1 liter capacity.
10 hand magnifiers, vulcanite or tripod form.
2 compound demonstration microscopes or 1 more expensive compound microscope.
300 insect pins, Klaeger, 3 sizes assorted.
10 feet rubber tubing to fit glass tubing, size $\frac{1}{8}$ inch.
1 chemical thermometer graduated to 100° C.
15 agate ware or tin trays about 350 mm. long by 100 wide.
1 gal. 95 per cent alcohol. (Do not use denatured alcohol.)

1 set gram weights, 1 mg. to 100 g.	2 books test paper, red and blue.
1 razor, for cutting sections.	10 Syracuse watch glasses.
1 box rubber bands, assorted sizes.	1 steam sterilizer (tin will do).
1 support stand with rings.	1 spool fine copper wire.

1 test tube rack.	1 alcohol lamp.	6 oz. nitric acid.
5 test tube brushes.	1 gross slides.	6 oz. ammonium hydrate.
10 pairs scissors.	100 cover slips No. 2.	6 oz. benzole or xylol.
10 pairs forceps.	1 mortar and pestle.	6 oz. chloroform.
20 needles in handles.	2 bulb pipettes.	$\frac{1}{2}$ lb. copper sulphate.
10 scapels.	1 liter formol.	$\frac{1}{2}$ lb. sodium hydroxide.
12 mason jars, pints.	1 oz. iodine cryst.	$\frac{1}{2}$ lb. rochelle salts.
12 mason jars, quarts.	1 oz. potassium iodide.	6 oz. glycerine.

The materials for Pasteur's solution Sach's nutrient solution can best be obtained from a druggist at the time needed and in very small and accurately measured quantities.

The agar or gelatine cultures in Petri dishes may be obtained from the local Board of Health or from any good druggist. These cultures are not difficult to make, but take a number of hours' consecutive work, often difficult for the average teacher to obtain. Full directions how to prepare these cultures will be found in Hunter's *Laboratory Problems in Civic Biology.*

INDEX